Artificial Intelligence Tools

DECISION SUPPORT SYSTEMS IN CONDITION MONITORING AND DIAGNOSIS

Artificial Intelligence Tools

DECISION SUPPORT SYSTEMS IN CONDITION MONITORING AND DIAGNOSIS

DIEGO GALAR PASCUAL

CRC Press is an imprint of the
Taylor & Francis Group, an **informa** business

CRC Press
Taylor & Francis Group
6000 Broken Sound Parkway NW, Suite 300
Boca Raton, FL 33487-2742

First issued in paperback 2020

© 2015 by Taylor & Francis Group, LLC
CRC Press is an imprint of Taylor & Francis Group, an Informa business

No claim to original U.S. Government works

ISBN-13: 978-1-4665-8405-1 (hbk)
ISBN-13: 978-0-367-73835-8 (pbk)

Visit the Taylor & Francis Web site at
http://www.taylorandfrancis.com

and the CRC Press Web site at
http://www.crcpress.com

To my father, who devoted his entire life to maintenance—
trying to keep the machines up and running

Contents

Preface

A serious concern in the industry is the extension of the useful life of critical systems by identifying ongoing problems to mitigate potential risks. If maintenance is included in risk mitigation, the useful life of these systems can be extended, their life-cycle costs reduced, and their availability and reliability improved.

In the past, repairs were performed only after a failure was noticed, but today, maintenance is done depending on estimations of a machine's condition. Thus, maintenance technology has shifted from "at-failure" maintenance to "condition-based" maintenance. A combination of diagnosis and prognosis technology is used to model the asset degradation process and predict its remaining life with the help of machine condition data. Condition-based maintenance (CBM) facilitates the effective life-cycle management system using results from the diagnostics and prognostics data. A well-modeled condition-monitoring (CM) technology guides maintenance personnel to perform the necessary support actions at the appropriate time, with lower maintenance and life-cycle costs, reduced system downtime, and minimal risk of unexpected catastrophic failures. Even though sensor and computer technology for obtaining CM data has made considerable progress, prognostics and health management (PHM) technology is relatively recent and still quite difficult to implement.

Two words have driven my life: diagnosis and prognosis. This book is the result of my reflection on and compilation of techniques used in many applications, especially in CM. There are various definitions of diagnosis and prognosis, but all definitions recognize diagnosis and prognosis as the most critical part of the CBM program. They are a key component of today's maintenance strategies, as it is now generally accepted that identification of current failures and the estimation of the remaining useful life (RUL) are essential. The process is simple. First, we must diagnose the state of the item, component, or system failure; hence, the purpose of this book. After determining the state of the item, prognosis comes into the picture to predict the RUL or remaining "system's lifetime" before a functional failure occurs, given the current machine condition and past operating profile.

Many methods have been used to monitor the condition of machinery. Most of these methodologies are model based and data driven. Each has its own strengths and weaknesses; consequently, they are often used in combination. Approaches range in fidelity from simple historical failure rate models to high-fidelity physics-based models. Depending on the type of approach, the required information includes the engineering model and data, failure history, past operating conditions, current conditions, and so on.

Model-based diagnosis usually detects, isolates, and identifies faults. The damage state of a system is estimated and its future degradation rate is predicted, thus determining the RUL of the system. This requires a deep knowledge of system functioning, often using an *a priori* approach. Model-based methods work well for components or simple subsystems that may be described in terms of behavior; in such cases, the derived equations can show degradation. However, if the item to be diagnosed is a complex system with many subsystems and components, the model must be very complex, making the mission impossible. Complexity and knowledge of the whole system are tremendous barriers to diagnose aircraft, vehicles, production machines, and other assets with many elements.

Therefore, model-based diagnosis is commonly used for diagnosis and prognosis even though it seems natural. The model-based approach is often called a "white box approach" since it requires in-depth knowledge of an asset's behavior and degradation mechanisms.

Unfortunately, knowledge may be unavailable or computations may be too complex to derive accurate models. In this case, diagnosis and prognosis are based on observations, the so-called "black-box approach."

This book discusses the merits of these techniques and the challenges associated with their applications in real life. On the one hand, methods based on a data-driven approach cannot identify situations that have not previously occurred. On the other, with adequate data, they can detect anomalies or abnormal behavior in assets that usually indicate an incipient failure.

Data-driven methods are based on simplicity. Usually, all data available for a particular machine are collected; features are extracted from the data and investigated to determine if they are normal (healthy condition) or are symptoms of failure. If the latter is true, the failure must be classified and categorized to identify the fault and determine its severity.

Data-driven algorithms obviously rely on large amounts of good-quality data. Chapter 1 introduces the topic of maintenance data and notes today's pursuit of the "holy grail" of self-diagnostics. Many balanced scorecards and key performance indicator solutions are available in the market, and all of them make similar claims—their product will make a manufacturing process run better, faster, more efficiently, and with greater returns. Yet, their efficacy is questionable as the necessary information is often scattered across disconnected silos of data in each department of an industry, and it is difficult to integrate these silos. For example, control system data are real-time data measured in terms of seconds, whereas maintenance cycle data are generally measured in terms of calendar-based maintenance (e.g., days, weeks, months, quarters, semiannual, and annual), and financial cycle data are measured by fiscal periods.

In summary, maintenance data sources are disparate but they offer rich information for diagnosis purposes if they can be correlated. However, the large size of the data and the complexity of contextual engines for correlating information create barriers.

Maintenance information is mostly the fusion of management information stored in CMMS/EAM (computerized maintenance management system/engineering asset management) software and CM records. These records comprise the readings of the sensors. A CM program can track a variety of measurements, including vibration, oil condition, temperature, operating and static motor characteristics, pump flow, and pressure output. These measurements are taken by monitoring tools such as ferrographic wear particle analysis, proximity probes, triaxial vibration sensors, accelerometers, lasers, and multichannel spectrum analyzers.

Chapter 2 describes the most common techniques used and the data acquired when these techniques are applied, paying special attention to the granularity of such data and the decision process associated with them to perform diagnosis. The chapter also describes information collected by automation systems via a PLC (programmable logic controller) that allows the maintainer to contextualize fault detection and identification.

Fault detection is discussed in Chapter 3. After collecting the data with different properties from disparate data sources, the maintainer must determine whether anything strange is happening. A failure will be revealed by the abnormal behavior of an asset, but this is only useful if the maintainer can detect the anomaly. This does not mean all anomalies mean failures but all failures are represented by some form of anomalies. The real challenge is that many failures occur infrequently; thus, while detection is difficult, identification is even more problematic. Events that seldom occur are explained by the metaphor of the black swan. In both nature and maintenance, a black swan is a highly improbable event with three principal characteristics: it is unpredictable; it has a massive impact; and finally, after the fact, we concoct an explanation that makes it appear less random and more predictable.

Fortunately, not all events for failure detection are rare and databases of past behaviors can be used to detect patterns even in the very early stages of fault progression. Chapter 4 describes three types of detection depending on the available data: supervised, semisupervised, and unsupervised. The user may have data of all possible failures and behaviors; so, supervised learning is an option. Or, maybe just healthy data are available due to the age of the asset, making semisupervised learning the only affordable option. In the worst-case scenario, in unsupervised learning, the asset is already running and we know nothing about its condition. During fault detection, failures must be classified according to certain characteristics; for example, failure may be an individual anomaly or associated with certain boundary conditions or contexts. Chapter 4 classifies failures as individual, collective, and contextual.

Once failure is detected, it must be classified. This is possible if the user has a closed catalog of failures or combinations of them where data can be projected and classified. Chapter 5 describes techniques of classification that have been successfully applied to many domains including CM. Some are old techniques (neural networks) and others are more recently used with excellent results in the CM field (Bayesian or support vector machines); all are very popular but have different pros and cons, depending on the case.

Classification is based on the concept of distance; instances are classified according to "failure distance." Chapter 6 discusses distance and pays special attention to outliers. Outliers represent a challenge in failure identification; maintainers may find abnormal behavior that is difficult to identify or cannot be assimilated to data on known failures. Here, supervised, unsupervised, and semisupervised learning are applicable.

Classification is an option when potential failures have been previously identified and no outliers are expected. In fact, if we identify many outliers during classification, this means our catalog of failures is out of date. Clustering permits the categorization of failures, namely, the creation of different categories of failures when these categories are not defined *a priori* and the user does not know if there are three, four, or dozens of potential failures. Chapter 7 describes these techniques and emphasizes the differences between classification and categorization. Both are valid and useful but they must be used as a function of the available information on the end user's system.

Outliers are a major issue in data analysis. In general, statistical techniques such as probability distribution models and stochastic process models are used for identification and removal of outliers from the data sets. Chapter 8 discusses the use of these techniques for two main purposes: first, to identify outliers, assuming data sets follow a certain distribution; second, to identify outliers and remove them for data cleaning, thus improving the quality of the processed data sets.

All techniques aim for fault identification and detection in CM data. These may be vibration data or other physical variables previously processed. Healthy signals are more chaotic and their entropy is higher than faulty signals. In faulty signals, the information contained due to failure reduces the chaos dramatically and entropy drops. These methods are described in Chapter 9. The chapter adds that techniques used in other domains have been proposed for CM to detect faults based on the classic information theories of Shannon.

Last but not least, we must deal with certainty/uncertainty. Diagnosis is a complex process with a high degree of uncertainty. Thus, the decision-making process involves a measure of risk. Chapter 10 proposes fuzzy logic to explain the uncertainties associated with diagnostic processes.

Acknowledgments

It is difficult to list all the people who should be thanked for this book but I will do my best.

First of all, I would like to thank my mother, Marisol Pascual. Without her continuous support and encouragement, I would never have been able to achieve my goals. This one is for you mom!

Second, a special thank you to my girlfriend, Ana Val. Words cannot describe how lucky I am to have her in my life. She has selflessly given more to me than I could have ever asked for. I look forward to our lifelong journey.

Thanks to the people who directly or indirectly participated in this book; none of this would have been possible without them. Their dedication and perseverance made the work enjoyable. Especially remarkable is the contribution of Professor Uday Kumar. His vision in maintenance engineering and management has always inspired my research. Kumar's ideas, especially in context-driven maintenance, were key for me in the decision to write this book.

I want to thank all my colleagues at the Division of Operation and Maintenance, University of Lulea, Sweden, for encouraging me to produce this document as a compilation of techniques widely used in our field, but especially to Aditya Thaduri who worked with me for many long hours to give coherence to a manuscript that addresses many different topics. It was a challenge to create a reader-friendly book that would be useful for maintainers!

Also, thanks to Numan, Angel, Victor, Roberto, Carl, Madhav, and others for your contribution and hours of reading and correcting. Without all of you, this book would not be a reality.

Author

Professor Diego Galar Pascual earned an MSc in telecommunications and a PhD degree in manufacturing from the Saragossa University in Zaragoza, Spain. He has served as a professor in several universities, including Saragossa University and the European University of Madrid. He was also a senior researcher in I3A, Aragon Institute for Engineering Research of Saragossa University, director of academic innovation, director of international relations, and subsequently the vice-chancellor of the university.

In industry, Professor Pascual has much experience in maintenance management and engineering. He has been the technological director and CBM manager of international firms such as VOLVO, SAAB, BOLIDEN, SCANIA, TETRAPAK, HEINZ, ATLAS COPCO, where he deployed maintenance techniques and methodologies all over the world.

Currently, he is professor of condition monitoring in the Division of Operation and Maintenance in Luleå University of Technology, Sweden, where he is coordinating several EU-FP7 projects related to different maintenance aspects and is also involved with the LTU-SKF University Technology Center, located in Luleå, focusing on SMART bearings. He has authored more than 200 journal and conference papers, books, and technical reports in the field of maintenance.

He is also a visiting professor at the University of Valencia (Spain), Polytechnic of Braganza (Portugal), Valley University (Mexico), Sunderland University (UK), Maryland University, and Northern Illinois University (NIU) (USA).

1

Massive Field Data Collection: Issues and Challenges

Fault diagnosis is as old as the use of machines. With simple machines, manufacturers and users relied on simple protections to ensure safe and reliable operation, but with the increasing complexity of tasks and machines, improvements were required in fault diagnosis. It is now critical to diagnose faults at their inception, as unscheduled machine downtime can upset deadlines and cause heavy financial losses. Diagnostic methods to identify faults involve many fields of science and technology and include the following: (a) electromagnetic field monitoring, search coils, coils wound around motor shafts (axial flux-related detection), (b) temperature measurements, (c) infrared recognition, (d) radio frequency (RF) emissions monitoring, (e) noise and vibration monitoring, (f) chemical analysis, (g) acoustic noise measurements, (h) motor current signature analysis, (i) modeling, artificial intelligence (AI), and neural network-based techniques (Li, 1994).

1.1 An Introduction to Systems

The word system (from the Greek συστημα) is used in many contexts and has thus assumed a variety of meanings. The most common refers to a group of parts linked by some kind of interaction. In System Theory, a system's evolution through time affects a certain number of measurable attributes; real-life systems include a machine tool, an electric motor, a computer, an artificial satellite, or the economy of a nation (Guidorzi, 2003).

A *measurable attribute* is a characteristic that can be correlated with one or more numbers, either an integer, real or complex, or simply a set of symbols. Examples include the rotation of a shaft (a real number), the voltage or impedance between two given points of an electric circuit (a real or complex number), any color belonging to a set of eight well-defined colors (an element of a set of eight symbols; for instance, digits ranging from 1 through 8 or letters from *a* through *h*), the position of a push button (a symbol equal to 0 or 1, depending on whether it is released or pressed), and so on.

In systems, we are interested, on the one hand, in internal functional relationships, and, on the other hand, in external relationships with the

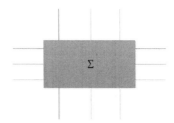

FIGURE 1.1
Schematic representation of a system. (From Basile, G. and Marro, G., 2002. *Controlled and Conditioned Invariants in Linear Systems Theory*. Italy: University of Bologna.)

environment. The former constitute the *structure* of the system, the latter its *behavior*. Behavior refers to the dependence of responses on stimuli. Structure refers to the manner of the arrangement, that is, organization, the mutual coupling between the elements of the system and the behavior of these elements.

A system can be represented as a block and its external relations as connections with the *environment* or other systems, as shown by the simple diagram in Figure 1.1. When quantities are separated into those produced by the environment and those produced by the system, we can distinguish inputs (the former) from outputs (the latter). If a system performs this type of separation, we call it an *oriented* or *controlled* system. If a separation between inputs and outputs is not given, we talk about *nonoriented* or *neutral* systems (Welden and Danny, 1999).

Note: The distinction between causes and effects, in some cases, is anything but immediate. Consider, for instance, the simple electric circuit shown in Figure 1.2a, whose variables are v and i. It can be oriented as in Figure 1.2b, that is, with v as input and i as output; this is the most natural choice if the circuit is supplied by a voltage generator. But the same system may be supplied by a current generator, in which case i would be the cause and v the

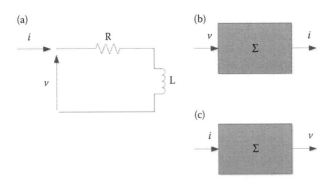

FIGURE 1.2
An electric system with two possible orientations. (a) An electric circuit; (b) voltage generator; and (c) current generator.

FIGURE 1.3
An oriented system.

effect, and the corresponding oriented block diagram would be as shown in Figure 1.2c (Basile and Marro, 2002).

In this case, a restriction is made toward an oriented system (i.e., a relatively closed physical system); see Figure 1.3. For the oriented system shown in Figure 1.3, $U(t)$ and $Y(t)$ are the external quantities and $X(t)$ is the internal quantity. The input is represented by $U(t)$ and the output by $Y(t)$.

Measuring the attributes of the system introduces the problem of establishing quantitative relationships, that is, constructing abstract (mathematical) models.

Since mathematical models are themselves systems, albeit abstract, it is customary to denote both the object of the study and its mathematical model by the word "system" (Basile and Marro, 2002). *System Theory* pertains to the derivation of mathematical models for systems, their classification, the investigation of their properties, and their use to solve engineering problems.

1.1.1 Evolution of Mathematical Models

The etymological roots of the word model are in the Latin *modus* and its diminutive *modulus*; both mean "measure."

The initial use of modeling in science and technology was in scaled representations created by architects to reproduce the shape of a building before its construction. Such models are physical systems that roughly approximate the aesthetic properties of other physical systems before their realization.

An important advancement occurred with the introduction of models which were still small-scale reproductions of physical systems but which were used to investigate the behavior of these systems before construction or under impossible or overly expensive operating conditions.

The next innovation was reproducing the behavior of a system on another system by taking advantage of the physical laws inherent in formally equal relations. A typical example is analog computers, structured as flexible electrical networks that when properly interconnected (programmed) reproduce the behaviors of other systems (mechanical, hydraulic, economic, etc.) less suitable for direct experiments. These models can be defined as *analog*. To avoid any confusion with the common use of this term (i.e., quantities whose measurement can be performed with continuity versus digital denotation of quantizations), we can refer to these as *models based on analogy laws*. These models offer greater flexibility than the others cited above, where the only degree of freedom is the scale, especially since the physical nature of the model is a matter of choice. However, the use of models based on analogy requires us to know the laws describing the system under study. More specifically, we can only select, construct, or configure a system using analog laws if we have comparable initial conditions. The ability to construct a complete mathematical model and use it to determine system behavior is not required, however.

The last step in the evolution of models was the development of abstract models, that is, mathematical models describing the links established by the system between its measurable attributes. Today, the widespread use of mathematical models in science and technology reflects the availability of abstract tools offered by computers, combined with System Theory, which allows their effective use (Guidorzi, 2003).

1.1.2 Models as Approximations of Reality

It has already been noted that the mathematical models limit their description to the quantitative links between their measurable attributes established by real systems. These constitute only partial descriptions. The asymptotic evolution of science has removed any illusions of achieving exact descriptions of reality (Guidorzi, 2003). Even the so-called laws of nature can, at most, be considered models that have not yet been falsified. On the one hand, Newton's law of motion is a good example of the ability of simple model to describe a wide range of situations; on the other hand, it is an equally good example of the widespread acceptance of a mathematical relation as an absolute description of a phenomenon before its falsification.

Many phenomena are simply too complex to be described in detail by tractable models and/or are not governed by any definite law of nature (e.g., national economies). The construction of mathematical models must, therefore, be ruled by criteria of *usefulness* rather than by (always relative) criteria of *truth*.

Different models of the same system may be utilized for different purposes (prediction, interpretation, simulation, diagnosis, filtering, synthesis, classification, etc.). The goal is to optimize the model's performance in its tasks.

Accordingly, the criteria for selecting and comparing models have both practical and philosophical importance. A well-known criterion is the "razor of Occam," from William of Occam (1290–1350): simply stated, among the models accounting for the same phenomenon, the simpler must be preferred. This principle certainly helped the acceptance of the model proposed by Copernicus for the solar system; he emphasized that his heliocentric model should be considered an exercise to obtain the results of the officially accepted Ptolemaic model in a simpler fashion.

Popper (1963) provided a different description of the principle of parsimony. According to him, among the models explaining available observations, the one explaining as little else as possible (i.e., the most powerful unfalsified model) is preferable.

The principle of parsimony is supported by philosophical (and by common sense) and mathematical arguments that when models are deduced from uncertain data, the resulting increased complexity leads to corresponding increases in the uncertainty of their parameters.

1.1.3 Modeling Classification Based on Purpose

Mathematical models are defined as sets of relationships among the measurable attributes of a system; they describe the links established by the system among these attributes. This means mathematical models can only describe those attributes that can be expressed by numbers, making these models approximate descriptions of reality. The approximation performed by models requires the use of classification; this, in turn, depends on what is being modeled (Guidorzi, 2003).

1.1.3.1 Interpretative Models

These models are designed to satisfy scientific curiosity and rationalize observed behavior. They can replace large amounts of data with a data-generating mechanism and thereby extract the essential information from complex experiments. The increased understanding of the reality behind observed phenomena is the purpose of interpretive models. They do not generally have the capability to generate other sets of data, but they should "interpret" sets of collected data.

The majority of the physical laws are models of this type. Ptolemy observed that it is possible to describe the same observations with different models; in other words, interpretation rests on measurable attributes of a phenomenon, not necessarily on its actual nature.

Another important observation is the limited range of validity of interpretative models and/or their approximations. For example, by giving a simple relation between the force acting on a mass and its acceleration, Newton's law of motion leads to large errors for speeds approaching the speed of light.

Nevertheless, models of this kind describing the motion of physical objects have been developed by Ptolemy, Copernicus, Kepler, Galileo, Newton, and Halley. Interpretative models are used in a large number of disciplines, such as econometrics, ecology, life sciences, agriculture, and physics.

1.1.3.2 Predictive Models

Predictive models are used to forecast the future behavior of a system, or interpolating available observations into the future. Mathematical models see more frequent use than any other kind, with numerous fields of application, including population growth, the future state of an ecosystem or a plant, weather conditions, and demands for specific products.

The resulting predictions are frequently used to manipulate the inputs of a system to achieve specific objectives, such as positioning a robot arm, or determining the desired altitude of an aircraft or a missile, the rate of inflation, or the degree of purity of the output of a distillation apparatus. Other less obvious applications of predictive models are speech and image processing to reduce bandwidth requirements for transmission and recording. Note that some models are both interpretative and predictive (Guidorzi, 2003).

1.1.3.3 Models for Filtering and State Estimation

As their name suggests, these models are used to extract external variables (i.e., output) from otherwise noisy system measurements and/or to estimate internal variables (i.e., state) from the external ones.

Applications include receiving and processing radio signals (e.g., telemetry, pictures from a spacecraft), transmitting digital data over noisy channels (e.g., telephone lines), processing radar signals, analyzing electrocardiographic and electroencephalographic signals, processing geophysical data, monitoring industrial plants and natural systems, and studying demography.

1.1.3.4 Models for Diagnosis

Diagnostic models compare the behaviors found in a particular dataset with a previously established reference class of behaviors. The goal is to detect abnormal conditions. For example, in industry, these models may detect a sensor fault that affect production, or in medicine, they may suggest the nature of a patient's disease.

1.1.3.5 Models for Simulation

Simulation models act as substitutions for real systems, with a view to evaluate the latter's response to certain situations (inputs). Such substitutions or simulations may result in financial benefits to a company, as the process will highlight the best course of action; equally, it may point to risks and situations that should be avoided. Examples of simulated systems include pilot training, the responses in a national economy to changes in interest rates, demographic studies, and nuclear reactor accidents.

The utility of the simulations, of course, depends on the precision of the model in reproducing the behavior of the actual system; the etymology of simulation (Latin *simulare* = to pretend) suggests possible ambiguities.

1.1.4 Model Construction

The unavoidable use of approximation in model construction puts modeling into a gray area; it is not pure science but must be based on results and methodologies offered by the abstract science of mathematics.

In addition, as the models described above have limited ability to solve specific problems, there is a need to construct special-purpose models.

1.1.4.1 Approaches to Model Building

Keeping in mind these directives, we can isolate two approaches to model building based on two types of information for a general system. The types are at opposite ends of the system's spectrum (see Table 1.1 and Figure 1.4) (University of Cauca, 2009) and can be stated as follows:

- Knowledge and understanding of the system (white box modeling)
- Experimental data of inputs and outputs of the system (black box modeling)

However, if we represent a model as a box containing the laws of mathematics that link the inputs (causes) with the outputs (effects), we can actually identify three boxes, not just two, as suggested by the above, till labeling them by "color": white, gray, and black.

White box modeling: This model considers the relationship between the system components, and is derived directly from first principles. Typical examples include mechanical and electrical systems in which physical laws ($F = ma$) can be used to predict an effect, given the cause.

Rather than white, however, the box should likely be termed "transparent," as we know the internal structure of the system.

Gray box modeling: Sometimes because the value of a parameter is missing, the model obtained by invoking first principles is too comprehensive. For example, a planet may be subject to the law of gravity, but its mass is unknown. In this case, experimental data must be collected; the unknown parameters must then be attuned with the outputs predicted by the model to match the observed data. As the internal structure of the box is only partially known, there are gray zones, hence the term gray box modeling.

TABLE 1.1

Two Main Approaches to Construct a Model

	Knowledge-Based Approach	Data-Based Approach
Synonyms	Modeling	System identification
	Top-down modeling	Bottom-up approach
Reasoning	Deduction	Induction
Does what?	Encodes the (inner) structure of the system	Encodes the behavior of the system (via experimental data)
Problem type	Analysis	Synthesis

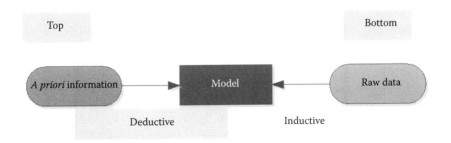

FIGURE 1.4
The two sources of information for model construction.

Black box modeling: When no first principles are available or the internal structure of the system is unknown, the only possibility is to collect data and use them to determine the links between inputs and outputs (a common situation in physiology and economics). Black box modeling is also useful for very complex systems, in cases when the white box approach would be long and expensive (e.g., modeling the dynamics of an internal combustion engine to develop the idle-speed controller).

1.1.4.2 Deducing Models from Other Models: Physical Modeling

A model constructed deductively can generally be considered a unique modeling solution. The top-down approach can and should be used if there is enough *a priori* knowledge and theory to characterize completely the mathematical equations. In such a context, model structure becomes important.

To build a physical model, a system is divided into subsystems, which can be described by known laws. The model joins these relations into a whole.

This approach requires a general knowledge of the "laws" describing the behavior of the system, as well as their structure or design. As physical laws are themselves models obtained from either observations or speculations, physical modeling constructs a model by combining several simple but well-established models.

The advantages of physical modeling include the ability to use *a priori* information about the system for model construction, including the physical meaning of model variables. Unfortunately, physical modeling cannot be used on systems whose internal structure is unknown, whose behavior does not comply with established laws, or whose complexity may result in an unmanageable model with unsuitable parameters (i.e., the parameters would not necessarily induce the system behavior that we want to reproduce).

The deductive or top-down approach is always preferable, if, of course, it is possible. Because deduction involves one-to-one mapping (a process in which no new knowledge is produced), it is a physical principle. In contrast, bottom-up modeling involves a one-to-many mapping (a process in which knowledge is induced). This process is especially important in machine learning.

1.1.4.3 Inductive Modeling and System Identification Methodology

We must treat the system as a black (or gray) box and attempt to infer or induce a model through data analysis of input and output signals, whenever deductive modeling is impossible. This treatment of the system is based on identification derived through experimentation. The *Concise Encyclopaedia of Modeling and Simulation* defines identification as the following:

> [It is] the search for a definition of a model showing the behavior of a process evolving under given conditions. It is usually realised by means of a set of measurements and by observing the behavior of a system during a given time interval. The model obtained by identification must enable the evolution of the identified process to be predicted for a given horizon, when the evolution of the inputs and various external influences acting on this system during this time interval are known (Atherton and Borne, 1992, p. 139).

The bottom-up approach tries to infer structural information from experimental data (i.e., using an experimental frame [EF]). Unfortunately, this approach can generate an infinite number of models that satisfy the observed input/output (I/O) relations.

Briefly stated, then, there is no simple method of determining the structure of a model. Inferring structure from data requires a set of guiding principles and quantitative procedures. It is also desirable to have additional assumptions or restrictions if we are to select the "optimal" model.

System observations can be obtained actively or passively. In the former case, the modeler isolates certain interesting inputs, applies them to the system under study, and observes the outputs. In the latter case, the modeler cannot specify inputs and must simply accept whatever I/O data are available.

To sum up, identification consists of the selection of a specific model in a specified class on the basis of observations performed on the system to be described and a selection criterion. The procedure makes no reference to either the physical nature of the modeled system or the *a priori* knowledge of the modeler. Only the data speak.

The parameters of models based on identification may lack physical meaning; obviously, this will also be true for the models' parameters. Nevertheless, such models can be both simple and accurate. In the end, we may extract relevant information from complex frameworks.

Physical laws are often obtained from identification procedures; the data collected by Galileo in his experiments on falling bodies led him to see that a simple model could explain all his experiments and could be considered a law (University of Cauca, 2009).

1.1.5 Modeling and Simulation

A model is a workable substitute for a system. It is sometimes called a *real system* because it can simulate and then solve problems in the original system. The following section discusses the basic concepts of modeling and simulation. Figure 1.5 shows these concepts, as they were introduced by Zeigler (1984).

- *Object* is an entity in the real world. Depending on the context in which it is studied and the aspects of behavior under study, the object's behavior can be highly variable.

- *Base model* is an abstract, hypothetical representation of an object's properties, especially its behavior. It describes all possible facets of the object in all possible contexts. A base model is hypothetical, as we will never—in practice—be able to construct/represent a "total" model.

- *System* is a well-defined object in the real world under specific conditions, considering certain specific aspects of its structure and behavior.

- *Experimental frame:* For systems in the real world, the EF describes experimental conditions (context) or aspects within which that system and corresponding models will be used. As such, the EF reflects the objectives of the experimenter who performs experiments either on a real system or, through simulation, on a model.

In its most basic form (see Figure 1.6), an EF has two sets of variables: input and output variables. These variables match the system or model terminals. An EF also has a *generator* and a *transducer*.

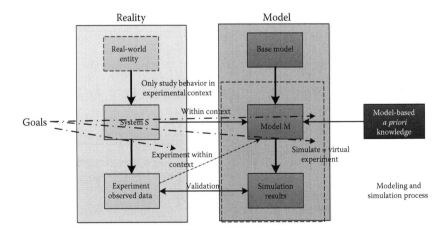

FIGURE 1.5
Modeling and simulation.

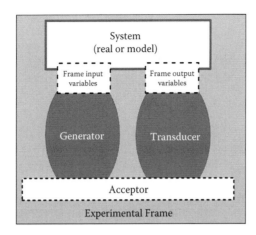

FIGURE 1.6
System versus EF.

Generator refers to the input or stimulus acting on a real system or model during an experiment; for example, a unit step could be a stimulus. Transducer refers to the transformation in a system's outputs during either experimentation or modeling. For example, a transducer may point to the values of some of the output variables. [*Note*: Here *output* refers to both physical system outputs and synthetic outputs observed in a model, such as state variables or parameters (University of Cauca, 2009).]

As well as I/O variables, a generator and a transducer, an EF may also include an *acceptor*, which compares the features of generator inputs with the features of the transduced outputs and determines whether the system (real or model) "fits" into this EF and meets the experimenter's objectives.

- *(Lumped) Model* accurately describes a system within the context of a certain EF. Although seemingly counterintuitive, the term "accurate description" requires precise definition.

  More precisely, certain properties of the system's structure should be modeled using an acceptable range of accuracy. [*Note*: A lumped model is not necessarily a lumped parameter model. Given the myriad applications of both modeling and simulation, however, overlapping terminology is perhaps inevitable (Cellier, 1991).]

- *Experimentation* is the physical act of performing an experiment. As an experiment may interfere with system operation (by influencing its input and parameters), the experimentation environment is a system in its own right (and may be modeled by a lumped model). Experimentation involves observation and this yields measurements.

- *Simulation* of a lumped model uses certain formalisms (such as Petri net, bond graph, or differential–algebraic equations [DAE]) to

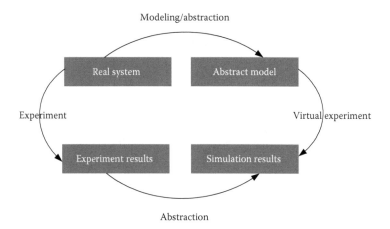

FIGURE 1.7
Modeling–simulation morphism.

produce results in the form of dynamic I/O behavior. Simulation, which mimics the real world, can be seen as virtual experimentation, allowing us to answer questions about (the behavior of) a system (Vangheluwe, 2001).

Crucial to the system experiment/model-virtual experiment scheme is a *homomorphic* relationship between model and system: building a model of a real system and then simulating its behavior should produce the same results as performing a real experiment, followed by observing and codifying the experimental results (see Figure 1.7).

A simulation model is a tool that we can use to achieve a goal (i.e., design, analysis, optimization). We must be able to accept with confidence any inferences drawn from modeling and simulation (Birta and Ozmizrak, 1996).

Confidence is guaranteed through verification and validation.

- *Verification* involves checking the consistency of a simulation with the lumped model from which it is derived. The transformation from an abstract representation (the conceptual model) to the program code (the simulation model) must be accurate, in that the program code must reflect the behavior of the conceptual model.

- *Validation* involves comparing the experimental measurements with the simulation results in the context of a particular EF.

A model cannot correspond to a real system if the comparison indicates differences. Moreover, even though the measurements and simulation results may match well, thereby increasing our confidence, this does not necessarily prove that the model is valid. Popper has introduced the concept of

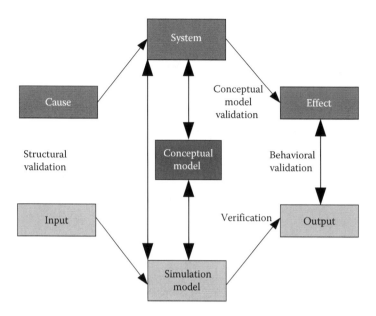

FIGURE 1.8
Verification and validation activities.

falsification (see Magee, 1985) to falsify or disprove a model, while methods of validation include conceptual model validation, structural validation, and behavioral validation.

- *Conceptual validation*: The evaluation of the realism of the conceptual model with respect to the goals of the study; alternatively, the evaluation of a conceptual model with respect to the system.
- *Structural validation*: The evaluation of the structure of a simulation model with respect to the perceived structure of the system.
- *Behavioral validation*: The evaluation of the simulation model behavior.

Figure 1.8 provides a general overview of verification and validation.

Note: The correspondence in behavior between a model and a system applies within the context of the EF. Thus, when we use models to exchange information, we must always match a model with an EF. Moreover, we should never develop a model without developing its EF (Vangheluwe, 2001).

1.1.6 Modeling and Simulation Process

To fully understand how an enterprise operates, we must analyze the processes of its various activities. What entities are involved? What are the causal relationships determining activity order and concurrency? As these

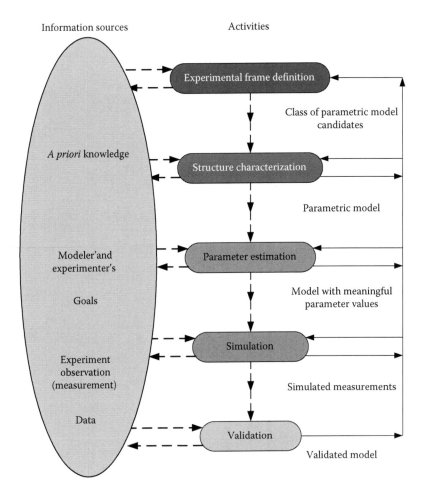

FIGURE 1.9
Model-based systems analysis.

questions suggest, simulation is part of a larger model-based systems analysis. Figure 1.9 presents a basic view of this type of analysis.

The example of a simple mass–spring experiment can illustrate the process. In Figure 1.10, we see a mass sliding without friction over a horizontal surface connected via a spring to a wall; the mass is pulled away from the rest position and let go.

A number of sources of information, whether *explicit* in the form of data/ model/knowledge bases or *implicit* in the mind of user, are used during the process:

1. *A priori knowledge*: In deductive modeling, we start from general principles such as energy, mass, momentum conservation laws,

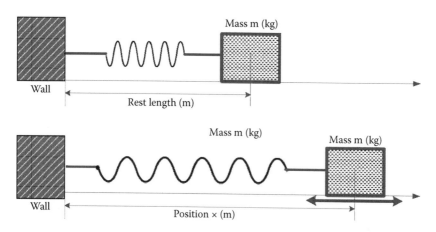

FIGURE 1.10
Mass–spring example.

constraints, and we use these to deduce specific information. Deduction is used during system design.

In the example, *a priori* knowledge is composed of the second law of Newton of movement, along with our knowledge about the behavior of an ideal spring.

2. *Objectives and intentions*: Methods employed, formalisms used, level of abstraction, and so on, are all determined by the type of questions we want to answer.

   In the example, possible questions are: "What is the spring constant?" "What is a suitable model for the behavior of a spring for which we have position measurements?" "Given performance criteria, how is an optimal spring built?" "Taking into account a suitable model and initial conditions, can we predict the behavior of spring?"

3. *Measurement data*: In inductive modeling, we start from data and try to extract structure.

   In turn, the resulting model can be used in a deductive fashion. Such iterative progression is typical of systems analysis.

Continuing with the example, the noisy measured position of the mass as a function of time is plotted in Figure 1.11.

The process begins when we identify an EF. Simply stated, the frame represents the experimental conditions with which we wish to investigate a system. Thus, it expresses our goals and our questions. As noted above, at the most basic level, the EF consists of a generator describing possible system inputs, a transducer describing outputs, and an acceptor describing the conditions which the system matches.

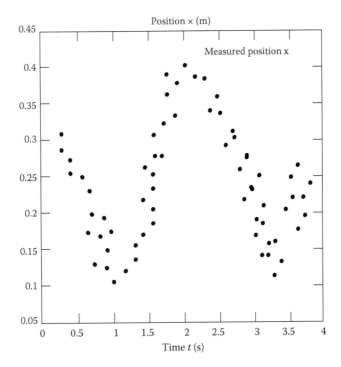

FIGURE 1.11
Measurement data.

In our example, the EF could choose to specify that the deviation from the rest position should never be greater than the rest length. It could also specify the environmental factors, such as room temperature or humidity, if those were relevant.

A class of matching models can be identified on the basis of a frame. The appropriate model can be selected using *a priori* knowledge and measurement data. In our example of the spring, a feature of an ideal spring (one connected to a frictionless mass) is that the position amplitude will remain constant. However, in a spring that is not ideal or in the case of friction, the amplitude will decrease over time. If we rely on the measured data, we will conclude that this is an ideal spring, and during model calibration, we will estimate the optimal parameters to produce a set of measurement data.

1.1.7 Simulation Model

From the model, we can build a simulator. Given the contradictory aims of modeling (meaningful model representation for understanding and reuse) and simulation (accuracy and speed), this process may involve many steps.

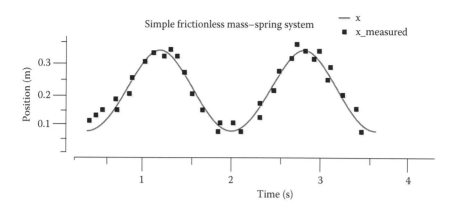

FIGURE 1.12
Fitted simulation results.

Figure 1.12 shows that by using the model and the identified parameters, simulation allows one to mimic a system's behavior (virtual experimentation). The resulting simulator can be embedded in optimizers, trainers, or tutoring tools.

Even if a model has predictive validity, a question remains: Is it possible to reproduce not only the data used to choose the model and parameter identification, but also to predict new behavior? This question should be asked to every user of the simulator.

1.2 System Identification Problem

The term identification was introduced by Zadeh (1956, p. 1) as a generic expression for the problem of "determining the I/O relationships of a black box by experimental means." He cites various terminologies that are prevalent for the same problem, such as "characterization," "measurement," "evaluation," "Gedanken experiments," and so on, and notes that the term "identification" states "the crux of the problem with greater clarity than the more standard terms" (Gaines, 1978).

Zadeh (1956) formulates the general identification problem as

1. A black box, x, whose I/O relationship is not known *a priori*
2. The input space of x
3. A class of models for such black boxes, M, which on the basis of *a priori* information about x is known to contain a model for it. By observing the response of x to various inputs, this class can determine a member of M, which is equivalent to x, in the sense that its

responses to all time functions in the input space of x are identical to those of x

The identification problem has become an essential area of study in modern control theory. The major effort in control research has tended to be with systems modeled as linear and continuous in their state variables and either continuous or uniformly sampled with respect to time. Such work has found a wide range of practical applications in plant measurement and to a lesser but still significant extent in online adaptive control.

Techniques based on linearity and continuity usually begin to break down when applied to nonartificial systems, for example, biological and economic modeling, but significant practical use of the linear "describing function" has been made, for example, in the inclusion of the pilot in aircraft design. However, human motor control is known to be based on discontinuous decision making leading to discrete corrections, rather than the smooth, linear motion of classical servomechanisms. As the extremes of biological systems behavior are approached, for example, in animal ethological studies, where the data are often purely descriptive with no metrical structure, linear systems techniques become inapplicable. Here, the times series to be modeled are strings of arbitrary symbols, and we have crossed into the domain of automata theory and the problems of *grammatical inference* (Fu and Booth, 1975).

It is interesting to note that Zadeh recognized this spectrum of problems in his discussion on system "identification" some 50 years ago, although the main part of his paper was concerned with continuous system identification.

1.2.1 Key Features of Identification Problem

Zadeh's definition given above forms a convenient framework for the general problem of identification. It already exhibits two key features of the problem (Gaines, 1978):

1. *The class of possible models must be determined in advance.* A basic conflict lies at the heart of the problem of identification, that is, between pure epistemology on the one hand (knowledge is the raw material of our experience and is prior to all "metaphysical" speculation about being) and ontology on the other (we have *a priori* reasons to suppose the world has a certain nature independent of our knowledge of it).

   In general systems theory, we cannot avoid operating in this region of conflict. The ontological approach to many questions is very satisfying: we can hypothesize and create structures that are adequate, complete, and consistent. Yet invariably, the question will arise as to whether this comforting sense of closure is "real."

2. *Identification is an active process of testing hypotheses.* This implies interaction with a system—not a passive process of data acquisition and modeling. Much research neglects the role of action in data

acquisition; we are only concerned with the model that best fits the data when, very often, the conclusion should be that there is inherent ambiguity in the results. No best model is determined, only a class of models, selection among which requires further data. Often, specific potential exemplars of behaviors may be indicated whose existence, or nonexistence, is the key to a separation between possible models.

What Zadeh's original definition did not attempt to cover are the additional key features:

3. *Identification is not carried out in the abstract but generally for a purpose.* The purpose of identification is an essential part of the identification problem. For example, we frequently evaluate identification in terms of *prediction*, but in complex systems we rarely need or can achieve prediction of all aspects. We often use identification in order to *control*, rather than out of pure scientific interest, and many aspects of prediction may be irrelevant.

4. *Identification may conflict with other objectives.* In terms of the preceding discussion, it is clear that the various requirements of prediction may be in conflict. An identification scheme that is optimal for one class of prediction may not be suboptimal for another, but may actually be in conflict with it. However, there are deeper problems when, for example, the purpose of identification is control; the simplest illustration is the classical "two-armed bandit" problem in which the gains of knowledge acquisition must be set against the costs of suboptimal control.

5. *The identification problem as stated may have no well-defined solution.* The requirement for a solution is a model whose responses are "identical" to those observed. For nondeterministic systems, such identity is not meaningful, and since "noise" is significant in most real-world systems, practical applications of identification generally have to allow for nondeterminism; we may then talk in terms of the degree to which the model *approximates* the observed behavior, but this is now an order relation rather than a unique classification.

6. *The identification problem may have a number of possible solutions, the choice amongst which is dependent on other factors.* Even when the system is deterministic, there may be several models whose responses are identical to those observed. Generally, all models will not be of equal status and there will be a preference ordering among them such that if two are of equal validity, one is preferred to the other; it is convenient to call this preference ordering one of *simplicity*, or its converse, of *complexity*.

These six aspects of the identification problem take it out of the realm of passive data analysis and point to its rich philosophical foundations, requiring more subtle formulation in system theoretic terms than might be expected (Gaines, 1978).

1.2.2 Identification Steps

System identification refers to the problem of building mathematical models of dynamic systems based on measurements of I/O. The approaches to identification as described in the *Concise Encyclopaedia of Modeling & Simulation* are quite general (Atherton and Borne, 1992). They consist of the following (Welden and Danny, 1999):

- Collecting I/O system data (data usually recorded by sampling in discrete-time sampling)
- Settling on a set of candidate models (or model/type of paradigm)
- Choosing a particular member of the model set as the most representative, guided by the information in the data

From a logical standpoint, the identification of a system may be divided into the following steps (Guidorzi, 2003):

1. *Experimental design (observing the system)*: The result of the identification process can be no better than a correspondence to the information contained in the data. The experimental design covers the choice of inputs to be made, presampling filters, sampling rates, and so on, to yield the most informative data. Experimental design also considers the goal, the data, and the *a priori* information (see Figure 1.13).

   Identification exercises are intended to replace data collection with data-generating mechanisms (models). Since this procedure is entirely based on the information found in the data, it cannot work if the data are faulty. The first step of identification is to collect observations of the process variables and apply these to the system inputs for identification purposes.

2. *Selecting a model set*: It is often difficult to justify the choice of a model set or model paradigm. A specific model can be selected, on the basis of observations, from an assumed family of models. In practice, many candidate models are tried out and the process of identification becomes the process of evaluating and choosing between them.

   A priori knowledge and intuition or vision must be combined with the formal properties of a model and identification methods if we are to have good results.

3. *Choosing a selection criterion:* The choice of criterion of fit affects the method of evaluating the quality of a particular model. Any model, no

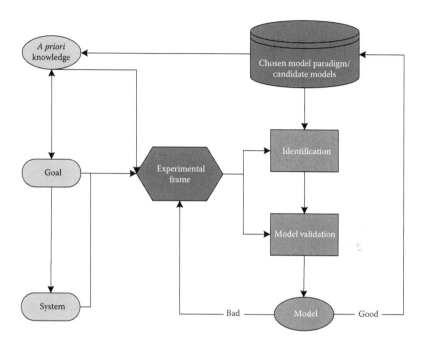

FIGURE 1.13
System identification.

matter how good it is, can only approximate the behavior of the real system, and the intended use of the model will determine the selection criteria. Many different models can be created from the same data, depending on the projected use.

4. *Computing model parameters*: Calculating the parameters of a model is an optimization problem (selecting the "best" model in the relevant class) or a way of "tuning" a model on the data.

5. *Validating the model*: Model validation is the process of examining the model, assessing its quality and possibly rejecting its use for the purpose in question. In a sense, this can be viewed as the "essential process of identification" (Lennart, 1994, p. 5). The estimation phase of model parameters is really just a means to provide candidate models that can be validated.

Model validation uses the criterion of fit to determine whether a model is good enough. Essentially, validation tries to falsify the model using collected data that differ from those data used during identification. If the model remains unfalsified, it is considered as validated and suitable for use.

These choices can always involve some *a priori* knowledge of the system, such as the choice of the class of models and/or the design of the experiments to collect inputs and outputs.

If the model is not validated, we must reconsider the choices made and start the identification process from the beginning, or at the very least, from an intermediate step (University of Cauca, 2009).

1.2.3 Identifiability

Once a class of models and a penalty function for the model behavior have been selected, identification is reduced, for any set of available data, to an optimization problem whose solution consists of selecting the model associated with the minimal value of the penalty function.

The problem is well-defined if and only if the range of the penalty function contains a single absolute minimum; in this case, the considered process is *identifiable* (Guidorzi, 2003).

Identifiability derives from the class of models selected from the penalty function and from the data, not from the system to be identified.

1.2.4 Classes of Models for Identification

The essence of classification is grouping *equivalence classes* according to a number of criteria. A classification may help in choosing the most appropriate formalism when modeling a system. This, in turn, may help in the selection of the most appropriate modeling and simulation tool.

Many different classes of models can be considered. The most relevant, from an identification standpoint, are the following (Guidorzi, 2003):

- Oriented and nonoriented
- Static and dynamic
- Causal and noncausal
- Lumped and distributed
- Constant and time varying
- Linear and nonlinear
- Deterministic and stochastic
- Single-input single-output (SISO), multi-input multi-output (MIMO)
- Parametric and nonparametric
- Continuous and discrete time
- Continuous and discrete event

1.2.4.1 Oriented and Nonoriented Models

Once the measurable attributes of a system have been defined, it is usual to partition them into two classes, inputs and outputs, or as in econometrics, exogenous and endogenous variables. The inputs can often be seen as the action of the surrounding environment on the system and the outputs as the system's reaction. It is also possible to describe the outputs as the variables explained by the model and the inputs as those left unexplained by the model.

In some cases, it can be desirable to avoid any *a priori* orientation of the system and treat all variables in a symmetric way. It is important to note that the orientation can be imposed by the external environment instead of by the system itself.

1.2.4.2 Static and Dynamic Models

Algebraic or static systems establish an instantaneous link between their variables and are described by sets of algebraic equations. Dynamic systems establish a link between the values assumed by their attributes at different times and are described by sets of differential or difference equations. Algebraic systems can be treated in a comparatively simple way and may describe a very limited slice of the real world; system identification is, thus, almost implicitly considered as dynamic and refers to dynamic models. The identification of models for algebraic processes can, however, be trickier than the commonly assumed conditions of nonidentifiability would suggest.

1.2.4.3 Causal and Noncausal Models

An oriented model is defined as causal when its output at time t is not affected by future input values. While all real systems are causal and can be properly described by models of this kind, it is also possible, from a mathematical standpoint, to introduce noncausal models.

1.2.4.4 Purely Dynamic and Nonpurely Dynamic Models

An oriented dynamic model is defined as purely dynamic when its input at time t does not affect its output at t, that is, when the system does not establish any instantaneous (algebraic) link between its input and output. If this condition is not satisfied, the model is defined as nonpurely dynamic. This property interacts with other properties and with the planned use of the model.

Thus, a nonoriented model is necessarily nonpurely dynamic because a purely dynamic model is intrinsically oriented and would even become noncausal for other orientations.

In contrast, a predictive model must be purely dynamic; considering discrete systems, the output at time $t + 1$ must be predicted at time t based on measurement taken only until that time.

1.2.4.5 Lumped and Distributed Models

Most aspects of reality are concerned with the phenomena that do not occur at a single point in space but in areas or volumes (e.g., heat transmission, electromagnetic phenomena, energy exchanges, mechanical systems). Such phenomena require distributed models, given by sets of partial differential (or difference) equations for dynamic systems. Lumped models, given by sets of ordinary differential (difference) equations, refer to simplified schemes that assume constant values for the system attributes in some properly defined space regions.

1.2.4.6 Constant and Time-Varying Models

Time-varying models can describe systems whose behavior changes over time; their parameters are generally functions of time. Time-invariant models feature sets of constant parameters and can describe constant systems. Time-invariant systems can appear time varying if there is a lack of knowledge on some of their inputs.

1.2.4.7 Linear and Nonlinear Models

Linear models describe systems where the superposition principle is valid. Most real systems are nonlinear but can be described quite accurately with a linear model near a working condition.

1.2.4.8 Deterministic and Stochastic Models

Real systems are always affected by disturbances (noise entering the system and/or affecting the measures, unknown inputs, quantization errors, etc.). These disturbances or their global effect can be described as noise acting on the input, state, and output of the model, which, in this case, is stochastic.

Often the global effect of disturbances is modeled as the output of a filter driven by white noise; when it is added to the output of the deterministic part of the model, the model is decomposed into a deterministic and a stochastic part. Depending on the application, it may be sufficient to identify the deterministic part of the model (e.g., diagnosis) or it may be necessary to identify both parts (e.g., prediction).

1.2.4.9 SISO and MIMO Models

SISO, multi-input single-output (MISO), and MIMO models are self-explanatory. While SISO models have limited usefulness (the world is multivariable), a multivariable model can be decomposed into a collection

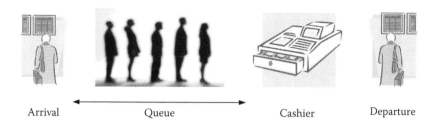

Arrival   Queue   Cashier   Departure

FIGURE 1.14
A discrete event system: A queuing system.

of MISO models. An approach of this kind has many conceptual and practical limits and is frequently followed to avoid the more complex tools required by truly multivariable models.

1.2.4.10 Parametric and Nonparametric Models

Some classes of models are given by sets of equations described by a certain number of parameters (parametric models), while other models are given without assigning parameters; for instance, in a graphical form (e.g., impulse or step responses and frequency responses for linear systems).

1.2.4.11 Continuous and Discrete-Time Models

Continuous models describe systems whose measurable attributes evolve with continuity in time while discrete models establish quantitative links between the values assumed by the variables at discrete (sampling) times. While the intrinsic nature of all natural systems and many technological systems is continuous, the widespread introduction of digital systems requires the use of discrete models that can describe continuous systems accurately when the variables are properly sampled (see Figure 1.14).

1.2.4.12 Continuous and Discrete-Event Models

Models belong to the discrete-event category when the time base is continuous and a finite number of events occur in a bounded time interval; only at those event times does the discrete state of the system change.

1.2.4.13 Free and Nonfree Models

In some instances, the environment does not act on a system, or at least, no action can be observed. Systems and models lacking input are called free systems and models; their outputs are termed time series.

1.3 Introduction to the Concept of Diagnostics

1.3.1 Meaning and Impact of Diagnostics

The word diagnostics comes from the medical field and refers to the identification of the nature of a health problem and its classification by examination and evaluation. Diagnostic criteria are the combination of signs, symptoms, and test results to be used in an attempt to define a problem. Diagnosis is the result of a diagnostics process (Czichos, 2013). Diagnostics is the examination of symptoms and syndromes to determine the nature of faults or failures of technical objects (ISO 13372, 2004).

- A symptom is a perception, made by means of human observations and measurements, which may indicate the presence of an abnormal condition with a certain probability.
- A syndrome is a group of symptoms that collectively indicate or characterize an abnormal condition.

The terms "fault" and "failure" are defined as follows (ISO 13372, 2004):

- *Fault*: The condition of an item that occurs when one of its components or assemblies degrades or exhibits abnormal behavior.
- *Failure*: The termination of the ability of an item to perform a required function. (Failure is an event, as distinguished from fault, which is a state.)

The failure mode is the phenomenon by which a failure is observed. After a failure, the systematic examination of an item to identify the failure mode and determine the failure mechanism and its basic cause is called root cause failure analysis.

The identification of faults and failures is an important task of technical diagnostics. Numerous failure characteristics are defined in international standards in various areas of technology and industry. To be detected by technical diagnostics, a failure can include the following (Czichos, 2013):

- Termination of the ability of a structure to perform its required function when one or more of its components is in a defective condition, either at a service or ultimate limit state, for example, mechanical vibration and shock (ISO 16587).
- Loss of the ability of a building or its parts (i.e., constructed assets) to perform a specified function (ISO 15686).
- Premature malfunction or breakdown of a function or a component or the whole engine, for example, internal combustion engines (ISO 2710).

- Sudden and unexpected ending of the ability of a component or equipment to fulfill its function, for example, gas turbines (ISO 3977).
- Actual condition of an item which does not perform its specified function under the specified condition, for example, earth-moving machinery (ISO 8927).
- An event causing the loss or reduction of nominal serviceability, for example, cranes (ISO 11994).
- Loss of structural integrity and/or transmission of fluid through the wall of a component or a joint, for example, petroleum and natural gas industries (ISO 14692).
- A state at which a component reaches its threshold level or terminates its ability to perform a required function, for example, hydraulic fluid power (ISO/TR 19972).
- Occurrence of bursting, leaking, weeping, or pressure loss, for example, plastics piping systems (ISO 7509).
- Any leakage or joint separation, unless otherwise determined, may be due to a pipe or fitting defect, for example, ships and marine technology (ISO 15837).
- Termination of the ability of an item to perform a required function, for example, space systems (ISO 14620).
- A system state that results in nonperformance or impaired performance as a result of a hardware or software malfunction, for example, road vehicles (ISO 17287).
- Insufficient load-bearing capacity or inadequate serviceability of a structure or structural element, for example, reliability for structures (ISO 2394).

These examples from industry and technology show that various faults and failures may detrimentally influence technical items. Damage identification by technical diagnostics generally considers four basic aspects:

- Existence of damage
- Damage location
- Damage type
- Damage severity

The probability that a technical item will perform its required functions without failure for a specified time period (lifetime) when used under specified conditions is called reliability (Hanselka and Nuffer, 2011). Risk is the combination of the probability of an event and its consequence (ISO Guide 73, 2002). The term "risk" is generally used only when there is at least the possibility of negative consequences. Safety is freedom from unacceptable risk

(ISO Guide 51, 1999). For the selection and application of technical diagnostic methods, the character of the item under consideration and the length and timescales associated with damage initiation and evolution must be considered. The length scale of technical items, subject to the application of technical diagnostics, can range across more than 10 dimensional decades. Today, technical diagnostics, conventionally applied to macrotechnology, must also consider microtechnology and nanotechnology. Consequently, reliability considerations have to be extended to micro–nano reliability.

The following aspects are of general importance to the timescales of the occurrence of faults and failures in technical items:

- The fault progression time, indicating the change in severity of a fault over time.
- The duration of a failure event; this may be very short (e.g., brittle fracture) or may extend over a long period of time (e.g., loading time until fatigue failure occurs). A catastrophic failure is a sudden, unexpected failure of an item resulting in considerable damage to the item and/or its associated components.
- The time-to-failure is the total operating time of an item from the instant it is first put in operation until failure, or from the instant of restoration until next failure. The detection and collection of information and data indicating the state of an item constitute condition monitoring (CM).

The application of CM to technical structures and systems allows actions to be taken in order to avoid the consequences of failure, before the failure occurs. The process of CM consists of the following main phases (Czichos, 2013):

- Detection of problems, that is, deviations from normal conditions
- Diagnosis of the faults and their causes
- Prognosis of future fault progression
- Recommendation of actions

1.3.2 Concepts, Methods, and Techniques of Diagnostics

The basic methods of technical diagnostics are structural health monitoring (SHM) and nondestructive evaluation (NDE), in combination with inductive and deductive concepts (Vesely, 2002):

- The inductive conceptual approach consists of assuming particular failed states for components and then analyzing the effects on the system. Inductive approaches start with a possible basic cause and go on to analyze the resulting effects. A basic inductive method is failure modes and effects analysis (FMEA).

- The deductive conceptual approach postulates that the system itself has failed in a certain way; an attempt is made to determine what modes of system or subsystem (component) behaviors contribute to this failure. A basic deductive method is the fault tree analysis (FTA).

1.3.2.1 Failure Modes and Effects Analysis

FMEA is a structured procedure to determine equipment and functional failures, with each failure mode assessed as to the cause of the failure and the effects of the failure on the system. Failure modes are any errors or defects in a process, design, or item, and can be potential or actual. The technique may be applied to a new system based on analysis or an existing system based on historical data. The FMEA method helps in identifying potential failure modes based on past experiences with similar products or processes.

An FMEA can identify, with reasonable certainty, those component failures with "noncritical" effects, but the number of possible component failure modes that can realistically be considered is limited. The objectives of this analysis are to identify single failure modes and to quantify these modes. An FMEA table for a component of a system contains the following information (Vesely, 2002):

- Component designation
- Failure probability
- Component failure modes
- Percent of total failures attributable to each mode
- Effects on overall system, classified into various categories; the two simplest are "critical" and "noncritical." Effects analysis refers to studying the consequences of those failures

1.3.2.2 Fault Tree Analysis

FTA attempts to model and analyze failure processes (Vesely, 2002). As a deductive approach, FTA starts with an undesired event, such as the failure of an engine, and then determines (deduces) its failure causes using a systematic, backward-stepping process.

- A fault tree (FT) is constructed as a logical illustration of the events and their relationships that are necessary and sufficient to result in the undesired event. FTA uses graphical design techniques to construct these diagrams. To do so, it looks at all types of events: including hardware problems and material failure or malfunction. It also considers combinations of factors contributing to an event. Ultimately, failure rates are derived from well-substantiated historical data, including mean time between failure of the component, unit, sub-system, and/or function.

- A success tree (ST) is the logical complement into which an FT can be transformed (Vesely, 2002). An ST shows the specific ways the undesired event can be prevented from occurring. It provides conditions that, if assured, guarantee the undesired event will not occur.

1.3.2.3 Structural Health Monitoring

SHM is the process of implementing a damage detection system for engineering structures (ISIS Canada, 2001). The objective of SHM is to monitor the *in situ* behavior of a structure accurately and efficiently, to assess its performance under various service loads, to detect damage or deterioration, and to determine the health or condition of the structure. SHM observes a system over time using periodically sampled dynamic response measurements from a set of sensors. It extracts damage-sensitive features from these measurements, and performs a statistical analysis to determine current system's health. The SHM system should be able to provide, on demand, reliable information on the safety and integrity of a structure. The information can then be incorporated into maintenance and management strategies and used to improve design guidelines. If it is immediate and sensitive, SHM can allow short-term verification of innovative designs, early detection of problems, avoidance of catastrophic failures, effective allocation of resources, and reduced service disruptions and maintenance costs. The physical diagnostic tool of SHM is the comprehensive integration of various sensing devices and auxiliary systems, including a sensor system, a data acquisition system, a data processing system, a communication system, and a damage detection and modeling system.

On the basis of the extensive literature now available on SHM, we can argue with some confidence that this field has matured to the point where we now have fundamental axioms and general principles. The axioms include the following (Worden et al., 2007):

- All materials have inherent flaws or defects; according to materials science, the following lattice defects (deviations of an ideal crystal structure) can be distinguished: (1) point defects or missing atoms: vacancies, interstitial, or substituted atoms; (2) line defects or rows of missing atoms: dislocations; (3) area defects: grain boundaries, phase boundaries, twins; and (4) volume defects: cavities, precipitates.
- Damage assessment requires comparing the status of two systems.
- Identifying damage and its location can be done in an unsupervised learning mode; identifying the type of damage and its severity usually requires a supervised learning mode.

- Sensors are unable to measure damage; feature extraction using signal processing and statistical classification is required for sensor data to be converted into information on damage.
- Unless we use intelligent feature extraction, the more sensitive a measurement is to damage, the more sensitive it will be to changing conditions, both operational and environmental.
- The required properties of the SHM sensing system rest on the length and time scales of damage initiation and its evolution.
- There is a trade-off between an algorithm's sensitivity to damage and its noise rejection capability.
- The output of long-term SHM is periodically updated with information on the ability of the structure to perform its intended function over time and with information on the degradation resulting from an operating environment.

1.3.2.4 Nondestructive Evaluation

NDE is the umbrella term for noninvasive methods of testing, evaluation, and characterization based on physical principles of sensing and assessment. NDE is an important method for performance control and CM. In engineering systems, flaws, especially cracks in the materials of structural systems' components, can be very detrimental. For this reason, the detection of defects and macro/micro/nano root cause analysis are essential elements of the quality control of engineering structures and systems and their safe and successful use.

The established NDE methods for technical diagnostics include (Erhard, 2011): radiography, ultrasound, eddy current, magnetic particle, liquid penetration, thermography, and visual inspection techniques. With the rapid advances in sensors, instrumentation, and robotics, coupled with the development of new materials and reduced margins of safety through stringent codal specifications, NDE has diversified with a broad spectrum of methods and techniques now being used for technical diagnostics in plants and structures. Industrial applications of sensors and noninvasive NDE methods are as wide-ranging as the technologies themselves and include mechanical engineering, aerospace, civil engineering, oil industry, electric power industry, and so on. The operation of NDE techniques in many industries has now become a standard practice, for example, to support CM for the proper functioning of the daily use of electricity, gas, or liquids in which pressure vessels or pipes are employed and where the correct operation of components under applied stress plays a big role in safety and reliability.

An overview of the concepts, methods, and techniques of technical diagnostics is given in Figure 1.15.

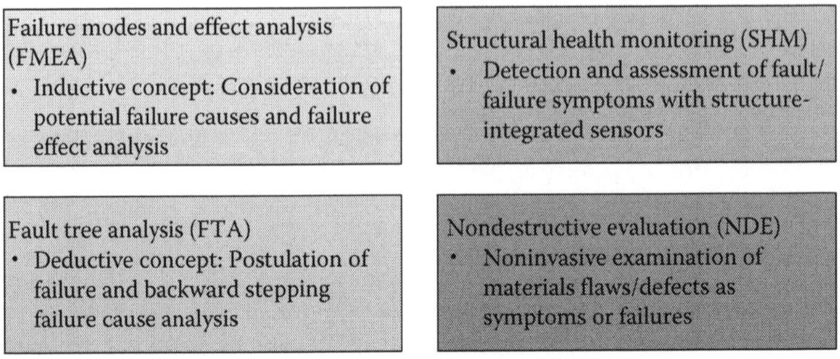

Failure modes and effect analysis (FMEA)
• Inductive concept: Consideration of potential failure causes and failure effect analysis
Structural health monitoring (SHM)
• Detection and assessment of fault/failure symptoms with structure-integrated sensors
Fault tree analysis (FTA)
• Deductive concept: Postulation of failure and backward stepping failure cause analysis
Nondestructive evaluation (NDE)
• Noninvasive examination of materials flaws/defects as symptoms or failures

FIGURE 1.15
Overview of the concepts, methods, and techniques of technical diagnostics.

1.3.3 Application of Technical Diagnostics

Technical diagnostics can be illustrated by the lifecycle of all man-made technical items: from raw materials to engineering materials and via design and production to structures and systems, and finally, to deposition or recycling; see Figure 1.16 (Czichos, 2013).

Technical diagnostics can be applied in almost all areas of technology and industry to ensure product quality, economical and efficient processes and, most importantly, to assure safety and reliability. Indeed, the economical–technological development of industrialized countries is affected by two major trends.

First, new technical products are lighter to increase speed, height, or distance, that is, to improve the global quality of life. This is achieved with the use of highly innovative materials, multimaterial designs, nanostructuring materials, and additional concepts for lightweight design. The following

Raw materials
Processing
Engineering materials
Design
Matter
Recycling
Structures and systems
The earth
Deposition
Technical objects

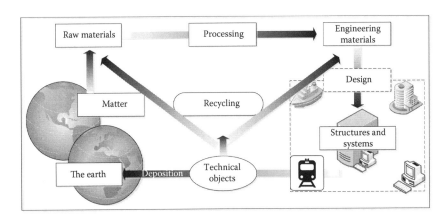

FIGURE 1.16
The product cycle of technical items and the potential of technical diagnostics.

aspects have to be considered in new technical products and their applied materials:

a. Materials are frequently designed for a single application purpose and, thus, material behavior and properties cannot be always transferred to other technical products. Nanostructured materials may be subjected to modern production and manufacturing processes that could significantly change their properties before the component is brought into service, for example, subjecting metallurgically nanostructured steels to laser hybrid welding in thick-walled pipeline sections.

b. Safety factors are frequently reduced to achieve the application goals; the materials in such components might be loaded to capacity.

c. The character of the loads may change, because materials in modern technical components are generally subjected to coupled loading. This means different types of loads are acting at the same time in the same process zone on a material in a specifically designed component. For instance, hip endoprostheses in the human body environment are subjected to corrosion fatigue, that is, to the corrosion process caused by the hip and leg bones and the dynamic mechanical load introduced by the leg movement acting simultaneously at the adjustable shaft. However, conventional threshold values have usually been determined for a single type of load.

d. The interaction and the load transfer capabilities from one material type to another in multimaterial-designed components are sometimes not fully understood, due to a lack of practical experience and a lack of scientific knowledge. This particularly applies to components containing joints between polymeric and metallic materials.

Second, the service lifetime of existing technical products is frequently extended over the original intended usage period, for instance, to save replacement costs. This is especially important with large-scale products belonging to the technical infrastructure, such as power stations, transportation systems, buildings, and so on, and is often achieved by lifetime elongation assessments, assisted by SHM, risk-based inspection, and preventive repair and restoration of critical parts.

The following aspects have to be considered for existing technical products and applied materials:

a. The material behavior within the later parts of the product life is not always completely understood, thereby precluding consistent forecasts of further safe use. This is especially concerned with components loaded mechanically at high frequencies within the giga-cycle range.

b. The loads during the elongated service life part might become significantly different, in particular, if the service purpose is changed.

c. Hazardous operations during service affecting the material properties are frequently not recorded, but can tremendously affect the residual loading capacities for lifetime extension.

d. Any repair can introduce additional loads and material property changes that cannot always be foreseen, but can significantly reduce the lifetime.

1.3.4 Management of Failure Analysis

While the design of technical products to ensure undisturbed function and operation for the whole lifecycle is an important topic in the field of component safety, the above-mentioned aspects increase the failure potential.

In the case of a failure event, engineers need reliable concepts and procedures so they can conduct failure analyses and make a fast and precise identification of the failure origins and the measures required for future failure avoidance. However, as outlined in the system approach, failure is defined as any disturbance or reduction of the function of a technical system. Thus, a failure does not necessarily lead to the breakdown or loss of a technical system or even to a technical catastrophe. For these, normally a sequence of events is necessary. This is becoming a topic of investigation during failure analysis (ISO 13373-1:(E), 2002).

Failure can pertain to one or more components (elements) of a technical system, especially in cases of faulty component interactions. Failures can occur during different parts of its lifecycle, that is, during production, service, or even during repair, replacement, or recycling of the various components.

1.4 Process of Diagnosis

The notion that computer technology can help solve problems has generated much interest, even enthusiasm; computers are now used in many fields of machine health, including maintenance. The technology continues to expand and evolve; in the past decade, a new branch of information technology has emerged, e-maintenance or maintenance informatics.

e-Maintenance is a branch of information technology (IT) or computer science that deals with information and software technology related to maintenance and machine health. e-Maintenance includes and is related to areas such as computers, computer science, IT, maintenance AI, maintenance guidelines, maintenance information, maintenance research data, maintenance technologies, maintenance processes and practices, maintenance terminology, maintenance law, and maintenance ethics. Diagnosis is the process of finding or identifying the cause and nature of a machine problem by carefully studying the symptoms and signs, evaluating history, and performing examinations. A diagnosis is the most important part of maintenance system, and a good diagnosis is vital for proper maintenance of decision support systems.

Diagnosis is still largely a manual process. It requires extensive knowledge, good interviewing skills, meticulous examination techniques, very high levels of analysis, and synthesizing skills. Unfortunately, all maintainers do not have the same level of expertise and skill; even more experienced ones sometimes fail to diagnose a condition correctly. Diagnostic mistakes are the major cause of maintenance errors. Therefore, any tool or technology with the potential to provide correct and timely maintenance diagnosis is worth serious consideration. Computer experts and scientists are now trying to develop a computer system that helps in complex diagnosis by eliminating human errors and misdiagnosis; such systems are called clinical diagnosis support systems (Douglas et al., 2005; Houghton and Gray, 2010).

Diagnosis process support systems are recent and important additions in determining machine's health. They guide maintainers in diagnosing conditions correctly and making the correct decisions. The systems analyze and process the machine data and make a diagnosis. This could be a multistage or a one-step process; the systems may request more data or further laboratory examinations based on the input.

Many research organizations and companies have developed clinical diagnosis support systems using a number of software and computer technologies, but the current generation of diagnosis support systems is both cumbersome and unsuccessful. Even though some companies claim their systems are more than 90% accurate, in our investigation we found this to be a gross exaggeration.

There are various reasons for the failure of these systems. First, they are not mature enough to use as life-critical systems; second, analysts and developers do not adequately understand the clinical diagnosis process; and finally, although the current systems can mimic human reasoning, the maintainers lack faith in them.

The main aim of this book is to study the current diagnosis support systems and determine a better way to develop reliable diagnosis support systems, taking the following steps:

1. Analyze diagnosis support systems.
2. Investigate current clinical diagnosis support systems and their impact on machine health.
3. Find a better way of developing diagnosis support systems.
4. Predict the future of diagnosis support systems.

Diagnosis systems are computer-based programs that help a clinician diagnose conditions. As they can guide a maintainer for correct diagnosis, they have the potential to reduce the rate of diagnostic errors. As noted above, many research organizations and companies have developed clinical diagnosis support systems using various technologies and providing various levels of functionality. Even though the systems use different technologies, however, they all work similarly. Maintainers enter the condition finding, and the system processes the input and comes up with a probable diagnosis. These diagnostic support systems can also be used as a teaching aid for engineering students in their maintenance study. Surprisingly, given their range of applicability, most have not gained widespread acceptance in maintenance.

It is important to understand the diagnosis and the diagnostic process, as well as the human reasoning behind clinical diagnosis, before plunging into an in-depth analysis of the relevant support systems.

Simply stated, diagnosis is a process of finding and establishing the characteristics and type of a machine problem based on signs, symptoms, and laboratory findings. More formal definitions of diagnosis include:

1. "The placing of an interpretive, higher level label on a set of raw, more primitive observations" (Ball and Berner, 2006, p. 100).
2. "The process of determining by examination the nature and circumstances of a faulty condition" (Ball and Berner, 2006).

Diagnosis is a complex, loosely defined, and multistep process. It establishes what the problem is, when it started, how it has manifested, and how it has affected a machine's normal function. Diagnosis involves the following series of individual steps:

1. Taking a machine's history
2. Performing a physical examination and systemic examination
3. Analyzing the machine's data
4. Performing a differential diagnosis (DD) and provisional diagnosis
5. Carrying out further examinations, including laboratory examination
6. Confirming or refuting the diagnosis
7. Starting maintenance actions

The diagnosis process starts as soon as the maintainer sees the asset. He/she immediately makes a general assessment of the machine, based on appearance, status, and so on. Next, the maintainer engages with the machine, gets the machine's details, and starts taking the history. To make a diagnosis, experienced maintainers recognize symptom patterns. After obtaining the history, the maintainer enquires about the other systems of the machine; this uncovers symptoms that might have been ignored. After taking the history, the maintainer will have a DD in his/her mind. At this point, the maintainer will examine the asset, looking for signs confirming or refuting the diagnosis.

After obtaining the details and assembling the relevant information, the maintainer should be able to produce a provisional or a confirmatory diagnosis.

After completing the general and systemic examination, based on his/her diagnosis, if necessary, the maintainer sends the machine to the laboratory for examination to confirm or refute the diagnosis and to decide what type of therapy can be applied. Laboratory research in the area is becoming increasingly important because of advancements in laboratory investigation technology (CM & NDT) (Douglas et al., 2005; Houghton and Gray, 2010; Walter et al., 2007).

The steps in diagnosis may not be identical in every case. Two maintainers may take very different steps; in fact, the same maintainer may approach two similar cases differently. As expertise and skill vary among maintainers, different maintainers encounter different diagnostic problems even though they are diagnosing similar cases. Circumstances such as the availability of a laboratory investigation facility in the maintenance center or the emergency of the case change the steps of the diagnosis process as well (Ball and Berner, 2006; Douglas et al., 2005).

Understanding diagnosis processes and patterns is very important in developing diagnosis support systems. For example, studying clinicians' information requirements will help us understand variability in clinical diagnosis process among clinicians.

For the same machine, depending on the maintainer's expertise and knowledge, different engineers will highlight different diagnostic problems. Because of the varying knowledge and expertise of the maintainers, diagnoses vary depending on the circumstances. Humans also have many limitations, such as the number of things they can remember at one time. Diagnosis support systems do not have human limitations such as short-term memory and can help maintainers overcome the problems of manual diagnosis processes.

At the same time, diagnosis support systems have their own problems. They are developed by different research organizations and companies using various algorithms and techniques. Their usage and output are not always the same. Like any other information storage, maintenance and condition data are stored in various ways depending on the maintenance center. A busy maintenance crew member may write down a few things on a paper record, leaving the administrative staff to convert the comments into text and save the record in a computer system. In fact, some maintenance centers still use paper-based records. There are many maintenance information management systems in the market, with a wide range of functionality. Some only provide information management, while others have much more integrated functionality. If a diagnosis software system is integrated into the latter systems and if the machine's findings are already saved in a medical information management system, the maintainer does not need to retype findings to get a diagnosis.

Understanding the human reasoning behind diagnosis is equally important. Human diagnostic reasoning is not based on precise logic. In terms of diagnosis, 2 + 2 does not always equal 4, leading many doctors to think computers cannot be useful in clinical diagnosis. Diagnostic reasoning involves a number of diverse, complex, and related activities, including pattern recognition, making provisional judgments under given circumstances, compiling a patient's data, solving the problem through trial and error, making decisions under uncertainty, performing further research on the basis of previous work, and comparing and combining data. Sophisticated skills are required in this complex but loosely structured area, including good organization and in-depth knowledge.

Diagnostic reasoning is based on generic human psychological experiences. It has some similarities with but differs from the game of chess, meteorological judgments, crypto-arithmetic patterns, and so on; hence, such patterns have been compared to diagnostic reasoning.

The psychological experiments looking at judgments made under uncertainty show the reasoning behind an individual's imperfect and partially logical reasoning skills (Ball and Berner, 2006; Greenes, 2006). In fact, judgment under uncertain data is one of the most important and complex aspects of human reasoning in diagnosis. To study the complex clinical diagnosis process, research look at patterns of human behavior when they combine introspection with diagnostic procedures. Researchers follow the thinking process, including all activities from the beginning of the diagnostic process to the end and interpret the process in depth, looking at knowledge, skill, motive, reasoning, hypothesis, logic, and strategies. However, every maintainer's thinking process and reasoning are likely to be different, so the studies remain general (Ball and Berner, 2006; Elstein et al., 1978; Kassirer and Gorry, 1978).

The main elements of diagnostic reasoning include developing a working hypothesis, testing the hypothesis, getting and analyzing additional information, and either accepting the hypothesis, rejecting it, or adding a new hypothesis based on further analysis.

In medicine, a working hypothesis is first developed during the process of information gathering when few facts are known about a patient's case. Because of limited human memory and analytic capacity, fewer than five such hypotheses are developed simultaneously; these hypotheses are basically developed from machine recognition using existing experiences and knowledge. Experts are usually better able to apply the knowledge gathered than novices, and experts rarely use causal reasoning (Ball and Berner, 2006; Elstein et al., 1978; Kassirer and Gorry, 1978).

Pople (1982) has noted the similarities between diagnosis reasoning and Simon's criteria (reasoning for ill-structured problems). According to Simon, an ill-structured problem can be divided into well-defined small tasks that are more easily solved than one big ill-defined task. Studies show that in medicine, physicians employ hypothetic-deductive methods after early hypothesis generation; usually early hypothesis reasoning produces results when there is a very high possibility of correct diagnosis. Medical researchers Kassirer and Gorry (1978, p. 2) describe it as a "process of case building" where the hypotheses are evaluated against data on diseases using Bayes' law, Boolean algebra, or pattern matching (some diagnosis support systems have been developed using this principle). Pople (1982) has also observed that separating complex differential diagnoses into problem areas allows engineers to apply very powerful additional reasoning heuristics.

Engineers can assume that the DD list is within the problem domain and consists of the mutual exclusion hypothesis. They can also assume that if the list is extensive, a correct diagnosis is always within it, and anything out of that list is incorrect.

Kassirer has recognized three abstract categories of human diagnostic reasoning strategies: probabilistic, causal, and deterministic. Bayesian algorithm logic is based on the probabilistic reasoning strategy that computes clinical finding statistics using mathematical models and results in optimal decisions. Other studies have shown that humans are not very good natural statisticians because they solve problems mostly based on judgmental heuristics (Greenes, 2006; Kassirer and Gorry, 1978; Pople, 1982).

Overall, researchers have observed that human reasoning for diagnosis is very complex and have suggested a plethora of human diagnosis-reasoning models (Reddy, 2009).

1.5 History of Diagnosis

In the 1980s and 1990s, several advanced techniques were integrated into existing diagnosis software systems and models, and more mathematical rigor was added to the models. However, mathematical approaches have a downside: they are dependent on the quality of the data. Therefore, many of the new systems were based on fuzzy set theory and Bayesian belief networks logic to overcome the limitations of heuristic approaches in the old models (Ball and Berner, 2006).

With the advent of artificial neural networks and AI, developers and researchers are taking a completely new approach to diagnosis decision support systems. Even though a simple neural network may be similar to Bayesian probabilities logic, in neural networks, generally, the technology is very complex and many data are required to train the network. Use of artificial data to train the neural network may not be realistic and may affect its performance on real machine data.

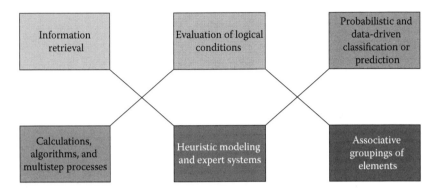

FIGURE 1.17
Methodologies and technologies for Collaborative Decision Support System (CDSS). Picture by Kaukuntla (author).

Important methodologies and technologies for decision support are information retrieval, evaluation of logical conditions, probabilistic and data-driven classification or prediction, heuristic modeling and expert systems, calculations, algorithms, and multistep processes, and associative groupings of elements. Figure 1.17 shows the methodologies and technologies for diagnosis.

- *Information retrieval*: Related basic information is sought for decision support systems. This is a basic search to retrieve maintenance data using keywords. Text search algorithms are also used to search for information in the machine's data (Greenes, 2006).

- *Evaluation of logical conditions*: This explores widely used logics for diagnosis support systems and various logical conditions. Decision tables refine and reduce the number of diagnostic possibilities, and Venn diagrams present clinical logic. Logical expressions with Boolean combinations of terms are used, along with comparison operators. Finally, alert reminders and other logical algorithms are developed (Greenes, 2006).

- *Probabilistic and data-driven classification or prediction*: As most decisions are precise, diagnosis support systems need to recognize the various types of maintenance data. Key developments are Bayes theorem based on a formula or decision theory, for example, whether a machine should be maintained or not. Data mining is a database technology to find hidden valuable data; an evidence-based system processes the data based on evidence. Artificial neural networks and AI are still evolving in the areas of rapid phase, belief networks, and meta-analysis (Ball and Berner, 2006; Combi et al., 2009; Greenes, 2006).

- *Heuristic modeling and expert systems*: These are used for diagnostic and therapeutic reasoning, given the uncertainty in human expertise. A key development is a rule-based systems model based on predefined rules and frame-based logic.

- *Algorithms, calculations, and multistep processes*: These are used in flowchart-based decision logic, interactive user interface control, computational process executions, imaging and image processing, and signaling. Key developments are workflow modeling and process flow, modeling languages, programming guidelines, and procedural and object-oriented concepts (Ball and Berner, 2006; Greenes, 2006; Taylor, 2006).

- *Associative groupings of elements*: This is used for structured reports, relational and structured data, order sets, presentations, business views, other specialized data views, and summaries. Key developments are business intelligence tools, document construction tools, report generators, document and report templates, document architectures, markup languages, tools, and ontology languages (Greenes, 2006).

Any successful software development requires many disciplined steps, and diagnosis support systems are no exception. Developers must start with a clear vision, not merely testing a single algorithm or a new technology. They should carefully define the scope and nature of the application and understand the manual process to be automated. They must know the limitations, including technical limitations, scope, boundaries, and data limitation, and ensure stakeholders are also aware.

Developers must analyze the requirements to determine the usage and scale of a proposed system. Algorithms must be studied in depth to find any possible condition in which they might fail. Developers and stakeholders must evaluate the automated system carefully outside the machine area; it should never be tried on functioning machines prematurely for safety reasons. It can be evaluated in the actual machine context, once its functionality is fully established and thoroughly evaluated. The practicality and use of the system must be demonstrated by the developers and analysts. They must show that it can be adopted by maintainers for productive daily use. No matter how wonderful the algorithm and technologies are, they are valueless if the system is not used by anyone.

Many critical phases impact both the development and the final product. These include understanding the diagnosis process, selecting the area, performing system analysis, developing and maintaining a knowledge base, developing algorithms and user interfaces, performing tests and quality control, testing user acceptance, and training.

Building a clinical diagnosis support system is a complex process; it requires good analytical skills, knowledge, and the ability to understand and follow a systemic procedure. Various methods, models, and approaches have been defined for software system analysis. In all cases, the functional requirements, technical requirements, and the nature of the problem need to be analyzed carefully, and a feasibility study has to be conducted. The cost-effectiveness must be analyzed and the required time effort should be estimated. All the alternative methods of building the system should be considered, and the knowledge base construction strategies and the data structure must be analyzed (ACM, 2009).

Every software system requires some kind of data to process or rely on. Knowledge base development and maintenance are vital for diagnosis support systems. Data gathering is crucial, but it depends on the type of application to be developed. Generally speaking, knowledge base development is not a onetime event; it evolves continuously. The data sources vary from machine statistics, to case records, to research data. Converting and encoding the data relevant to the application is also very important. Initial reports of new discoveries and inventions must await confirmation before their content is added to a database of maintenance knowledge. Knowledge base development should reflect the latest scientific knowledge in the terminology and mark-up used in diagnosis. The maintenance of such terminology and data is crucial, as both may change over time.

The development of the knowledge base must be scientifically reconstructible, with the ability to add further data from technical literature, statistics, and expert opinion. Ideally, the data should be verifiable by experts with a user interface. The long-term value and reliability of these systems depend on the accuracy, quality, and up-to-date nature of their knowledge base. In brief, the long-term success of diagnosis software depends on updating and maintaining the knowledge base.

Although CPU speed has increased and computer memory has become cheaper, a developer needs to analyze and plan for the use of complex algorithms and the amount of data they process. Analysts and developers need to convert theory-based models into practical implementations of specific models. The development of diagnostic systems involves balancing theory and practicality, while maintaining the knowledge database. Developers must design a way to store data access methods; broad-based diagnosis systems require robust and detailed designs. The resources required to construct and maintain a knowledge base are measured in dozens of person-years of complex effort; obviously, a large knowledge base may require more resources.

Even today, human-to-human interaction is more advanced than computer-to-human interaction; therefore, in medicine, physician interaction with a patient is more advanced than patient–computer interaction. A maintainer might not be able to express his/her full understanding of a machine to a computer system, and many researchers believe computers will never replace diagnosticians. However, computers may replace physicians if researchers and developers understand the evolution of medical knowledge and the immaturity and uncertainty of asset data and knowledge. Designs and algorithms must be realistic for the given circumstances.

1.6 Big Data in Maintenance

Manufacturers are hearing increasing technical and professional buzz about "Big Data," with little explanation of what it is. In fact, Big Data refers to the vast collection of detailed information and documentation gathered and stored as a result of computing processes, which can be processed further to provide valuable insights to optimize a process or an operation. For example, Facebook collects "likes" so that it can target advertisements to the right users (O'Brien, 2014).

According to the cloud computing software developer Asigra, 2.5, a quintillion bytes of data were created daily by business and consumers worldwide in 2013. The total amount of stored data is expected to grow by 50 times by 2020. The data themselves are worthless unless information can be extracted from them and used to inform decision making. Therefore, organizations are examining ways to decipher data. The goal is to be able to use data to generate forward-looking insights that can be acted upon to improve the way they do business. Over the next decade, data will become as important to manufacturing as labor and capital.

In the past, systems left a factory never to be seen or heard from again. However, as the price of sensors, cloud computing, broadband, and data storage continues to fall, the ability of manufacturers to communicate directly with their equipment in the field is becoming more commonplace.

Beyond the references to "Big Data," there is increasing discussion and conjecture in IT about the "Internet of Things." Generally, this refers to the strategies pursued by capital goods manufacturers to build ecosystems in which their finished products host built-in sensors and monitoring software that continuously feed operating and performance data to the original equipment manufacturer's (OEM's) central data warehouse for processing.

Boeing is a good example. Boeing aircrafts are continually transmitting flight data back to Boeing headquarters (HQ) via satellite for analysis. Such data could be invaluable in explaining the disappearance of Malaysia Airlines MH370 in March 2014.

By collecting data like these, manufacturers can spot potential equipment failure in advance by identifying early signs of potential downtime and component issues. Issues could be addressed proactively to optimize maintenance schedules, reduce warranty repair costs, and improve customer satisfaction.

The data can also be used to identify equipment that is behaving differently from the rest of the fleet, or to determine how certain failure events will affect the life expectancy of the asset or its reliability going forward. This information can help organizations maintain and optimize their assets for improved availability, utilization, and performance.

Big Data can also be used to influence the next generation of products by identifying the issues that cause emergency downtime across the product fleet, feeding those insights back into the design process and improving the manufacturing process and product quality.

Today, manufacturers are collecting valuable information from their capital assets either directly from connected sensors, or indirectly in their computer maintenance management systems (CMMS) by recording meter readings or logging work orders. The data comprise two types: structured data, such as sensor readings, real-time monitoring, failure codes, and parts consumption on work orders, and unstructured data such as maintenance reports and service logs. These data offer a treasure trove of information that can improve the way we manage our assets.

Exactly how can Big Data improve our management of physical assets? Organizations can improve asset reliability by using CMMS data to predict component and system failure through preventive maintenance (PM), link failures that occur close together, and monitor for conditions that require further investigation. Asset managers can better plan production runs, downtime, and expected maintenance-related costs based on historical data. The data can also be used to improve decision making, such as repair versus replacement based on the expected life of the asset in its current condition.

It is well documented that when cutting maintenance-related costs, optimizing spare parts inventory can deliver the biggest savings. Maintenance managers must strike a balance between reducing maintenance-related costs and maintaining equipment availability. CMMS data can be used to help predict asset failure, optimize quality and supply-chain processes, and limit the number of parts onsite to what will be needed in the near future. By analyzing the rate of consumption, failure rates, and lead times, parts can be procured at the right time, mirroring just-in-time manufacturing methodology.

There is no doubt that Big Data will improve how we manage our assets. OEMs will use the data to manufacture more reliable systems while CMMS data will help maintenance technicians to work smarter. The Internet of Things may be a few years off for many organizations, but they can extract valuable information from CMMS data even today.

The good news is many organizations are already mining valuable information from their CMMS data and making decisions that drive revenue. CMMS software has become more intuitive and easier to use, which is critically important for organizations that need to control maintenance-related costs and improve system reliability today.

1.7 Maintenance Data: Different Sources and Disparate Nature

Maintenance teams are responsible for maintaining the operation of a system based on observations of users of the system or messages originating from test devices built into the equipment to establish an equipment failure diagnostic, identify and locate system equipment failures, and if required, replace a failed item of equipment, in part or in full. The gathering of messages and observations to track down equipment failures is called maintenance management. The architecture of complex systems is, in general, based on information transfers between electronic equipment linked by a communication network. The location of the equipment in the system sometimes complies with strict specifications on dimensions; moreover, the equipment can be difficult to access, from 10 cm to several hundred meters away (Horenbeek and Pintelon, 2012).

Various management information systems are able to monitor and record messages and observations in connection with industrial processes. These systems often are reactive in that they respond to present levels of monitored parameters or, at most, respond to present trends to control the generation of alarms and the like when a parameter exceeds preset values or threatens to do so. A typical process control system monitors sensed parameters to ensure they remain within preset limits defined by the programmer of the system. Often, the present levels can be displayed graphically to highlight trends.

Another form of management information system involves the scheduling of maintenance procedures. By defining a useful life for each article of equipment among a number of related or interdependent articles, it is possible to schedule repair, replacement, or PM operations more efficiently so as to minimize downtime. The idea is to plan replacement or repair of equipment articles for as late as practicable before an actual failure, preferably using intelligent scheduling procedures to minimize downtime by taking maximum advantage of any downtime. The scheduling system prompts or warns plant personnel to attend to each of the articles that may need attention at or soon after the time at which its maintenance becomes critically important.

It would be advantageous to provide an integrated system that not only monitors various assets of plant equipment, but also accounts for the interdependence of the subsystems, makes decisions or predictions in view of stored design criteria, and makes all this information generally available to plant personnel. Therefore, the interrelations of the articles or subsystems, their design specifications, their history, and their current conditions should all be taken into account when assessing operational conditions and maintenance needs, or when evaluating operations on an engineering level.

It is generally advisable for maintenance personnel to collect any available data on the subsystems operating in a plant or in an area of the plant to coordinate maintenance and repair activities. In this manner, downtime for work on one or more articles or subsystems can be used for simultaneous work on others. However, the comprehensive calculation and analysis of relevant plant conditions can be lengthy and costly. In a monitoring system where information on operational conditions is only immediately available to the operators and maintenance technicians, the engineers and managers must collect and analyze much of the same information to plan their own activities. Each group tends to collect and analyze data in a manner best suited to its specific area of concern.

An integrated arrangement is certainly more efficient and useful than one in which the various departments operate substantially on independent information systems. To accomplish this objective, the maintenance management system and the process control system can be integrated with instrument data collection from a variety of sources. These condition parameters can be factored together in integrated diagnostics systems with technical specifications and historical data for condition-based maintenance (CBM) and aging management. However, this integration requires the creation of a common taxonomy to select the right information, create links among data sources, and extract further knowledge; if this is done, operations and maintenance decisions can be made more effectively from a greater base of knowledge.

The need for improved asset performance through appropriate diagnosis and prognosis is considerable. A barrier has been the lack of a performance management solution that includes the widely diverse divisions of maintenance, operations, and finance; for example, if each division uses its own performance metrics, it is hard to make optimal decisions such as balancing reliability objectives against those of asset utilization.

Many people are seeking an ideal combination of self-diagnostics and prognosis. As a result, there are numerous versions of balanced scorecards and key performance indicator solutions available in the market today. They all say the same thing: their product will make a manufacturing process run better, faster, more efficiently, and with greater returns, but what they do not address, however, is one of the greatest challenges in improving plant asset performance. Simply stated, the necessary information is scattered across disconnected silos of data in each department; hence, it is difficult to integrate these silos for several fundamental reasons. For example, control system data are real-time data measured in seconds, whereas maintenance cycle data are generally measured in calendar-based maintenance time (e.g., days, weeks, months, quarters, semiannual, and annual), and financial cycle data are measured in fiscal periods.

For maintenance information, CMMS and CM are the most popular repositories. Although both store information on the deployed technology, their use creates isolated information islands. While using a good version of either technology can assist a company to reach its defined goals of maintenance, the combination of the two (CMMS and CM) in a seamless system can have exponentially more positive effects on the performance and maintenance of assets that either system could achieve alone. By combining the strengths of a premier CMMS (PM programming, automatic work order generation, inventory control of maintenance, and data integrity) with those of a leading-edge CM system (multiple method CM, trend monitoring, and expert system diagnosis), work orders could be generated

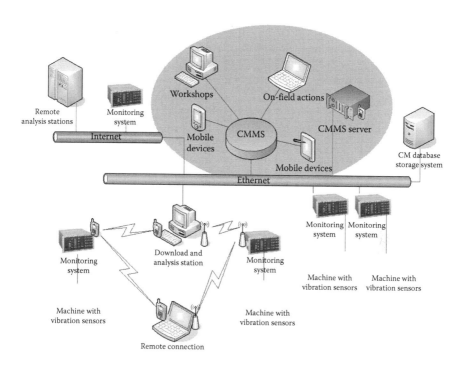

FIGURE 1.18
The ICT architecture for the integration of CMMS and CM systems in maintenance and asset management.

automatically based on information provided by diagnostic and prognostic abilities of CM. A few years ago, linking CMMS and CM technology was dismissed as impossible or too expensive, making it difficult to justify research. Now, however, the technology in CMMS and CM makes it possible to achieve this "impossible" link relatively easy and inexpensively too (Figure 1.18).

In general, a good CMMS will perform a variety of functions to improve maintenance performance and can be considered a central organizational tool to achieve world class maintenance (WCM). Among other things, it is designed to shift emphasis from reactive maintenance to preventative maintenance. For example, it will allow a maintenance professional to set up automatic preventative maintenance work order generation.

A CMMS is able to provide historical information which, in turn, can be used to adjust preventative maintenance over time, thus minimizing unnecessary maintenance or repairs, while avoiding run-to-failure repairs. In such a system, preventative maintenance for a piece of equipment can follow a calendar schedule or use meter readings.

A fully featured CMMS includes inventory tracking, workforce management, and purchasing. It ensures database integrity to safeguard information.

Optimized equipment downtime, lower maintenance costs, and improved plant efficiency are only a few of the resulting benefits.

A CM system must alert the maintenance professional to any change in performance trends and accurately monitor equipment performance in real time. A CM package may be designed to track a wide variety of measurements: vibration, oil condition, temperature, operating and static motor characteristics, pump flow, and pressure output, to name a few. These measurements are taken by special monitoring tools, including ferrographic wear particle analysis, proximity probes, triaxial vibration sensors, accelerometers, lasers, and multichannel spectrum analyzers. Top-notch CM systems can analyze measurements such as vibration and diagnose machine faults. This type of expert system analysis postpones maintenance procedures until they are absolutely necessary, thereby creating maximum equipment uptime. Finally, the best systems offer diagnostic fault trending, whereby an individual machine fault severity can be monitored over time.

CMMS and CM systems are indispensable to any organization seeking to improve its maintenance operations. CMMS is a good organizational tool; unfortunately, it cannot directly monitor the condition of equipment. In contrast, CM is able to monitor condition but cannot organize overall maintenance operations. Obviously, the two technologies should be combined to create a seamless system, one that both avoids catastrophic breakdowns and eliminates needless repairs.

Maintenance staff generally sense that the use of IT has a dramatic impact on machine reliability and maintenance efficiency. Yet few can actually explain or demonstrate the benefits of applying information technologies. Technology developers continue to deliver increasingly advanced tools, leaving maintenance departments to implement, integrate, and operate these systems. When users combine their experience and heuristics to define maintenance policies and use CM systems, the resulting maintenance systems are a heterogeneous combination of methods and systems; the integrating factor is the maintenance staff. The human mind is an organizational information system, and the inauguration of any new maintenance program relies on the expertise of maintenance personnel. The literature offers many models that could be used to support maintenance decisions, but the majority are too simple to accurately represent real life. Therefore, they are not widely used in industry.

With the increased use of information technology and communication (ICT) in organizations and the emergence of intelligent sensors to measure and monitor the health status of components, the conceptualization and implementation of e-maintenance is becoming a reality. Although e-maintenance shows great promise, however, the seamless integration of ICT into the industrial environment continues to be a challenge. It is essential to address the needs and constraints of maintenance and the capabilities of ICT simultaneously.

1.8 Required Data for Diagnosis and Prognosis

Maintenance can be considered an information processing system that produces a vast amount of data. Having data is not synonymous with information; rather, data must be processed using analytical tools to extract information. In maintenance, IT and AI tools are supporting an unprecedented transformation from the industrial age to the information age; the existing and emerging technologies are analyzing real-time asset systems data to make predictions and determine maintenance capability. Several technological advances at various levels have moved toward making CBM a reality for industry. The benefits of these technologies have already been proven.

The transition to CBM requires collaboration on a large scale and is contingent on the identification and incorporation of new and improved technologies into existing and future production systems. This will require new tools, test equipment, and embedded onboard diagnosis systems. More importantly, the transition to CBM will require the construction of data-centric, platform operating capabilities built on carefully developed and robust algorithms. As a result, maintenance personnel, support analysts, and engineers will simultaneously—and in real time—be able to translate conditional data and proactively respond to maintenance needs based on the actual condition of equipment.

To reiterate the above discussion, two main systems are implemented in most maintenance departments today. CMMS is the core of traditional maintenance record-keeping practices and can facilitate the use of textual descriptions of faults and actions performed on an asset, while CM systems are able to directly monitor the parameters of the active components. To this point, however, attempts to link CMMS observed events to CM sensor measurements remain relatively limited in both scope and scalability.

CM systems are able to directly monitor the parameters of asset components, but as noted above, the attempts to link observed CMMS events to CM sensor measurements remain limited with respect to both scope and scalability. A CBM strategy that estimates the optimal time for a service visit, based on the present condition of equipment, could be a way to increase efficiency and reduce costs over the lifecycle. But predictive maintenance approaches are frequently hampered, first, by the lack of knowledge of the features indicating the equipment's condition, and second, by the enormous processing power required to create prediction algorithms to forecast the evolution of the selected features, especially when large measurements are collected. To avoid these issues, we propose data mining.

The development of future maintenance information systems to improve automatic CM systems enabled by embedded electronics and software in industrial machines is an extremely important research problem. Further, understanding of the requirements and limitations from both maintenance AI and IT perspectives is necessary if we are to reach conclusions that are relevant to different end users.

1.8.1 Existing Data in the Maintenance Function

Maintenance documentation systems for recording and conveying information are an essential operational component of the maintenance management process. Maintenance documentation can be defined as any catalog, record, drawing, computer file, or manual which contains information that may be needed to facilitate the maintenance work. An information system of maintenance may be defined as the formal mechanism to store, collect, examine, analyze, and report maintenance information (Galar et al., 2012).

How a maintenance documentation system generally functions is shown in Figure 1.19. This model has evolved over a number of years through maintenance information base for plant units extensive study of both paper-based and computerized systems. It illustrates the principal features of both types,

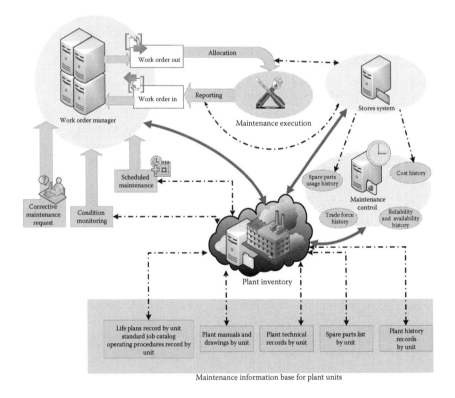

Maintenance information base for plant units

FIGURE 1.19
Functional model for maintenance documentation system.

including those they have in common. The system is composed of the following interrelated modules:

1. Plant inventory
2. Maintenance information base
3. Maintenance schedule
4. CM
5. Maintenance control

The plant inventory comprises a coded list of the plant units. It offers a way into the system. In the plant inventory, asset data are collected in an organized and structured way. The major data categories for equipment are the following:

- Classification data, for example, industry, plant, location, and system.
- Equipment attributes, for example, manufacturer's data and design characteristics.
- Operating data, for example, operating mode, operating power, and environment.

These data categories apply to all equipment classes, so data specific to an equipment class (e.g., number of stages for a compressor) are also needed. The classification of equipment into technical, operational, safety related, and environmental parameters is the basis for collecting asset data according to the nature of devices (safety instrumented systems, productive assets, maintenance tools, CM systems, etc.). This information is necessary to determine whether the data are suitable for various applications.

There are two kinds of maintenance actions:

- Actions to correct an item after it has failed (corrective maintenance or CM).
- Actions to prevent an item from failing (PM). A part of this may only be checks.

In both cases, the information is recorded to supply the following additional information:

- The total resources used for maintenance (man–hours, spare parts).
- The full story of an item's life (all failures and maintenance).
- The total downtime and, by extension, total equipment availability, both operational and technical.
- The balance between preventive and corrective maintenance (inspections, tests) to verify the condition of the equipment and decide if any PM is required.

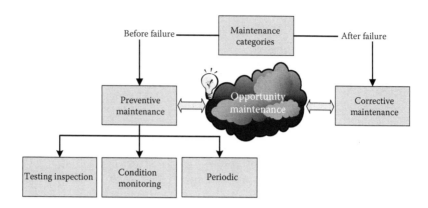

FIGURE 1.20
Maintenance categorization.

Figure 1.20 shows the common maintenance actions. To this, we can add another maintenance type: maintenance is either deferred or advanced in time when an unplanned opportunity becomes available.

Maintenance actions are the result of a specific maintenance program. The choice of methodology, that is, the ratio of preventive to corrective maintenance, and so on, is always up to the maintenance manager who uses the plant inventory and maintenance information base to construct the model to be used. As the name suggests, the maintenance information base is a database of maintenance information, for example, unit life plans, job catalog, and so on, for each of the units. These data include:

- *Identification data*: For example, maintenance record number, related failure, and/or equipment record.
- *Maintenance data*: Parameters characterizing a particular maintenance event, for example, date of maintenance, maintenance category, maintenance activity, impact of maintenance, and items maintained.
- *Maintenance resources*: Maintenance man–hours per discipline and total utility equipment/resources applied.
- *Maintenance times*: Active maintenance time and downtime.

A common report should be used for all equipment classes. For some equipment classes, minor adaptations may be needed. Table 1.2 shows the minimum data needed to meet international standards, maintenance association standards, and CMMS manufacturers' recommendations.

Recording maintenance actions is crucial for successful knowledge extraction; therefore, all actions should be recorded. PM records are useful for the maintenance engineer, but will be helpful for the maintenance engineer

TABLE 1.2

Maintenance Records Meeting International Standards and Recommendations

Category	Data to be recorded	Category	Data to be recorded
Identification	Maintenance record	Maintenance resources	Maintenance man-hours, per discipline
	Equipment location		Maintenance man-hours, total
	Failure record		Maintenance equipment resources
Maintenance data	Date of maintenance	Maintenance times	Active maintenance time
	Maintenance category		Down time
	Maintenance priority		
	Interval (planned)		Maintenance delays/problems
	Maintenance activity	Remarks	Additional information
	Maintenance impact on plant operations		
	Subunit maintained		
	Component/maintainable item(s) maintained		
	Spare part location		

Data relevant for reliability prediction and estimation

If user can add handwritten comments or documents, further data mining becomes more difficult

Accuracy in data feeding process is required to warrant a proper *maintenance control* and trustable results in *data mining*

who wants to record or estimate the availability of equipment and do a lifetime analysis, taking failures into account, as well as maintenance actions intended to restore the item to "as-good-as-new" condition. PMs are often performed on a higher indenture level (e.g., "package" level); hence, there may not be data available that can be related to the items on the lower indenture level. This restriction must be considered when defining, reporting, and analyzing PM data.

During the execution of PM actions, impending failures may be discovered and corrected. In this case, the failure(s) is (are) recorded as any other failure with the subsequent corrective action done even though it was initially considered as a PM-type activity. In this case, the failure detection method is referred to as the type of PM being done. However, some failures, generally minor ones, may be corrected as part of the PM and not recorded. The practice may vary between companies and should be addressed by the data collector(s) to reveal the type and number of failures included in the PM program.

A final option is to record the planned PM program as well. In this case, it is possible to additionally record the differences between the planned and the actual performed PM (backlog), (EN 15341, 2007). An increasing backlog will indicate that control of the conditions of the plant is being jeopardized and may, in adverse circumstances, lead to equipment damage, pollution, or personnel injury.

For corrective maintenance, failure records are especially relevant to knowledge extraction, so failure data must be recorded in way to facilitate further computation. A uniform definition of failure and a method of classifying failures are essential when data from different sources (plants and operators) need to be combined in a common maintenance database.

These failure data are characterized as:

- *Identification data*: For example, failure record number and related equipment that has failed.
- *Failure data for characterizing a failure*: For example, failure date, items failed, failure impact, failure mode, failure, cause, and failure detection method.

The type of failure and resulting maintenance data are normally common to all equipment classes, with exceptions where specific data types need to be collected. Corrective maintenance events should be recorded to describe the corrective action following a failure.

Finally, the combination of plant inventory and maintenance-based information leads to the production of a maintenance schedule. This schedule is a mixture of available techniques to fulfill stakeholders' needs and achieve company goals. This mixture is usually composed of both scheduled maintenance and CM. The maintenance schedule includes the preventive maintenance jobs (over a year and longer) listed against each of the units in the life plans. The CM schedule includes the CM tasks, for example, vibration monitoring, listed against each unit. PM records must contain the complete lifetime history of an equipment unit.

The system has to plan and schedule preventive jobs (arising from the maintenance schedule), corrective jobs (of all priorities) and, where necessary, modification jobs. The jobs are carried out via hard copy or electronic work orders. Information coming back on the work orders (and other documents) is used to update the planning systems and provide information for maintenance control. The maintenance control system uses information from a number of sources, including work orders, stores, shift records, and so on, to provide reports for cost control, plant reliability control, and so on.

A main issue is the integration of these data with the rest of company records, such as health and safety, finances, and so on. Up until about 10 years ago, most CMMS were stand-alone; that is, they had no electronic linkage with other company software. The most recent computerized maintenance systems are integrated electronically (in the same database) with stores, purchasing, invoicing, company costing, and payroll and can have electronic links to project management and CM software.

1.8.2 In Search of a Comprehensive Data Format

All mentioned data become a database record, for example, failure events, and are identified in the database by a number of attributes. Each attribute describes one piece of information, for example, the failure mode. Each piece of information should be coded where possible. The advantages of this approach versus free text are:

- Facilitation of queries and analysis of data
- Ease of data input
- Consistency check at input by having predefined code lists
- Minimized database size and shorter response time to queries

The range of predefined codes should be optimized. On the one hand, a short range of codes will be too general to be useful. On the other hand, a wide range will give a more precise description, but will slow the input process and may not be used fully by the data collector. If possible, selected codes should be mutually exclusive.

The disadvantage of a predefined list of codes versus free text is that some detailed information may be lost. It is recommended that free text be included to provide supplementary information. A free-text field with additional information is also useful for quality control of data. This free-text box is extremely risky in further data-mining processes due to difficulties of text recognition and interpretation (see Table 1.2). Different employees have different skills to describe failures, events, and actions, and expert systems are not good at distinguishing these variations. For all mentioned categories, the inclusion of additional free text is recommended to give more information if available and deemed relevant, for example, a more detailed verbal description of the occurrence leading to a failure event. This would assist in quality checking the information and browsing through single records to extract more detailed information. However, users should be aware of the risk in the automatic processing of these records.

1.8.3 Database Structure

The data collected should be organized and linked in a database to provide easy access for updates, queries, and analysis. Several commercial databases are available for use as main building blocks in designing a reliability database. Two relevant aspects of data structure are:

- *Logical structure*: This requires a logical link between the main data categories in the database. The model represents an application-oriented view of the database. The example in Figure 1.21 shows a hierarchical structure, with failure and maintenance records linked to the classification/equipment description (inventory). Records describing PM are linked to the inventory description in a many-to-one relation. The same applies for failures, which also have related corrective maintenance records linked to each failure. Each record (e.g., failure) may consist of several attributes (e.g., failure date and failure mode).
- *Database architecture*: This refers to the design of the database, specifically how the individual data elements are linked and addressed.

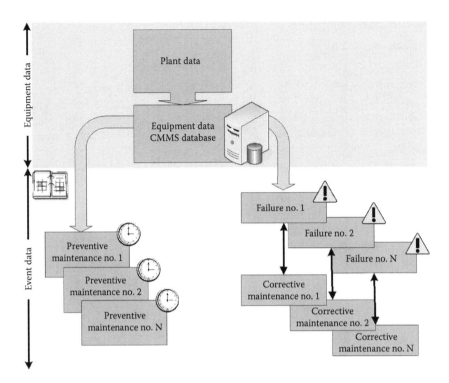

FIGURE 1.21
Logical data structure.

Four model categories are available, ranked here in ascending order of complexity and versatility:

- *Hierarchical model*: Data fields within records are related by a "family tree-like" relationship. Each level represents a particular attribute of data.

- *Network model*: This is similar to the hierarchical model, but each attribute can have more than one parent.

- *Relational model*: The model is constructed from tables of data elements, called relations. No access path is defined beforehand; the manipulation of data in tabular form is possible. The majority of database designs use this concept.

- *Object model*: Software is considered a collection of objects, each with a structure and an interface. The structure is fixed within each object, while the interface is the visible part providing the link address between the objects. Object modeling enables the database design to be flexible, extendable, reusable, and easy to maintain. This model is popular in new database concepts.

1.8.4 CM Data and Automatic Asset Data Collection

CM involves comparing online and offline data with the expected values; if necessary, it should be able to generate alerts based on preset operational limits. Health assessment determines if the health of the monitored component or system has degraded, and conducts fault diagnostics. The primary tasks of prognostics involve calculating the future health and estimating the remaining useful life (RUL). In reality, however, a reliable and effective CBM faces challenges. For one thing, initiating CBM is costly. Often the cost of instrumentation can be quite large, especially if the goal is to monitor equipment that is already installed. It is, therefore, important to decide whether the equipment is important enough.

Implementing CBM requires setting an information system to meet the following basic requirements:

- Collecting and processing a large quantity of information not previously available, on the condition of each part of a machine.
- Initiating corrective maintenance actions within the lead time (the period of time between the off-limits condition and an emergency shutdown). There are two possible situations:
  - The condition of machine is not yet close to breakdown. In this case, the normal procedure through the maintenance planning section is followed.
  - The condition of machine is already well within the lead time (near to breakdown). In this situation, the information must be directly passed on to the maintenance supervision for emergency corrective maintenance actions.

To operate the CBM program correctly, the maintenance personnel should introduce the following into the system:

- Condition of machine
- Part of machine probably defective
- Probable defect
- Time during which failure must be repaired

By scrutinizing and correlating the diagnosis to actual findings during repair work, it will be possible to:

- Control the examiner training
- Improve the correlation between parameters chosen for condition measurement and actual defects found
- Obtain severity curves specific to each machine

Making the potential of CM a reality requires large amounts of data to be collected, monitored, filtered, and turned into actionable information. The cheaper and more ubiquitous the computerized monitoring hardware becomes, the greater the volume of data and the more challenging it becomes to manage and interpret. The vast amount of diagnostic data produced by today's smart field devices can be a very important source for accurate documentation of maintenance activities. But the sheer volume and complexity of such information can be daunting and difficult for maintenance personnel to manage. What is needed is an effective means of compiling and organizing the data for day-to-day use by staff, while preserving and recording significant events for future reference.

Data are becoming more and more available. However, in most cases, these data may not be used due to their bad quality or improper storage:

- Project managers do not have sufficient time to analyze the computerized data and do not care about proper storage.
- The complexity of the data analysis process is beyond the capabilities of the relatively simple maintenance systems commonly used.
- There is no well-defined automated mechanism to extract, preprocess and analyze the data, and summarize the results; so stored data are not reliable.

Maintenance personnel not only cope with large amounts of field-generated data, they turn that information to their advantage in a number of ways. Real-time condition monitoring (RTCM) systems produce numerous warnings, alarms, and reports that can be used by maintenance people for many purposes. This allows the most important issues to be identified and handled quickly.

The ultimate goal is to fully integrate RTCM data with CMMS to generate work orders as needed. This will provide true automation, from the time a field device begins to show signs of reduced performance to the time a work order is printed in the maintenance department and a technician is dispatched to the scene. Figure 1.22 shows the automation of work-order dispatching. This level of integration of CMMS and CM is feasible due to today's rapid IT evolution. With the development of open-communication protocols, the information accumulated by smart field devices can be captured by asset management software. It is no longer necessary for technicians to carry handheld communicators or laptops into the plant to evaluate the condition of instruments, some of which are quite inaccessible or in hazardous areas, to be followed by manually documenting test results and current device status.

Current applications compile databases of every smart instrument used for process control, including design parameters, original configuration, maintenance history, and current operating condition. With these online tools, technicians can obtain up-to-date information on any device and never have

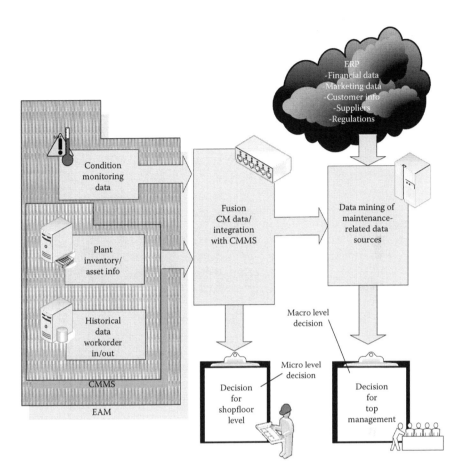

FIGURE 1.22
Two-step integration of RTCM and CMMS databases.

to make manual entries back into a system. Every event is recognized and recorded, whether initiated by a technician or caused by an external force such as an equipment breakdown or power failure. This process produces an immediate result for the shop floor level: work orders can be opened and closed with the help of devices that collect automatically information and send a warning if something wrong happens. Users can refer to recorded alerts to identify any devices that have been problematic over time and determine what corrective steps may have been taken previously. Automated documentation provides a seamless record of events in a given production area, including communication failures, device malfunctions, and process variables that are out of range. Armed with this information, maintenance personnel are better equipped to understand and resolve nagging repetitive issues to improve the process. If there is an issue, or if maintenance personnel are experiencing a rash of issues, they can go back into the records and get a sense of what has been going on over time, searching by a specific device or by location.

Since all records are date and time stamped, users can easily determine when and by whom a particular device was changed or tested, including "as found/as left" notations. With this information in a database that cannot be edited, it should never be necessary for technicians to spend time searching for historical information on a device. Since events can also be recorded manually, users can document unusual occurrences affecting the entire plant, such as a lightning strike or power outage, or individual events like device inspections. This decision level is extremely useful for technicians to take immediate actions.

A vast amount of available information can produce new knowledge but must be properly exploited. Modern CMMS information are stored in large relational or tabular databases. This format is appropriate for integrated research, as many software tools are available to investigate the tables. As historical analysis requires only certain fields, sensitive data can be removed or filtered. Despite this filtering process, the database subset will retain a full history of a component's faults and any related maintenance actions, thus creating a comprehensive maintenance profile while alleviating security concerns.

Importing CM data into a relational database can be challenging. Each type of sensor generates different data classes, sampling rates, and numbers of compiled indicators. Manufacturers store collected data in unique proprietary formats, each of which requires platform-specific importation software. However, most CM software allows CM data to be exported from the original interface so that it can be expanded and generalized.

Although CMMS and CM data can now coexist within a single database where they can be queried and explored, automating the discovery of linked events requires additional processing. Data on given maintenance faults are textual; sensor data represent an arbitrary data class. Relating them can only be done by compiling overlapping metadata (Tianhao et al., 1992). The fields generated characterize the location and significance of events, creating a quantified set of parameters that can be used to compare the disparate data.

The generation of metadata for CM records depends on the data class. One dimensional and dimensionless data classes can be assigned rarity parameters using statistical distribution analysis. Higher dimensional data classes require neural networks to identify anomalies. Determining rarity is often accomplished through simple single variable statistical analysis, while severity is typically derived from developers' recommended threshold values. More complex domain types require more advanced but well-understood analyses such as neural networks that can isolate anomalous points from multidimensional data.

It is predicted that through the integration process, more advanced metrics and indicators will be discovered to implement previously unexplored relationships in the data such as multiparameter trending. This new knowledge will help maintenance personnel determine the RUL of a system and schedule operation and maintenance processes accordingly. The information affects the replacement of assets, plant shutdowns, overhauls, and so on. It constitutes the second decision level displayed in Figure 1.4; strongly related to business goals, it is useless for immediate interventions.

References

ACM, 2009. [Online] Available at: http://www.acm.org/.

Atherton, D. and Borne, P., 1992. *Concise Encyclopedia of Modeling & Simulation (Advances in Systems Control and Information Engineering)*. 1st edition. England: Pergamon.

Ball, M. and Berner, E. S., 2006. *Clinical Decision Support Systems: Theory and Practice (Health Informatics)*. 2nd edition. New York: Springer.

Basile, G. and Marro, G., 2002. *Controlled and Conditioned Invariants in Linear Systems Theory*. Italy: University of Bologna.

Birta, L. G. and Ozmizrak, F. N., 1996. A knowledge-based approach for the validation of simulation models: The foundation. *ACM Transactions on Modeling and Computer Simulation*, 6(1), 76–98.

Cellier, F. E., 1991. *Continuous System Modeling*. New York: Springer-Verlag.

Combi, C., Shahar, Y., and Abu-Hanna, A., 2009. *Artificial Intelligence in Medicine*. Berlin: Springer.

Czichos, H., 2013. *Handbook of Technical Diagnostics—Fundamentals and Application to Structures and Systems*. Heidelberg, Berlin: Springer-Verlag.

Douglas, G., Nicol, F., and Robertson, C., 2005. *Macleod's Clinical Examination*. 11th edition. Edinburgh: Churchill Livingstone.

Elstein, A., Shulman, L., and Sprafka, S., 1978. *Medical Problem Solving: An Analysis of Clinical Reasoning*. Cambridge, MA: Harvard University Press.

EN 15341, 2007. *Maintenance Key Performance Indicators*. Brussels, Belgium: European Committee for Standardization (CEN).

Erhard, A., 2011. Nondestructive evaluation. In: Czichos, H., Saito, T., and Smith, L. E. (Eds.), *Springer Handbook of Metrology and Testing*. Heidelberg, Berlin: Springer-Verlag, pp. 888–900.

Fu, K. and Booth, T., 1975. Grammatical inference: Introduction and survey. *IEEE Transactions Systems Man and Cybernetics*, SMC-5, 95–111.

Gaines, B., 1978. General system identification—Fundamentals and results. *Applied General Systems Research: NATO Conference Series*, pp. 91–104.

Galar, D., Gustafson, A., Tormos, B., and Berges, I., 2012. Maintenance decision making based on different types of data fusion. *Maintenance and Reliability*, 14(2), 135–144.

Greenes, R. A., 2006. *Clinical Decision Support: The Road Ahead*. San Diego, CA: Academic Press.

Guidorzi, R., 2003. *Multivariable System Identification: From Observations to Models*. Bologna: Bononia University Press.

Hanselka, H. and Nuffer, J., 2011. Characterization of reliability. In: Czichos, H., Saito, T., and Smith, L. E. (Eds.), *Springer Handbook of Metrology and Testing*. Heidelberg, Berlin: Springer-Verlag, pp. 949–967.

Horenbeek, A. V. and Pintelon, L. G. D., 2012. Integration of disparate data sources to perform maintenance prognosis and optimal decision making. *Insight (Northampton)*, 54(8), 440–445.

Houghton, A. R. and Gray, D., 2010. *Chamberlain's Symptoms and Signs in Clinical Medicine: An Introduction to Medical Diagnosis*. Boca Raton, FL: CRC Press.

ISIS Canada, 2001. *Guidelines for Structural Health Monitoring*. [Online] Available at: www.isiscanada.com/index.html.

ISO 13372, 2004. *Condition Monitoring and Diagnostics of Machines-Vocabulary*.

ISO 13373-1:(E), 2002. *Condition Monitoring and Diagnostics of Machines—Vibration Condition Monitoring*.

ISO Guide 51, 1999. *Safety Aspects—Guidelines for Their Inclusion in Standards*.

ISO Guide 73, 2002. *Risk Management—Vocabulary*.

Kassirer, J. and Gorry, G., 1978. Clinical problem-solving a behavioral analysis. *Annals of Internal Medicine*, 89, 245.

Lennart, L., 1994. *From Data to Model: A Guided Tour of System Identification*. Sweden: Linkoping.

Li, E. Y. 1994. Artificial neural networks and their business applications. *Information & Management*, 27(5), 303–313.

Magee, B., 1985. *Popper*. London: Fontana Press.

O'Brien, J., 2014. *Understanding 'Big Data' for Manufacturing, Maintenance*. [Online] Available at: http://americanmachinist.com/enterprise-software/understanding-big-data-manufacturing-maintenance.

Pople, H. J., 1982. Heuristic methods for imposing structure on ill-structured problems: The structuring of medical diagnostics. In: Czichos, H., Saito, T., and Smith, L. E. (Eds.), *Artificial Intelligence in Medicine*. Boulder, CO: Westview Press, pp. 119–190.

Popper, K., 1963. *Conjectures and Refutations: The Growth of Scientific Knowledge*. London: Routledge.

Reddy, K., 2009. *Developing Reliable Clinical Diagnosis Support System*. [Online] Available at: http://www.kiranreddys.com/articles/clinicaldiagnosissupportsystems.pdf.

Taylor, P., 2006. *From Patient Data to Medical Knowledge: The Principles and Practice of Health Informatics*. London: Wiley/Blackwell.

Tianhao, W. et al., 1992. *Posting Act Tagging Using Transformation-Based Learning*.

University of Cauca, 2009. *Identificacion De Sistemas*. [Online] Available at: ftp://ftp.unicauca.edu.co/Facultades/FIET/DEIC/Materias/Identificacion/2005a/parte%20I/clase%2001%20ident/capitulo_01.pdf.

Vangheluwe, H., 2001. *Modeling and Simulation Concepts*. [Online] Available at: http://www.cs.mcgill.ca/~hv/classes/MS/MSconcepts.pdf.

Vesely, W., 2002. *Fault Tree Handbook*. Washington, DC: NASA.

Walter, M., Siegenthaler, A., Aeschlimann, E. B., and Bassetti, C., 2007. *Differential Diagnosis in Internal Medicine: From Symptom to Diagnosis*. New York: Thieme Medical Publications.

Welden, V. and Danny, F., 1999. *Induction of Predictive Models for Dynamical Systems Via Datamining*.

Worden, K., Charles, R., Graeme, M., and Gyuhae, P., 2007. The fundamental axioms of structural health monitoring. *Proceedings of the Royal Society A: Mathematical, Physical & Engineering Sciences*, 463(2082), 1639–1664.

Zadeh, L., 1956. On the identification problem. *IRE Transactions on Circuit Theory*, 3, 277–281.

Zeigler, B. P., 1984. *Theory of Modeling and Simulation*. Malabar, FL: Robert E. Krieger, p. 1.

2

Condition Monitoring: Available Techniques

2.1 Role of Condition Monitoring in Condition-Based Maintenance and Predictive Maintenance

Currently, to know when a maintenance action is needed, it is necessary to monitor (follow) the health status of the machine (its condition) by means of different parameters that reflect the health status, for example, the evolution of the vibration level over time. This action is called condition monitoring (CM), which is the fundamental tool of condition-based maintenance (CBM). By following, numerical or graphically, the evolution of the measured parameters over time, the health status of the machine can be detected at any point in time. Together with preset alarm levels, it will let us know when it is time to intervene on the machine to prevent a failure/problem. Extrapolating historical evolution (the trend), numerically or graphically, we can calculate or estimate how long it will take until the alarm level is reached. This means this type of maintenance has predictive features that make it possible to anticipate the failure with sufficient time to avoid it. This technique is called predictive maintenance (PdM).

In summary, PdM and CBM have the same meaning and are essentially the same thing. But it should be clear that CBM is just that, maintenance actions based on the condition of the equipment, in contrast to maintenance actions based on time or failure, while PdM is based on the equipment's condition history to establish predictions (any CM could be used to develop a history or record). In other words, CM is the basic tool used by both CBM and PdM to assess the condition of the machine, and detect potential failure as soon as the first signs of failure appear. CBM is the maintenance strategy and CM is the tool used to achieve it.

2.2 Difference between CM and Nondestructive Testing

Nondestructive testing (NDT) is an umbrella term for a wide range of analysis techniques used in science and industry to evaluate the properties of a material, component, or system without causing damage (Mobley, 2001).

In engineering, NDT includes all methods of detecting and evaluating flaws in materials. Flaws can include cracks or inclusions in welds and castings, or variations in structural properties, all of which can affect the ability of the material or structure to perform its functions. In such cases, using NDT will ensure both safe operation and quality control (including environmental concerns).

NDT can be used for in-service inspections and for CM in an operating plant. It can measure components and spacings or physical properties such as hardness and internal stress.

The test process itself has no deleterious effects on the material or structure being tested. In addition, NDT has no clearly defined boundaries; it ranges from simple techniques such as visual examination of surfaces, to well-established methods of radiography, ultrasonic testing, magnetic particle crack detection, to new and highly specialized methods such as the measurement of Barkhausen noise and positron annihilation. Finally, NDT methods can be coupled with automated production processes or with the inspection of localized problem areas.

Traditionally, NDT was used to detect material defects (i.e., cracks and voids) using x-ray, ultrasonic, or other similar techniques. However, computerized signal processing, data interpretation, and processing play an increasingly important role. Certain nondestructive methods, formerly used for final product inspection, such as infrared thermography (IT), vibration analysis (VA), and acoustic emission (AE), are now being used for check-ups during the lifespan of a component or a machine, along with the accompanying diagnostics and prognostics. This latter usage is called CM (BINDT, 2014).

CM originally used mainly vibration and tribology analysis, but now includes a number of nondestructive techniques: thermal imaging, AE, and so on. These diagnostic and prognostic elements have increasingly sophisticated signal processing; in addition, they are using trends from repeated measurements in time intervals of days and weeks. Smart systems, an even newer concept, incorporate measuring elements directly into structures.

In other words, we are seeing a considerable overlap between NDT and CM. Both disciplines will benefit from close collaboration.

2.2.1 What Is NDT?

NDT is a test on or an evaluation of any type of test object without the need to change or alter that object in any way, to determine the presence or absence of conditions or discontinuities that may affect the utility or serviceability of that object. NDT can also measure test object characteristics, such as size, dimension, configuration, or structure, including alloy content, hardness, grain size, and so on. The most basic definition is a test performed on an object of any type, size, shape, or material to determine the presence or absence of discontinuities, or to evaluate other material characteristics. Nondestructive examination (NDE), nondestructive inspection (NDI), and nondestructive evaluation (NDEX) are also terms used to describe this technology. Although the technology has been around for years, it remains generally unknown to the average person, who takes it for granted that buildings will not collapse, planes will not crash, and products will not fail (Marwan, 2004).

Although NDT cannot ensure that failure is not going to occur, it plays an important role in reducing the chances of failure. That being said, other variables, such as inadequate design and improper application of the object, can contribute to failure even when NDT is appropriately applied.

Over the past 25 years, NDT technology has experienced significant innovation and growth. It is, in fact, considered to be one of the fastest-growing technologies from the standpoint of uniqueness and innovation. Recent equipment improvements and modifications, as well as a more complete understanding of materials and the use of diverse products and systems, have all contributed to creating a technology with widespread use and acceptance throughout many industries.

This technology can also affect our daily lives, especially in the area of safety, where it has accomplished more than any other technology, including medicine. If it were not for the effective use of NDT, it is hard to imagine the number of accidents that might occur, not to mention the unplanned power outages. NDT is now an integral part of practically all industrial processes where product failure can result in accidents or bodily injury. Virtually every major industry in existence today depends on it, to some extent or other.

In addition to everyday safety, NDT is a process performed on a daily basis by the average individual, who is not aware that it is taking place. For example, when a coin is deposited in the slot of a vending machine and the selection is made, whether for candy or a soft drink, that coin is actually subjected to a series of nondestructive tests. It is checked for size, weight, shape, and metallurgical properties very quickly, and if it passes all tests satisfactorily, the product being purchased will make its way through the dispenser. To cite another example, it is common to use sonic energy to determine the location of a stud behind a wallboard. These examples, in a very broad sense, meet the definition of NDT—an object is evaluated without changing it or altering it in any fashion.

Finally, the sense of sight is employed regularly by individuals to evaluate characteristics such as color, shape, movement, and distance, as well as for identification purposes (Hellier, 2003). In fact, the human body has been described as one of the most unique NDT instruments ever created. Heat can be sensed by placing a hand in close proximity to a hot object and, without touching it, determining there is a relatively higher temperature present in that object. With the sense of smell, a determination can be made that there is an unpleasant substance present based simply on the odor. Without visibly observing an object, it is possible to determine roughness, configuration, size, and shape simply through the sense of touch. The sense of hearing allows the analysis of various sounds and noises and, based on this analysis, judgments and decisions relating to the source of those sounds can be made. For example, before crossing a street, we may hear a truck approaching. The obvious decision is not to step out in front of this large, moving object. But of all the human senses, the sense of sight provides us with the most versatile and unique NDT approach. When we consider the wide application of the sense of sight and the ultimate information that can be determined by mere visual observation, it becomes apparent that visual testing (VT) is a widely used form of NDT.

NDT, in fact, can be considered an extension of the human senses, often through the use of sophisticated electronic instrumentation and other unique equipment.

In industry, NDT can do the following:

1. Examine raw materials before processing
2. Evaluate materials during processing for process control
3. Examine finished products
4. Evaluate products and structures once in service

It is possible to increase the sensitivity and application of the human senses when used in conjunction with these instruments and equipment. But the misuse or improper application of a nondestructive test can cause catastrophic results. If the test is not properly conducted or if the interpretation of the results is incorrect, disastrous results can occur. It is essential that the proper nondestructive test method and technique be employed by qualified personnel to minimize these problems. Conditions for effective NDT will be covered later in this chapter.

To summarize, NDT is a valuable technology that can provide useful information on the condition of an object if it tests essential elements, follows approved procedures, and is conducted by qualified personnel.

2.2.2 Concerns about NDT

Certain misconceptions and misunderstandings should be addressed. One widespread misconception is that the use of NDT will ensure, to a degree, that a part will not fail or malfunction. This is not necessarily true. Every nondestructive test method has limitations. A nondestructive test by itself is not a panacea. In most cases, a thorough examination will require a minimum of two methods: one for conditions that would exist internally in the part and another method more sensitive to conditions at the surface of the part. It is essential that the limitations of each method be known prior to use. For example, certain discontinuities may be unfavorably oriented toward detection by a specific nondestructive test method. The threshold of detectability is another major variable that must be understood and addressed for each method. It is true that there are standards and codes describing the type and size of discontinuities considered acceptable or not acceptable, but if the examination method is not capable of capturing these conditions, the codes and standards are basically meaningless.

Another misconception involves the nature and characteristics of the part or object being examined. It is essential that as much information as possible be known and understood as a prerequisite to establishing test techniques. Important attributes such as the processes that the part has undergone and the intended use of the part, as well as applicable codes and standards, must be thoroughly understood as a prerequisite to performing a nondestructive test. The nature of the discontinuities anticipated for the particular test object should also be well known and understood.

At times, the erroneous assumption is made that if a part has been examined using an NDT method or technique, some magical transformation guarantees the part is sound. Codes and standards establish minimum requirements and are not a source of assurance that discontinuities will not be present. Acceptable and unacceptable discontinuities are identified by these standards, but there is no guarantee that all acceptable discontinuities will not cause some type of problem after the part is in service. Again, this illustrates the need for some type of monitoring or evaluation of the part or structure once it is operational.

A final widespread misunderstanding is related to the personnel performing these examinations. NDT is a "hands-on" technology, but not necessarily an easy one to apply. In fact, the most sophisticated equipment and the most thoroughly developed techniques and procedures can result in potentially unsatisfactory results when applied by an unqualified examiner. A major ingredient in the effectiveness of a nondestructive test is the personnel conducting it and their qualifications (Hellier, 2003).

2.2.3 Conditions for Effective NDT

Many variables associated with NDT must be controlled and optimized. The following major factors must be considered for an NDT to be effective (Hellier, 2003):

1. The product must be "testable." There are inherent limitations with each nondestructive test method, and it is essential that these limitations be known so the appropriate method is applied based on the variables associated with the test object. For example, it would be very difficult to provide a meaningful ultrasonic test on a small casting with very complex shapes and rough surfaces. In this case, it would be much more appropriate to consider radiography. In another case, the object may be extremely thick and high in density, making radiography impractical. Ultrasonic testing, on the other hand, may be very effective. In addition to the test object being "testable," it must also be accessible.

2. Approved procedures must be followed. It is essential that all NDEs be performed following procedures developed in accordance with the applicable requirements or specifications. In addition, it is necessary to qualify or "prove" the procedure to assure it will detect the applicable discontinuities or conditions and the part can be examined in a manner that will satisfy the requirements.

3. Once the procedure has been qualified, a certified NDT Level III individual or other quality assurance person who is suitably qualified to properly assess the adequacy of the procedure should approve it.

4. Equipment must operate properly. All equipment to be used must be in good operating condition and properly calibrated. In addition, control checks should be performed periodically to ensure the equipment and accessory items are functioning properly.

5. Annual calibrations are usually required but a "functional" check is necessary as a prerequisite to actual test performance.

6. Documentation must be complete. It is essential that proper test documentation be completed at the conclusion of the examination. This should address all key elements of the examination, including calibration data, equipment and part description, procedure used, identification of discontinuities if detected, and so on. In addition, the test documentation should be legible. There have been cases where the examination was performed properly and yet the documentation was so difficult to interpret that it cast doubt on the results and led to concerns regarding the validity of the entire process.

7. Personnel must be qualified. Since NDT is a "hands-on" technology and depends on the capabilities of the individuals performing the examinations, personnel must not only be qualified, but also properly certified. Qualification involves formalized planned training, testing, and defined experience.

2.2.4 Qualification as a Main Difference

The effectiveness of a nondestructive test primarily depends on the qualifications of the individuals performing the examinations. Most nondestructive tests require thorough control of the many variables associated with these examinations. The subject of personnel qualification has been an issue of much discussion, debate, and controversy over recent decades. There are many different positions on what constitutes qualification. The most common approach is to utilize some form of certification, but there are many very different certification programs.

The term "qualification" generally refers to the skills, characteristics, and abilities of the individual performing the examinations which are achieved through a balanced blend of training and experience. "Certification" is defined as some form of documentation or testimony that attests to an individual's qualification. Therefore, the obvious process involved in the attainment of a level of certification necessitates that the individual satisfactorily completes certain levels of qualification (training combined with experience) as a prerequisite to certification. In fact, a simple way to explain this system would be to consider the steps involved in becoming a licensed driver. A candidate for a driver's license must go through a series of practical exercises in learning how to maneuver and control a motor vehicle and, in time, is required to review and understand the various regulations dealing with driving that vehicle. Once the regulations are studied and understood, and upon completion of actual practice driving a vehicle, the individual is then ready to take the "certification examination." Most states and countries require applicants to pass both written and vision examinations, as well as to demonstrate their ability to operate and maneuver the motor vehicle. Once those examinations are completed, the candidate is issued the "certification" in the form of a driver's license. The mere possession of a driver's license does not guarantee there will not be mistakes. It is obvious that there are individuals who carry driver's licenses but are not necessarily qualified to safely drive the vehicles. This is quite apparent during "rush-hour" traffic time.

Unfortunately, the same situation occurs in NDT. Since individuals by the thousands are certified by their employers, there are major variations within a given level of certification among NDT practitioners. Those countries that have adopted some form of centralized certification do not experience these variations to the same degree as those still using employer certification approaches.

One of the earliest references to any form of qualification program for NDT personnel appears in the 1945 spring issue of a journal titled *Industrial Radiography*, published by the American Industrial Radium and X-Ray Society. The name of this organization was eventually changed to the Society for Nondestructive Testing (SNT) and, ultimately, the American Society for Nondestructive Testing (ASNT). The original journal, *Industrial Radiography*, is now referred to as *Materials Evaluation*. An article in that 1945 issue titled "Qualifications of an Industrial Radiographer" proposed that the society establish standards for the "registration" of radiographers by some type of examination, leading to a certification program. By the late 1950s, the subject of qualification or registration was being discussed more frequently. A 1961 issue of the journal, now called *Nondestructive Testing*, contained an article titled "Certification of Industrial Radiographers in Canada." Then, in 1963, at the Society for Nondestructive Testing's national conference, a newly formed task group presented a report titled "Recommended Examination Procedure" for personnel certification. Finally, in 1967, ASNT published the first edition of "Recommended Practice" for the qualification and certification of NDT personnel in five methods (applicable NDT methods are used as eddy current [ET], liquid penetrant [PT], magnetic particle [MT], radiography [RT], ultrasonic [UT], and visual testing [VT]). This first edition, generally called the 1968 edition of SNT-TC-1A, was a set of recommendations designed to provide guidelines to assist employers in the development of a procedure referred to by the document as a "Written Practice." This Written Practice became the key procedure for the qualification and certification of NDT personnel.

Today, SNT-TC-1A continues to be used widely in the United States as well as in many other countries and, in fact, it is probably the most widely used program for NDT personnel certification in the world. Over the years, it has been revised, starting in 1975, then again in 1980, 1984, 1988, 1992, 1996, and 2000. With this pattern of revisions, it is anticipated that there should be a new revision every four years. The Canadian program was first made available in 1961.

With the development of a new training and certification standard for CM, namely, ISO18436, through the efforts of ISO/TC108/SC5, a new level of harmonization now exists between the various CM methods. Methods covered include VA, AE, lubrication management (LM), and IT. AE and IT are testing methods also specified in some NDT standards such as EN473 and ISO9712. With the acceptance of ISO18436, the world is now experiencing "growth pains" with a good deal of resulting confusion for some trainers and certifying bodies (CBs) issuing certificates of compliance to ISO18436. The issues of concern include the following:

1. Two different ISO committees (TC108 and TC135) work on training specification documents for CM and NDT, respectively.
2. People in the NDT community do not appreciate the differences between NDT and CM for the same method.
3. Companies are unfamiliar with ISO18436 and the 30+ "technical foundation" CM standards.
4. Not all CBs within the European Federation of Non-Destructive Testing (EFNDT) will extend their scope to include certification of CM under ISO18436 within 2006–2007, leaving implementation of the program to a few CBs.
5. The world obviously recognizes the growth in CM, but a significant number of "rogue" trainers are falsely claiming to be accredited trainers complying with ISO18436.
6. Some CBs prefer to stay with EN473, ISO9712, IAEA 628, and ASNT for training and certification of AE and IT personnel under an NDT umbrella, even though their duties are really CM-based, thus increasing the confusion among practitioners.
7. NDT training bodies see the establishment of CM training bodies as detrimental to their business, therefore offering resistance to CM scheme development in general.
8. Confusion exists within the CM industry, which has never had certification standards, as to who accredits or approves whom and who certifies whom.

These concerns being experienced by people in the CM community with the roll-out of ISO18436 are also topics of discussion at ISO/TC108.

With some large multinational companies looking to include ISO18436 into company policy in 2006 and many global CM training companies looking to supply the demand, the future for CM is bright. Resolving some of these issues and improving the relationship with NDT practitioners will ease the growth pains.

To begin discussion on these issues, it is pertinent to state that both NDT and CM methods have been serving industry, side by side, for many decades. The mature NDT methods are adequately specified in standards such as EN473 and ISO9712 and guidelines such as SNT-TC-1A. CM methods such as VA and LM (which includes tribology and wear debris analysis) are also mature methods, but have never been covered by a certification of competence. In CM, in 1996, the ISO technical committee (ISO/TC108/SC5) began work on standards for vibration and shock in machines. The scope of TC108 has since evolved to include AE, LM, and IT, with the most recent extension (December 2005) to include airborne UT, laser stereography (interferometry), and motor current analysis. The committee scope has now been extended to include monitoring of structures.

2.2.4.1 Commonality between NDT and CM

The commonality of methods between NDT and CM specified by ISO18436 exist mainly with AE and IT, but there is dissention among practitioners. More specifically, there is a genuine equivalence only in AE at level 3, as specified in ISO/DIS18436-6 and ISO9712:2005. Of added concern throughout the CM community is the inclusion of IT and AE into the NDT technical reference documents ISO/TR 25107 (Non-Destructive Testing—Guidelines for NDT Training Syllabuses), which enhances the similarity of these methods in NDT and CM. However, the member CBs of CEN/TC138 decided in 2005 not to adopt the ISO9712:2005 version, deciding instead to review EN473:2000 (Stephen, 2006).

Regardless of the commonality of methods between NDT and CM in areas of general theory and basic applications, there is a definite demarcation in areas such as CM program design, implementation and management, CM standards, and CM applications (in machines, structures, electrical, and other applications), which do not exist in NDT training. Any commonality is restricted by the differences in scope and applications in different industrial applications. That is, CM is primarily predictive in intent, as it applies test methods to monitor the condition (or health) of a machine or structure over time, generating diagnostic data that yield prognostic (predictive) output using CM prognostic standards, providing residual life determination and whole life cycle cost analysis of major assets.

NDT applications are restricted to reactive or scheduled testing within a defined maintenance or production protocol, normally on materials or components that are not part of a machine. The latter is also an application undertaken by a CM practitioner, suggesting that NDT methods are tools (a subset) of a CM program.

To harmonize NDT and CM, each CB could establish training modules that comply with both CM and NDT requirements, for example, theory and basic applications, which all NDT-qualified persons would satisfy. The remaining CM-specific modules could be CM-diagnostics/prognostics, CM-design/implementation and management, CM-standards and codes, and CM-sector-specific machine, structures, or electrical systems applications. Future development of IT modules may include medical and/or veterinary thermography for monitoring the condition (health) of patients (human or animal). That is, for example, a level 2 NDT-qualified AE person could have his/her NDT certificate accepted as proof of partly meeting the CM prerequisite, and needs only to attend a shortened training program consisting of selected CM-specific modules, together with meeting the CM experience requirement of testing in a CM environment.

2.3 Oil Analysis

Oil must be sampled and analyzed for various properties and materials to monitor wear and contamination in an engine, transmission, or hydraulic system. Regular sampling and analysis establishes a baseline of normal wear; thus, abnormal wear or contamination is more readily apparent (Bob is the Oil Guy, 2014).

Oil that has been inside any moving mechanical apparatus, including engines, for any length of time will reflect the exact condition of that apparatus. Over time and with increasing wear, tiny metallic trace particles enter the oil and remain in suspension. Many other products of the combustion process also become trapped in the circulating oil. In effect, the oil tells the machine's whole history.

The oil will contain particles caused by normal wear and operation, along with particles from externally caused contamination. By identifying and measuring the impurities, we can determine the rate of wear and highlight excessive contamination. At the same time, the oil analysis can suggest ways to reduce any accelerated wear or contamination.

Oil analysis (e.g., ferrography, particle counter testing) can be performed on different types of oils such as lubrication, hydraulic, or insulation oils. It can indicate machine degradation (e.g., wear), oil contamination, improper oil consistency (e.g., incorrect or improper amount of additives), and oil deterioration.

The science of oil analysis falls into four main areas:

1. Fluid physical properties (viscosity, appearance)

2. Fluid chemical properties (TBN, TAN, additives, contamination, water percent)
3. Fluid contamination (ISO cleanliness, ferrography, spectroscopy, dissolved gases [transformer])
4. Machine health (wear metals associated with plant components)

In addition, oil analysis can be divided into a number of different categories (Randall, 2011):

1. *Chip detectors:* Filters and magnetic plugs are designed to retain chips and other debris in circulating lubricant systems; these are analyzed for quantity, type, shape, size, and so on. Alternatively, suspended particles can be detected in flow past a window.
2. *Spectrographic oil analysis procedures (SOAP):* Here, the lubricant is sampled at regular intervals and subjected to spectrographic chemical analysis. Detection of trace elements can tell of wear of special materials such as alloying elements in special steels, white metal or bronze bearings, and so on. Another case applies to oil from engine crankcases, where the presence of water leaks can be indicated by a growth in NaCl or other chemicals coming from the cooling water. Oil analysis includes analysis of wear debris, contaminants and additives, and measurement of viscosity and degradation. Simpler devices measure total iron content.
3. *Ferrography:* This represents microscopic investigation and analysis of debris retained magnetically (hence the name), but which can contain nonmagnetic particles caught up with the magnetic ones. Quantity, shape, and size of the wear particles are all important factors in isolating the type and location of failure.

Oil analysis typically tests for a number of different materials to determine sources of wear, find dirt, and other contamination. This analysis can even check if lubricants are appropriate. To sum up, this analysis can detect the following:

1. Fuel dilution of lubrication oil
2. Dirt contamination in the oil
3. Antifreeze in the oil
4. Excessive bearing wear
5. Misapplication of lubricants

Of course, wear is to be expected, but abnormal levels of wear in a particular material can give early warning of a potential problem and perhaps prevent a major breakdown. Early detection, in turn, permits corrective action such as repairing an air intake leak before major damage. In fact, the advantage of an oil analysis program is the ability to anticipate problems and schedule repair work to avoid downtime during a critical time of use. Early detection offers the following benefits to an organization:

1. Repair bills are reduced
2. Catastrophic failures are reduced
3. Machinery life is increased
4. Nonscheduled downtime is decreased

2.3.1 Looking Inside

Oil analysis allows maintainers to predict possible impending failure without taking the equipment apart. Using this method, a maintainer can look inside anything—an engine, a transmission, a hydraulic system—without taking it apart.

2.3.2 Physical Tests

Oil sample analysis generally tests for the following physical properties (Grisso and Melvin, 1995):

1. *Antifreeze* forms a gummy substance that may reduce oil flow, leading to oxidation, oil thickening, acidity, and, ultimately, engine failure.
2. *Fuel dilution* thins oil, lowers lubricating ability, and may drop oil pressure causing higher wear.
3. *Oxidation* measures gums, varnishes, and oxidation products. The use of oil that is too hot or the use of oil for too long a period leaves sludge and varnish deposits, thickens the oil, and causes high oxidation.
4. *Total base* number indicates the remaining acid-neutralizing capacity of the lubricant.
5. *Total solids* include ash, carbon, lead salts (gasoline engines), and oil oxidation.
6. *Viscosity* is measures oil's resistance to flow. On the one hand, oil may thin if it is sheared in multiviscosity oils or diluted with fuel. On the other hand, oil may thicken from oxidation when run too long or too hot. Oil may also thicken from contamination by antifreeze, sugar, and other materials.

2.3.3 Metal Tests

Metals tested for and generally included in oil sample analysis are:

1. *Aluminum (Al)*: Used in construction of thrust washers, bearings, and pistons. High readings of aluminum in oil can reflect piston skirt scuffing, excessive ring groove wear, broken thrust washers, and so on.
2. *Boron, magnesium, calcium, barium, phosphorus, and zinc*: Generally due to lubricating oil additive packages, that is, detergents, dispersants, extreme-pressure additives, and so on.
3. *Chromium (Cr)*: Usually associated with piston rings. High levels of chromium are found in oil when dirt enters by way of the air intake or broken rings.
4. *Copper (Cu) and tin*: Normally from bearings or bushings, and valve guides, but oil coolers and some oil additives can also contribute to high readings. Copper readings will normally be high at first usage of a new engine but will decline over time, that is, in a few hundred hours.
5. *Iron (Fe)*: Is found in many places in the engine: liners, camshafts, crankshaft, valve train, timing gears, and so on.
6. *Lead (Pb)*: Very high lead readings result from use of regular gasoline, with bearing wear having some influence as well. Generally speaking, however, fuel source (leaded gasoline) and sampling contamination (use of galvanized containers for sampling) are critical in lead tests.
7. *Silicon (Si)*: Dirt or fine sand contamination from a leaking air intake system generally cause high readings of silicon. Dirt and sand are an abrasive, causing excessive wear. *Note*: In some oils, silicon is used as an antifoam agent.
8. *Sodium (Na)*: Coolant leaks are generally the source of high sodium readings, but an oil additive package can be influential as well.

2.3.4 Oil Analysis Benefits

One of the benefits of oil analysis is that it detects problems in both the fluid and the machine. It can also detect some defects earlier than other technologies. Oil analysis is often referred to as the first line of defense as far as predictive technologies are concerned (Symphony Teleca, 2013). The oil sample reports will define the following items:

1. The presence of foreign fluids or destructive surface contaminants
2. The overall physical and chemical condition of the fluid
3. The presence of machine wear materials, how much and of what type and morphology

Successful use of oil analysis requires that oil sampling, changing, and top-up procedures are all well-defined and documented. It is much more difficult to apply lubricant analysis to the grease of lubricated machines, but grease sampling kits are now available to make the process more reliable.

2.4 Vibration Analysis

One of the many forms of CM utilizes the principles of VA. This system has been successfully used for many years to monitor equipment as a whole by measuring the vibration of the overall machine or by analyzing the individual components.

These vibrations are used to trend machinery and components, tracking any changes that may arise. These changes indicate possible problems associated with the machinery and that further monitoring is required until maintenance can be performed.

Vibration can be defined as the motion of a machine and its components from a resting position. External forces act to produce motion or movement inherent to a particular machine, its components, and use. It can be measured by specific equipment, for instance, an accelerometer, which converts the vibrating motion to an electrical signal in preparation for analysis. This motion takes a sinusoidal waveform, with characteristics of frequency, amplitude, wavelength, and phase. When machine vibration is measured, several sinusoidal waveforms or motions are usually found; they combine to give an overall time waveform. To improve the use of this waveform, a Fourier analysis is performed using specialized equipment to convert the time waveform to amplitude versus frequency spectrum. This equipment is known as a Fourier transform analyzer or a fast Fourier transform (FFT).

1. $x(t)$—Continuous-time signal
2. $X(f)$—Fourier transform

When $X(f) = \int_{-\infty}^{\infty} x(t)e^{-j2\pi ft}dt$, we find a continuous signal. If the transform is computed numerically using k samples, the expression is

$$X_k = \sum_{n=0}^{N-1} x_n e^{-j2\pi kn/N} \quad k = 1, 2, \dots, N$$

for the line spectrum at frequency $\omega_k = (2\pi)^k/T$.

The frequency spectrum identifies the frequency that gives an indication of possible components with problems. This frequency forms an essential

TABLE 2.1

Correlation of Lubricant Wear and Wear Particle Analysis with Vibration and Thermography

Technology	Correlative Method	Indication	Usage
Vibration	Time sequence	Wear particle build-up precedes significant vibration increase in most instances	Routinely (monthly)
Thermal analysis	Time coincident	Major wear particle production (near end of bearing life) occurs as the bearings fail	When bearing degradation is suspected
Advance filtration/ debris analysis	Time sequence/ coincident	Major bearing damage has occurred when significant amounts of material appear in the lubricating system filters	Routinely with each filter cleaning or change

Source: Data from Wenzel, R., 2011. *Condition-Based Maintenance: Tools to Prevent Equipment Failures.* [Online] Available at: http://www.machinerylubrication.com/Read/28522/condition-based-maintenance.

part of the analysis, in conjunction with the amplitude. The amplitude identifies the magnitude of the component vibration, giving an indication of its condition. It is to be noted that time waveform analysis is also very useful, as it identifies discrepancies and energy developed by vibrations.

Tables 2.1 and 2.2 show the strengths and weaknesses of VA compared to other methods.

2.4.1 Machine Vibration Causes

Almost all machine vibration is due to one or more of these causes (Vyas, 2013, GE Measurement & Control, 2013):

1. Repeating forces
2. Looseness
3. Resonance

1. Repeating forces in machines are mostly due to the rotation of imbalanced, misaligned, worn, or improperly driven machine components. Examples of these four types of repeating forces are shown below.

   Imbalance machine components contain "heavy spots," which, when rotating, exert a repeating force on the machine. Imbalance is often caused by machining errors, nonuniform material density, variations in bolt sizes, air cavities in cast parts, missing balance weights, incorrect balancing, uneven electric motor windings, and broken, deformed, corroded or dirty fan blades or dirt that has dropped from a fan blade, suddenly creating a big imbalance.

TABLE 2.2

Strengths and Weaknesses of VA Compared with Oil Analysis

Equipment Condition	Oil Analysis	Vibration Analysis	Correlation
Oil-lubricated antifriction bearings	Strength	Strength	Oil analysis can detect infant failure condition. Vibration analysis provides late failure state information
Oil-lubricated journal/thrust bearings	Strength	Mixed	Wear debris will generate in the oil prior to a rub or looseness condition
Imbalance	N/A	Strength	Vibration analysis can detect imbalance. Oil analysis will eventually detect the effect of increased bearing load
Water in oil	Strength	N/A	Oil analysis can detect water in oil. Vibration analysis is unlikely to detect this
Greased bearings	Mixed	Strength	Some labs do not have adequate experience with grease analysis. Vibration analysis can detect greasing problems
Shaft cracks	N/A	Strength	Vibration analysis is very effective in diagnosing a cracked shaft
Gear wear	Strength	Strength	Oil analysis can predict the failure mode. Vibration analysis can predict which gear is worn
Alignment	N/A	Strength	Vibration analysis can detect a misalignment condition. Oil analysis will eventually see the effect of increased load
Lubricant condition	Strength	N/A	Oil analysis can determine inadequate lubrication
Resonance	N/A	Strength	Vibration analysis can detect resonance. Oil analysis will eventually see the effect
Root cause	Strength	Strength	Need oil and vibration analysis to work best

Source: Data from Wenzel, R., 2011. *Condition-Based Maintenance: Tools to Prevent Equipment Failures.* [Online] Available at: http://www.machinerylubrication.com/Read/28522/condition-based-maintenance.

Misaligned machine components create "bending moments," which, when rotated, exert a repeating force on the machine. Misalignment is often caused in inaccurate assembly, uneven floors, thermal expansion, distortions due to fastening torque, and improper mounting of couplings.

Worn machine components exert a repeating force on the machine because of the rubbing of uneven worn surfaces. The wear in roller bearings, gears, and belts is often due to improper mounting, poor lubrication, manufacturing defects, and overloading.

Improperly driven machine components exert a repeating force on the machine because of intermittent power use. Examples include pumps receiving air in pulses, internal combustion engines with misfiring cylinders, and intermittent brush–commutator contact in DC motors.

One or several of those repeating forces in combination with looseness and/or resonance will create even more problems.

2. Looseness of machine parts causes a machine to vibrate. If parts become loose, vibration that is normally of tolerable levels may become unrestrained and excessive. Looseness can cause vibration in both rotating and nonrotating machinery. Looseness is often due to excessive bearing clearances, loose mounting bolts, mismatched parts, corrosion, and cracked structure.

3. Resonance is a state of operation wherein an excitation frequency is close to a natural frequency of the machine structure; at this frequency, the damping of the vibration is very low and the result is resonance. When resonance occurs, the resulting vibration levels can be very high and can cause damage very quickly.

Machines tend to vibrate at certain oscillation rates. The oscillation rate at which a machine tends to vibrate is called its natural oscillation rate or the resonant frequency (Lindley et al., 2008). The natural oscillation rate of a machine is the vibration rate most natural to the machine, that is, the rate at which the machine "prefers" to vibrate. A machine left to vibrate freely will tend to vibrate at its natural oscillation rate.

A repeating force, close to a natural frequency, causing resonance may be small and may originate from the motion of a good machine component. Such a mild repeating force is normally not a problem until it begins to cause resonance. Resonance, however, should always be avoided as it can cause rapid and severe damage. To avoid resonance, we must either change the frequency of the repeating force or change the resonant frequency.

Making the structure stiffer makes the resonant frequency higher, while increasing the weight makes the resonant frequency lower.

2.4.2 How Is Machine Vibration Described?

To analyze the condition of a machine, we must first accurately describe the behavior or symptoms of the machine (Commtest Instruments, 2014a,b).

1. How can vibration symptoms be described accurately?
2. How do vibration analysts describe the condition of a machine?

By watching, feeling, and listening to machine vibration, we can sometimes roughly determine the severity of the vibration. We may observe that certain kinds of machine vibration appear "rough," others "noticeable," and yet others "negligible." We can also touch a vibrating bearing house and feel that it is "hot," or hear that it is "noisy," and so conclude something is wrong.

Describing machine vibration with these general terms is, however, imprecise and depends on the person making the assessment. What appears "rough" to one person may appear acceptable to another. Verbal description is usually unreliable (Dennis, 1994).

To accurately analyze a vibration problem, it is necessary to describe the vibration in a consistent and reliable manner. Vibration analysts rely primarily on numerical descriptions rather than verbal descriptions to analyze vibration accurately and to communicate effectively. The three most important numerical descriptors of machine vibration are amplitude, frequency, and phase.

Amplitude describes the severity of vibration, and frequency describes the oscillation rate of vibration (how frequently an object vibrates). Phase describes the synchronizing in time compared to a reference. Together, they provide a basis for identifying the root cause of vibration.

2.4.2.1 What Is Amplitude?

The amplitude of vibration is the magnitude of vibration.

A machine with large vibration amplitude experiences large, fast, or forceful vibratory movements. Normally, the larger the amplitude, the more movement or stress is experienced by the machine, and the more prone the machine is to damage. A machine with a resonant problem may sometimes behave a bit differently than it normally does.

Vibration amplitude is usually an indication of the severity of vibration; sudden changes in phase may also be a sign of trouble (Dennis, 1994).

In general, the severity or amplitude of vibration relates to

1. The size of the vibratory movement (displacement)
2. The speed of the movement (velocity)
3. The force associated with the movement (acceleration)

The vibration amplitude can be expressed in displacement, velocity, or acceleration, Vd, Vv, and Va, respectively.

If the frequency is F, then $Vv = 2*\pi*F*Vd$

$$Va = 4*\pi^2*F^2*V_d$$

In most situations, it is the speed or velocity amplitude of a machine that gives the most useful information about the condition of the machine.

Amplitude can be expressed in terms of its peak value, or what is known as its root-mean-square value.

The peak velocity amplitude of a vibrating machine is simply the maximum (peak) vibration speed attained by the machine in a given time period, as shown in Figure 2.1.

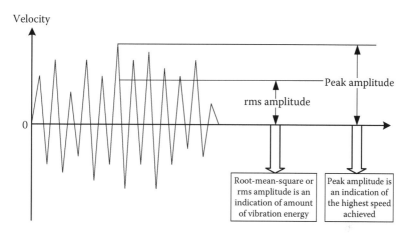

FIGURE 2.1
The maximum (peak) vibration speed attained by the machine in a given time period.

In contrast to the peak velocity amplitude, the root-mean-square velocity amplitude of a vibrating machine tells us the vibration energy in the machine. The higher the vibration energy, the higher the root-mean-square velocity amplitude.

The term "root-mean-square" is often shortened to "rms." It is useful to remember that the rms amplitude is always lower than the peak amplitude.

2.4.2.2 What Is Frequency?

A vibrating machine component oscillates; that is, it goes through repeated cycles of movement. Depending on the force causing the vibration, a machine component may oscillate rapidly or slowly. The rate at which a machine component oscillates is called its oscillation or vibration frequency. The higher the vibration frequency, the faster the oscillation.

The frequency of a vibrating component can be determined by counting the number of oscillation cycles completed every second. For example, a component going through five vibration cycles every second is said to be vibrating at a frequency of five cycles per second. As shown in Figure 2.2, one cycle of a signal is simply one complete sequence of the shortest pattern that characterizes the signal.

The vibration rate or frequency of a machine component is often a useful indicator of the root cause of the vibration.

Frequency, as with amplitude, is always expressed with a unit. Commonly used frequency units are cycles per second (cps), Hertz (Hz), and cycles per minute (cpm). Hertz is a unit equivalent to "cycles per second." One Hz is equal to one cps (one cycle per second) or 60 cpm (60 cycles per minute).

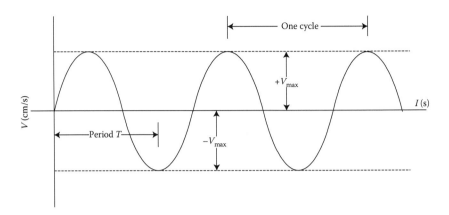

FIGURE 2.2
One cycle of a signal.

2.4.2.3 What Is Phase?

Phase is all about the relative timing of related events. Here are a few examples:

1. When balancing, we are interested in the timing between the heavy spot on the rotor and a reference point on the shaft. We need to determine where that heavy spot is located, and the amount of weight required to counteract the rotational forces.

2. When we look at fault conditions such as imbalance, misalignment, eccentricity, and foundation problems, we are interested in the dynamic forces inside the machine and, as a result, the movement of one point in relation to another point.

3. We can use phase to understand the motion of the machine or structure when we suspect a machine of structural resonance, where the whole machine may be swaying from side to side, twisting this way and that, or bouncing up and down.

In other words, phase is very helpful when balancing and when trying to understand the motion of a machine or structure. But it is also very useful when trying to diagnose machine fault conditions. Phase is the position of a rotating part at any instant with respect to a fixed point/reference. Phase describes the synchronizing in time compared to this reference. It gives us the vibration direction. A phase study is a collection of phase measurements made on a machine or structure and evaluated to reveal information about relative motion between components.

In VA, phase is measured using absolute or relative techniques. Absolute phase is measured with one sensor and one tachometer referencing a mark on the rotating shaft (Figure 2.3). At each measurement point, the analyzer

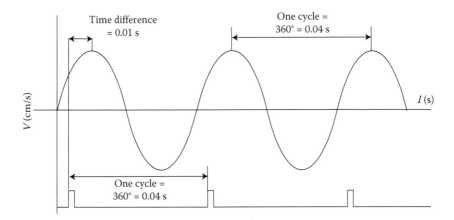

FIGURE 2.3
Absolute phase. (Redrawn from Mobius Institute, 2008.)

calculates the time between the tachometer trigger and the next positive waveform peak vibration. This time interval is converted to degrees and displayed as the absolute phase. Phase can be measured at shaft rotational frequency or any whole number multiple of shaft speed (synchronous frequencies). Absolute phase is required for rotor balancing.

Instead of a tachometer to generate a once-per-revolution signal, we can use a displacement (proximity) probe, which is aimed at a keyway or setscrew. Stroboscopes can also be used to collect phase readings. If the strobe is tuned to the running speed of the machine (so that the shaft or coupling appears to have stopped rotating), the output of the strobe can be connected to the tachometer input of the data collector. The data collector will treat the signal from the strobe as if it were a normal tachometer input. The strobe can also be used to trigger an optical tachometer connected to the input of the data collector. To compare phase readings from one time to another, the reference has to be the same every time. When FFT is used for analysis, the reference signal triggers the start of data collection.

Relative phase is measured on a multichannel vibration analyzer using two or more (similar type) vibration sensors (Figure 2.4). The analyzer must be able to measure cross-channel phase. One single-axis sensor serves as the fixed reference and is placed somewhere on the machine (typically on a bearing housing). Another single-axis or triaxial sensor is moved sequentially to all of the other test points. At each test point, the analyzer compares waveforms between the fixed and roving sensors. Relative phase is the time difference between the waveforms at a specific frequency converted to degrees. Relative phase does not require a tachometer so phase can be measured at any frequency.

Both types of phase measurements are easy to make. Relative phase is the most convenient way to measure phase on a machine because the machine

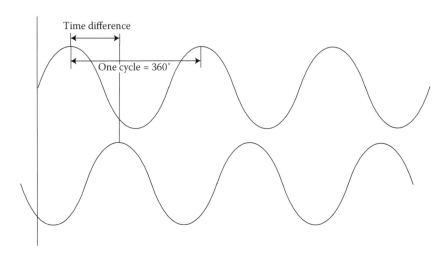

FIGURE 2.4
Relative phase. (Redrawn from Mobius Institute, 2008.)

does not need to be stopped to install reflective tape on the shaft. Phase can be measured at any frequency. Most single-channel vibration analyzers can measure absolute phase. Multichannel vibration analyzers normally have standard functions for measuring both absolute and relative phase.

2.4.2.4 What Is a Waveform?

The graphical display of electrical signals from a person's heart (electro-cardiogram or ECG) is useful for analyzing the heart's medical condition. In a similar way, graphical displays of vibratory motion are useful tools for analyzing the nature of vibration. We can often find clues to the cause and severity of vibration in the graphical display of vibratory motion. One display commonly used by vibration analysts is the waveform, a graphical representation of how the vibration level changes with time. Figures 2.5 and 2.6 show an example of a velocity waveform. A velocity waveform is simply a chart showing how the velocity of a vibrating component changes with time.

Sometimes we can also use the waveform of the signal to see a repeating pattern.

The amount of information a waveform contains depends on its duration and resolution. The duration of a waveform is the total time period over which information may be obtained from it. In most cases, a few seconds are sufficient. The resolution of a waveform is a measure of its level of detail and is determined by the number of data points or samples characterizing its shape. The more samples there are, the more detailed the waveform is.

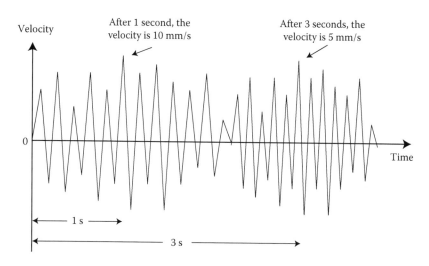

FIGURE 2.5
Velocity waveform.

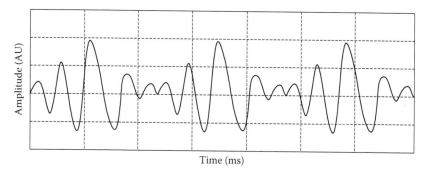

FIGURE 2.6
Waveform with repeating pattern.

2.4.2.5 What Is a Spectrum?

Another kind of display commonly used by vibration analysts is the spectrum. This is a graphical display of the frequencies at which a machine component is vibrating, together with the amplitudes of the component at these frequencies. Figure 2.7 shows a velocity spectrum.

A single machine component can be simultaneously vibrating at more than one frequency, because the machine vibration, as opposed to the simple oscillatory motion of a pendulum, does not usually consist of a single vibratory motion; rather, many take place simultaneously.

For example, the velocity spectrum of a vibrating bearing usually shows the bearing is vibrating at not just one frequency but at various frequencies.

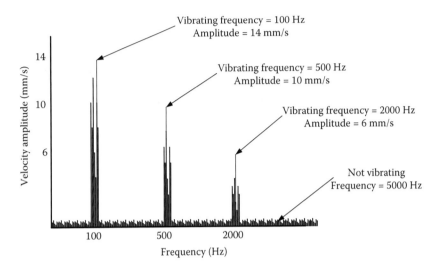

FIGURE 2.7
Velocity spectrum.

Vibration at some frequencies may be due to the movement of bearing elements, at other frequencies because of the interaction of gear teeth, and at yet other frequencies because of electrical problems in the motor.

Because a spectrum shows the frequencies at which vibration occurs, it is a very useful analytical tool. By studying the individual frequencies at which a machine component vibrates, as well as the amplitudes corresponding to those frequencies, we can infer a great deal about the cause of the vibration and the condition of the machine.

In contrast, a waveform does not clearly display the individual frequencies at which vibration occurs. Rather, a waveform displays only the overall effect. It is, thus, not easy to manually diagnose machine problems using waveforms. With the exception of a few specialized cases, spectra (not waveforms) are usually the primary tool for analyzing machine vibration.

The information contained by a spectrum depends on the F_{max} and resolution of the spectrum. The F_{max} of a spectrum is the frequency range over which information may be obtained from the spectrum. How high F_{max} needs to be depends on the operating speed of the machine. The higher the operating speed, the higher F_{max} must be. The resolution of a spectrum is a measure of the level of detail in the spectrum and is determined by the number of spectral lines characterizing the shape of the spectrum. The more spectral lines, the more detailed the spectrum.

2.4.3 Vibration Sensors–Transducers

The type of sensors and data acquisition techniques employed in a maintenance program are critical factors that can determine its success or failure.

Accuracy, correct application, and proper installation will determine whether the findings of these techniques are valid or not.

Three types of vibration transducers can be used to monitor the mechanical condition of plant equipment, each with specific applications and limitations:

1. Displacement probe (proximity probe)
2. Velocity transducer
3. Accelerometer

Figure 2.8 shows the different transducers' normal limitations. With different kinds of compensation techniques, their use can be extended.

2.4.3.1 Displacement Probe (Proximity Probe)

Displacement probes are normally either capacity or eddy current displacement sensors.

Eddy current displacement sensors use a magnetic field that engulfs the end of the probe. As a result, any metallic objects close to the probe will affect the sensor output.

Capacitive sensors use the electrical property of "capacitance" to make measurements. Capacitance is a property that exists between any two conductive surfaces within some reasonable proximity. Changes in the distance

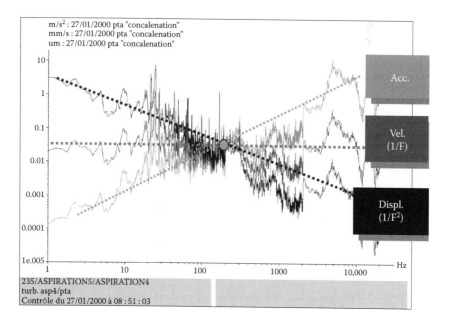

FIGURE 2.8

Normal limitation for different vibration transducers.

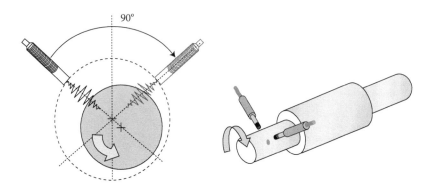

FIGURE 2.9
Displacement probe (proximity probe)—typical arrangement. Two channel measurement of the 0 to peak displacement. The S1 and S2 outputs are vector summed to produce a new time series called an "orbit," which is equal to one shaft rotation.

between the surfaces change the capacitance. It is this change of capacitance that capacitive sensors use to indicate changes in the position of a target.

Displacement probes are normally used to measure the absolute motion of a rotating machine axis with respect to the probe but can also be used to measure fixed parts. Therefore, the displacement sensor should be mounted on a rigid structure to ensure safe and repeatable data.

Turbines, compressors, and other heavy machines usually have displacement sensors mounted permanently in key measuring positions to supply data to the CM program. Two perpendicular proximity probes are often used to analyze journal bearing and shaft behavior with the so-called orbit analysis. The orbit represents the path of the shaft centerline within the bearing clearance. Two orthogonal probes are required to observe the complete motion of the shaft within. The dynamic motion of the shaft can be observed in real time by feeding the output of the two orthogonal probes to the X and Y of a dual-channel oscilloscope.

The useful frequency range for a displacement probe is normally 1–2000 Hz (60–120,000 rpm). The displacement data are usually recorded in mils peak to peak (when applied to the normative of an Anglo-Saxon source).

Laser displacement sensors can be used to measure displacement, but they are not common in industrial applications for vibration measurement in CM (Figure 2.9).

2.4.3.2 Velocity Transducers

The velocity transducer was one of the first vibration transducers to be built. It consists of a wire coil and magnet arranged so that if the housing is moved, the magnet tends to remain stationary due to its inertia. This can also be done the other way around. The relative motion between the magnetic field

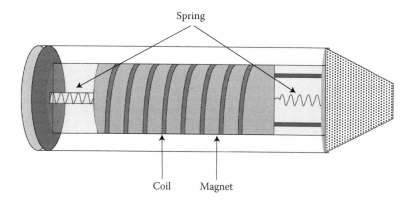

FIGURE 2.10
Velocity transducer. (Redrawn from Vibration Training Course Book Category II. 2013. Vibration school website: http://www.vibrationschool.com.)

and the coil induces a current proportional to the velocity of motion. The unit thus produces a signal directly proportional to vibration velocity. It is self-generating, needs no conditioning electronics to operate, and has a relatively low electrical output impedance, making it fairly insensitive to noise induction.

Velocity transducers commonly measure with a response range from 10 Hz to 2 kHz (Figure 2.10).

2.4.3.3 Accelerometers

Accelerometers can be based on the following technologies:

1. Piezoelectric
2. Piezoresistive
3. Capacitive

Piezoelectric accelerometers are the most widely used. These devices measure the vibration or acceleration of the motion of a structure. The force caused by vibration or a change in motion (acceleration) causes the mass to "squeeze" the piezoelectric material, which produces an electrical charge proportional to the force exerted upon it. Since the charge is proportional to the force, and the mass is a constant, the charge is also proportional to the acceleration.

There are two types of piezoelectric accelerometers (vibration sensors). The first is a "high impedance" charge output accelerometer. In this type of accelerometer, the piezoelectric crystal produces an electrical charge that is connected directly to the measurement instruments. This type of accelerometer is also used in high-temperature applications (>120°C), where low-impedance models cannot be used.

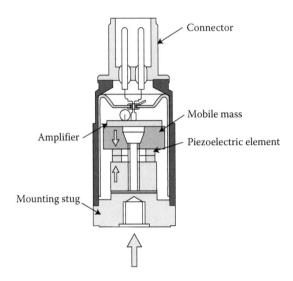

FIGURE 2.11
Typical accelerometer.

The second type is a low-impedance output accelerometer. It has a charge accelerometer as its front end, with a tiny built-in microcircuit and FET transistor able to convert that charge into a low-impedance voltage that can easily interface with standard instrumentation. The piezoelectric crystals are usually preloaded so that either an increase or decrease in acceleration causes a change in the charge they produce.

The output is normally 10–100 mV/G (1G = 9.8 m/s²), and the frequency range is from a few hertz (much lower with compensation) up to 20 kHz or even higher (Figure 2.11).

2.4.4 Mounting Techniques

PdM programs based on VA need accurate and repeatable data to determine the operating conditions of plant machinery. Apart from the transducers, three factors affect the quality of the data:

1. Location of the measurement point
2. Attachment and orientation of the transducer
3. Operational data for the machine (rotations per minute [RPM], power, etc.)

The location and orientation of the key measurement points of the machine are selected to provide the best possible detection of its incipient problems. Any deviation of exact point or orientation will affect the accuracy of the obtained data. Therefore, it is important that each measurement for the duration of the program be performed in exactly in the same place and direction.

The best method to ensure this is to

1. Drill and tap fixing holes for the transducer
2. Use cementing studs with fixing holes or quick connectors for the transducer
3. Use cementing studs with flat magnetic surfaces for permanent magnets

If we want permanent measuring points on a machine but do not want to drill and tap fixed holes, we can opt for cementing studs. These are attached to the measuring point with a hard glue, such as epoxy and cyanoacrylate types. Soft glues can considerably reduce the usable frequency range of the accelerometer. This fixing method gives good results and can be used for frequencies up to about 10–15 kHz (Figure 2.12).

A permanent magnet is a simple attachment method. The measuring point is a flat magnetic surface. If possible, a cemented stud with a flat magnetic surface should be used. This method can be used up to about 2 kHz. The magnet's holding force can usually handle vibration levels up to 1000–2000 m/s².

A hand-held probe with the accelerometer mounted on top is convenient for a fast survey. However, this can result in gross measuring errors because of low overall stiffness, and it may be difficult to get repeatable results. A low-pass filter can limit the measuring range at about 1000 Hz (Figure 2.13).

If measurements are taken with a three-axis accelerometer (a fast and convenient method), the orientation (x,y,z) must remain the same every time.

With single-axis transducers, if there is an angular alignment problem, the transducer should be positioned axially; if the underlying problem is looseness the transducer should be positioned vertically; the clearest indications of an unbalanced machine is normally given by transducers with a horizontal placement.

FIGURE 2.12

Typical studs and quick connectors for mounting of vibration transducers. (Redrawn from Farnell Electronics. http://www.farnell.com.)

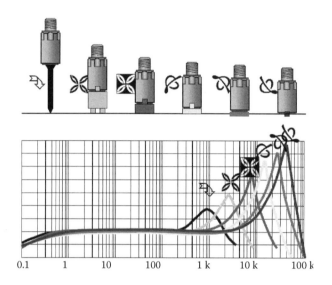

FIGURE 2.13
Different mounting techniques frequency response.

For each measurement, the actual operational condition for the machine should be recorded if it varies from time to time.

2.4.5 VA Applications

Vibration monitoring and analysis are the most common tools to prevent emerging mechanical problems related to manufacturing equipment in any production plant; they are not limited to rotating machines. Until recently, machines with low operating speeds were excluded from VA; however, at present, VA techniques are used in machines whose nominal speeds are of the order of 6 rpm or higher.

Keep in mind that all machines vibrate due to tolerances inherent to each of their construction elements and also due to their resonant frequencies and changes in operating conditions. In a new machine, these tolerances set the basic characteristics of that machine's vibration, which can be compared with future vibrations under similar operating conditions. Similar machines working in similar conditions will have similar characteristic vibrations that differ from each other mainly by their construction tolerances.

A change in the basic vibration of a machine, assuming it is operating under normal conditions, indicates an incipient fault in some of its elements, causing a change in the operating conditions of those elements. Different types of failures give rise to different types of changes in the characteristic vibration of the machine and can, therefore, help determine the source of the problem (Figure 2.14).

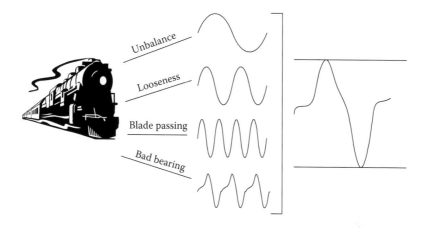

FIGURE 2.14
Different types of failures give rise to different types of changes in the characteristic vibration of the machine (in this picture the amplitude is in displacement).

2.4.5.1 How Does the Instrument Work?

Before taking a vibration measurement, we must attach a sensor that can detect the vibration behavior of the machine being measured. Various types of vibration sensors are available, but an accelerometer is normally used as it offers advantages over other sensors. An accelerometer produces an electrical signal proportional to the acceleration of the vibrating component to which it is attached.

The acceleration of a vibrating component is a measure of how quickly the velocity of the component is changing.

The acceleration signal produced by the accelerometer is passed on to the instrument that, in turn, converts the signal to a velocity signal. Depending on the user's choice, the signal can be displayed as a velocity waveform or a velocity spectrum. A velocity spectrum is derived from a velocity waveform by means of a mathematical calculation called the FFT.

Figure 2.15 is a very simplistic explanation of how vibration data are acquired.

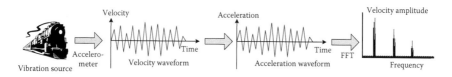

FIGURE 2.15
Acquisition of vibration data.

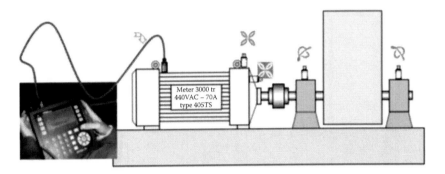

FIGURE 2.16
Vibration measurement with vibration meter.

2.4.6 Vibration Meters

To distinguish between vibration measurements, a portable device called a vibration meter can be used. This is a basic instrument for a PdM program based on vibrations. It is a small microprocessor specifically designed to collect, pack, and store vibration data in both the time and frequency domains (Figure 2.16).

This unit is used to check the mechanical condition of machines at regular intervals. As noted above, it includes a microprocessor with memory, allowing it to record and store the levels of vibration on all the machines selected in the plant. Scheduled messages appear on an LCD screen, guiding the operator to the correct measuring points. Additional information can also be entered using the front keypad. Measurements can be done easily and quickly; for example, the operator only needs to position the transducer against the point to be measured and press the "store" key to register the vibration level.

2.4.7 Vibration Analyzer

The function of a vibration analyzer is to determine the condition of critical machinery in a plant. When a failure is detected, the vibration meter is not able to pinpoint the specific problem or root cause. This is the function of a vibration analyzer.

Today, vibration analyzers are hand-held computers which combine in a lightweight unit the capabilities of a data collector and a vibration analyzer with the ability to not only obtain, store, and deal with collected data in both time and frequency domains, but also to simultaneously collect and store process variables, such as pressure, flow, or temperature. This capability provides the analyst with all the data required to detect incipient problems in the machine (Figure 2.17).

FIGURE 2.17
Vibration analyzer (SKF).

2.4.8 Periodic Monitoring or Online Monitoring

Normally, most of the rotating machines in a plant are condition monitored periodically with portable systems as described earlier. For more critical and essential assets, online monitoring is an option. The functionality of an online system is similar to that of the portable system but the functionalities for trending, alarm setting, phase measurement, and so on are normally more advanced and, of course, the measurement can be taken around the clock 24/7. The asset pyramid in Figure 2.18 shows the typical criticality distribution of rotating assets in any plant. Usually, only the *critical, essential, more expensive,* and *important* assets are considered for online monitoring.

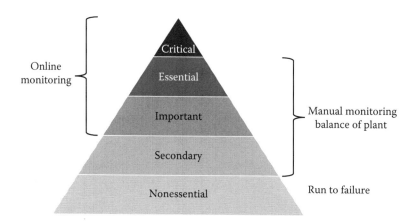

FIGURE 2.18
The asset pyramid. (Redrawn from Emerson. 2014. *Measurement Types in Machinery Monitoring.* White Paper. www.assetweb.com.)

2.5 Motor Circuit Analysis

Most electrically caused faults in electrical motors occur because of shorted turns with the windings. However, winding contamination and poor connections may lead to motor failure as well. Winding faults usually start in the end-turns of the winding because, at this location, the insulation is weakest, yet the stress is the highest. Most of the faults in electric motors are small in the beginning, but they accelerate over time and almost always end in a motor failure (Penrose, 2004).

Motor circuit analysis (MCA) is a technology that helps us observe the electrical health of a motor by evaluating the electrical properties of the motor windings. The purpose of applying MCA is to measure the electromagnetic properties of a motor; its condition is determined by seeing it as an electric circuit (Figure 2.19). The following are measured with the help of an MCA analyzer in each of the three winding phases:

1. Winding resistance, inductance, and impedance, with high-accuracy bridges of the analyzer
2. Insulation resistance to ground
3. Phase angle, by applying a low-voltage AC signal
4. Multiple-frequency current response test (I/F)
5. Phase imbalance
6. Relation current/frequency in function of the impedance
7. State of rotor rods
8. Uniformity of spark gap

These measurements are all balanced in a healthy motor. If even one is not balanced, a fault is likely to occur. For example, unbalanced resistance indicates loose connections and an unbalanced phase angle indicates shorted turns within the motor windings.

In the MCA instrument, a software program produces a report indicating the motor condition. The software program will either indicate a good

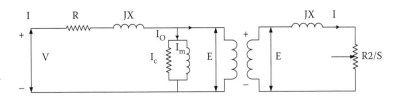

FIGURE 2.19
Electric circuit (a way of seeing the motor).

winding condition or specify the area of concern. MCA is nondestructive and takes less than 15 min to show grounded windings, burned windings, loose connections, phase-to-phase faults, or conversely, that a motor is perfectly healthy. MCA helps the analyst determine the ground insulation condition, phase angle, inductance, simple as well as complex resistance, and the condition of the winding of the electric motors. The MCA equipment can read the mutual inductance between the rotor and the stator, thereby helping the analyst to detect some of the defects in the rotor quickly and safely.

2.5.1 MCA Application

An electric motor has the following five sources of electrical faults (Bethel, 1998):

1. Power circuit
2. Stator windings
3. Rotor (failure rods)
4. Air gap
5. Isolation

The power quality is important as well. First, let us understand what we are really talking about when we speak of power quality problems. Voltage and current harmonic distortion, voltage spikes, voltage imbalance, and power factor are a few of the many concerns we may have when discussing power quality. Variable frequency drives (VFDs) and other nonlinear loads can significantly increase the distortion levels of voltage and current.

MCA will identify the following faults:

1. Contaminated winding, grease, dust, or moisture
2. Short circuit in loops, between turns, or between phases
3. Phase imbalance, impedance imbalance (increases the motor's energy consumption and reduces its lifetime)
4. Failure in conductivity in coils
5. Rotor failure, defective rods, eccentric rotor, or encrustations on the core
6. Insulation failure

The analysis is performed with the energized motor without any harm to the technician doing the analysis.

Data can be collected from different points on the power circuit, such as the junction box of the motor, circuit breakers, fuses, contactors, or the control board. At times, it may be necessary to measure at different points to find the root cause of the problem (Figure 2.20).

FIGURE 2.20
Control board.

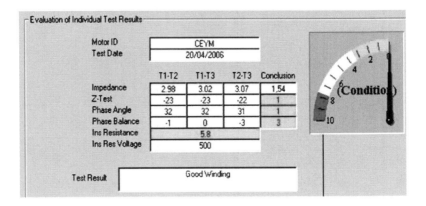

FIGURE 2.21
Special software. (From Vibratec, 2008. *Análisis de Circuito Eléctrico en Motores.* [Online] Available at: http://www.vibratec.net/pages/tecnico4_anacircelec.html.)

Once the field data are entered into the computer, special software (condition calculator) will give a report on the electrical condition of the engine (Figure 2.21).

The data will be stored in a database; with these data, it is possible to observe variations in the trend of the electrical variables in subsequent monitoring (Vibratec, 2008) (Figure 2.22).

2.6 Thermography

Thermal measurement technology measures the absolute or relative temperatures of key equipment parts or areas being monitored. Abnormal

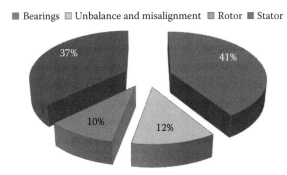

Bearings ▦ Unbalance and misalignment ▦ Rotor ▦ Stator

FIGURE 2.22
Typical failures in electric motors.

temperatures indicate developing problems. Temperature and thermal behavior of machine components are important factors in the maintenance of plant equipment. For this reason, temperature is frequently measured in good plant maintenance.

There are two types of equipment used in this technology, contact and noncontact. Contact methods of temperature measurement, using thermometers and thermocouples, are still commonly used for many applications. However, noncontact measurement using infrared sensors has become an increasingly desirable alternative to conventional methods (Kaplan, 2007).

2.6.1 Contact Temperature Measurement

Contact temperature measurement involves measuring surface or interior temperature by sensing conducted heat energy. In the past, mercury or alcohol thermometers were used to measure temperature. They consisted of a liquid glass bulb reservoir leading to a long narrow tube. As the bulb was heated, the fluid expanded and began to fill the tube, with the expansion of the liquid proportional to the temperature. A temperature scale was marked along the tube, allowing the temperature to be read by an observer.

Bimetallic thermometers operate on the principle of different thermal expansions of two metals (Kaplan, 2007). Two metal strips are connected at one end by soldering or welding. As the strips are heated, they expand. All metals expand at different rates. The expansion difference can be translated by a mechanical linkage turning an indicator needle. A temperature scale behind the needle allows an observer to read the temperature.

Resistance temperature detectors (also called thermistors) use sensors that are electrical conductors. When heated, the conductor's electrical resistance changes. Analysts determine the temperature by measuring the resistance and by knowing the resistance-to-temperature relationship.

A thermocouple also operates on electrical principles. Two different pieces of metal welded together at one end will produce a voltage proportional to the absolute temperature. The voltage produced is proportional to the heat sensed. Thermocouples provide very accurate temperature measurement in certain temperature ranges.

2.6.2 Noncontact Thermal Measurement

The four most commonly stated advantages of noncontact thermal infrared measurement over contact measurement are that it

1. Is nonintrusive
2. Is remote
3. Is faster than conventional methods
4. Provides a thermal distribution of the object surface

Any one, or a combination, of the following conditions warrants consideration of a noncontact sensor (Kaplan, 2007):

1. *Target is in motion*: When the target to be measured is moving, it is usually impractical to have a temperature sensor in contact with its surface. Bouncing, rolling, or friction can cause measurement errors and a sensor might interfere with the process.
2. *Target is electrically hot*: Current-conducting equipment and its components present a hazard to personnel and instruments.
3. *Target is fragile*: When thin webs or delicate materials are measured, a contacting sensor can often damage the product.
4. *Target is very small*: The mass of a contacting sensor that is large with respect to the target being measured will usually conduct thermal energy away from the target surface, thus reducing the temperature and producing erroneous results.
5. *Target is remote*: If a target is very far away from, or inaccessible to, contacting sensors, infrared measurement is the only option.
6. *Target temperature is changing*: Infrared sensors are much faster than thermocouples. Infrared radiant energy travels from the target to the sensor at the speed of light. A rapidly changing temperature can be monitored by infrared sensors with a millisecond response or faster.
7. *Target is destructive to thermocouples*: When the high mortality rate of thermocouples due to jarring, burning, or erosion becomes a serious factor, an infrared sensor is a more cost-effective alternative.
8. *Multiple measurements are required*: When many points on a target need to be measured, it is usually more practical to re-aim an infrared sensor than it is to reposition a thermocouple or deploy a great

number of thermocouples. The fast response of the infrared sensor is important.

Some basic concepts of noncontact thermal measurement are:

1. Electromagnetic spectrum
2. Infrared energy
3. Infrared thermography
4. Infrared image
5. Emissivity
6. Blackbody, graybody, and realbody

Each of these is discussed in detail below.

2.6.2.1 Electromagnetic Spectrum

Infrared radiation is a form of electromagnetic radiation that has a longer wavelength than visible light. Other types of electromagnetic radiation include x-rays, ultraviolet rays, radio waves, and so on.

Electromagnetic radiation is categorized by wavelength or frequency. For example, broadcast radio stations are identified by their frequency, usually in kilohertz or megahertz.

Figure 2.23 graphically illustrates where the electromagnetic spectrum and types of electromagnetic radiation fall within the wavelength ranges and the expanded infrared measurement region (Abbott, 2001).

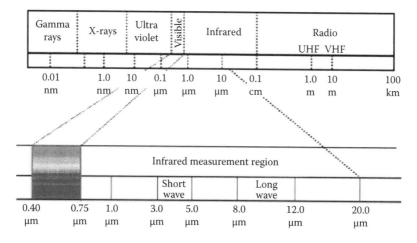

FIGURE 2.23
Electromagnetic spectrum.

As they are a form of electromagnetic radiation, infrared detectors or systems are categorized by their wavelength. The unit of measurement is the micrometer, or micron. A system able to detect radiation in the 8–12 μm band is usually called longwave. Alternately, one that detects radiation between 3 and 5 μm is termed shortwave. (A 3–5 μm system can also be classified as midband because some systems can detect radiation shorter than 3 μm.) The visible part of the electromagnetic spectrum falls between 0.4 and 0.75 μm. Different colors can be seen because they can be discriminated by their different wavelengths. For a laser pointer, the wavelength is usually about 650 nm. If we examine a chart of the electromagnetic spectrum at 650 nm (0.65 μm), we can see it is the radiation of red light.

2.6.2.2 Infrared Energy

All objects emit infrared radiation as a function of their temperature (Kastberger and Stachl, 2003). That means all objects emit infrared radiation. In fact, infrared energy is generated by the vibration and rotation of atoms and molecules. The higher the temperature of an object, the more these nuclear particles are in motion and, hence, the more infrared energy that is emitted. This is the energy detected by infrared cameras. The cameras do not "see" temperatures; they detect thermal radiation. At absolute zero (–273.16°C, –459.67°F), material is at its lowest energy state so infrared radiation is almost nonexistent.

Infrared energy is part of the electromagnetic spectrum and behaves similarly to visible light. It travels through space at the speed of light and can be reflected, refracted, absorbed, and emitted.

The wavelength of IR energy is about an order of magnitude longer than visible light, between 0.7 and 1000 μm. Other common forms of electromagnetic radiation include radio, ultraviolet, and x-ray.

2.6.2.3 Infrared Thermography

IT is based on measuring the distribution of radiant thermal energy (heat) emitted from a target surface and converting this to a surface temperature map or thermogram. The thermographer requires an understanding of heat, temperature, and the various types of heat transfer as an essential prerequisite to undertaking a program of IR thermography (ISO 18434-1:2008, 2008).

IT is the technique of producing an image of invisible infrared light emitted by objects due to their thermal condition. A typical thermography camera resembles a typical camcorder and produces a "live" TV picture of the heat radiation. More sophisticated cameras can actually measure the apparent temperature of any object or surface in the image. The cameras can also produce color images that make interpretation of thermal patterns easier. An image produced by an infrared camera is called a thermogram or sometimes a thermograph.

Used in a variety of commercial applications since the early 1970s, IT has become a well-recognized, cost-effective technique for use in energy management and equipment monitoring and diagnostics. This nonintrusive method readily represents thermal patterns and can be used to identify locations where heating is excessive. This method is also used to evaluate the severity of heating problems.

2.6.2.4 Infrared Image

The IR camera captures the radiosity of the target that it is viewing. Radiosity is defined as the infrared energy coming from a target modulated by the intervening atmosphere (Infrared Training Center, 2014).

Radiosity consists of emitted, reflected, and sometimes transmitted IR energy. An opaque target has a transmittance of zero. The colors on an IR image vary due to variations in radiosity. The radiosity of an opaque target can vary due to the target temperature, target emissivity, and reflected radiant energy variations.

Thermographers see targets exhibiting such contrasts in emissivity behavior every day. It could be an insulated electric cable with a bare metal-bolted connection, or a bare metal nameplate on a painted surface, such as an oil-filled circuit breaker or load tap changer. It could equally well be a piece of electrical tape placed by the thermographer on a bus bar to enable an accurate reading (Figure 2.24).

For opaque objects, the emissivity and reflectivity are complementary. High emissivity means low reflectivity and vice versa. As Kirchhoff showed, in thermal equilibrium, the absorptivity of an object equals its emissivity. Combining this with the law of conservation of energy results in an equation that quantifies these concepts (Monacelli, 2005):

$$\varepsilon + \rho + \tau$$

Note: Greek letters for e, r, and t are typically used where

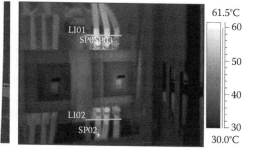

FIGURE 2.24
Imaging using thermographers.

1. ε = Emissivity
2. ρ = Reflectivity
3. τ = Transmissivity

For opaque targets, $\tau = 0$, and the equation reduces to

$$\varepsilon + \rho = 1$$

In simple terms, the above equation indicates that a high emissivity means a low reflectivity, and a low reflectivity means a high emissivity. Thermographers like the emissivity to be as high as possible. This allows them to obtain the most accurate readings because most of the radiosity is due to the radiant energy emitted by the target. Modern IR cameras correct for emissivity with limited user input; however, the uncertainty in the measurement increases with decreasing emissivity. Calculations show the measurement uncertainty gets unacceptably high for target emissivities below about 0.5.

2.6.2.4.1 *Emissivity*

Every target surface above absolute zero radiates energy in the infrared spectrum. The hotter the target, the more radiant infrared energy is emitted. Emissivity is a very important characteristic of a target surface and must be known in order to make accurate noncontact temperature measurements. The methods for estimating and measuring emissivity are discussed throughout the industry literature. Therefore, the emissivity setting that must be put into the instrument can usually be estimated from available tables and charts. The proper setting needed to make the instrument produce the correct temperature reading can be learned experimentally by using samples of the actual target material. This more practical setting value is called effective emissivity (Gorgulu and Yilmaz, 1994).

Emissivity tables exist, but establishing the exact emissivity of a target is sometimes difficult.

Emissivity has previously been discussed here as a material surface property, but it is more than a surface property. Surface properties are continually changing, and the shape of an object affects its emissivity. For semitransparent materials, thickness will affect emissivity. Other factors affecting emissivity include viewing angle, wavelength, and temperature. The wavelength dependence of emissivity means different IR cameras can get different values for the same object. And they would all be correct. It is recommended that the emissivity of key targets be measured under conditions in which they are likely to be monitored during routine surveys.

In general, dielectrics (electrically nonconducting materials) have relatively high emissivities, ranging from about 0.8 to 0.95, which includes painted metals. Unoxidized bare metals have emissivities below about 0.3 and should not be measured. Oxidized metals have emissivities ranging from about 0.5 to 0.9, and are considered a problematic category due to the large range of values. The degree of oxidation is a key ingredient in an object's emissivity: the higher the oxidation, the higher the emissivity.

For opaque objects, if the emissivity and the background (reflected) temperature are known, an IR camera with a temperature measurement feature can, in theory, give temperatures accurate to within a few percent. To get the temperature, the IR camera must extract just the fraction of the radiosity that is due to the energy emitted by the target. Modern IR cameras are capable of doing this; however, for emissivities below 0.5, the errors might be unacceptably large. The reflected component is subtracted and the result is scaled by the target emissivity. The resulting value can then be compared to a calibration table and the temperature extracted.

2.6.2.4.2 Blackbody, Graybody, and Realbody

A blackbody is a perfect radiator because it has zero transmittance and zero reflectance; according to the emissivity equation, the emissivity of a blackbody is 1. Blackbodies were first defined for visible light radiation. In visible light, something that does not reflect or transmit anything looks black; hence, the name blackbody (Infrared Training Center, 2014). A graybody has an emissivity less than 1, and this remains constant over the wavelength. A realbody has an emissivity that varies with wavelength.

IR cameras sense infrared radiant energy over a waveband. To obtain the temperature, they compare the results explained above with a calibration table (Table 2.3) generated using blackbody sources. The implicit assumption is that the target is a graybody. Most of the time, this is true or close enough to get meaningful results. For highly accurate measurements, the thermographer should understand the spectral (wavelength) nature of the target.

Max Karl Ernst Ludwig Planck is credited with developing the mathematical model for blackbody radiation curves. Figure 2.25 shows the magnitude of emitted radiation due to an object's temperature versus the wavelength for various temperatures. Note that the sun has a peak wavelength in the middle of our visible light spectrum.

Blackbody curves are nested and do not cross each other. This means that a blackbody at a higher temperature will emit more radiation at every wavelength than one at a lower temperature. As the temperature increases, the wavelength span of radiation widens, and the peak of radiation shifts to shorter and shorter wavelengths.

The peak of infrared radiation at 300 K is about 10 µm. In addition, an object at 300 K emits radiation only down to about 3 µm. Because human eyes are not sensitive beyond about 0.75 µm, this cannot be seen; however, if the object is warmed up to about 300°C, a faint red glow can begin to be seen.

2.6.3 Infrared Inspection Techniques

This section is divided into two subsections dealing with the common problems encountered when using IT in a plant or industrial environment, along

TABLE 2.3

Correlation of Thermal Analysis with Other Technologies

Technology	Correlation Method	Indication	Usage
Vibration	Time coincident	Increasing or already high vibration at the same time as increasing temperature	Suspected bearing or coupling problem
Lubrication	Time sequence	High and/or increasing large wear particles	Suspected bearing problem
Debris analysis	Time sequence	Damage has occurred or it is occurring	After abnormal temperature rise in bearing or lubricant from the bearing
Leak detection	Time coincident	Abnormal temperature coincident with acoustic signals indicating internal leak of a fluid or gas	Suspected leak
Electrical circuit testing	Time coincident	High resistance in energized circuit radiating abnormal heat	Suspected circuit problem
Visual inspection	Time sequence	Signs of overheated parts of equipment such as electrical insulation on wires or corrosion/oxidation of normally shiny terminal connections	Suspected problem not previously repaired

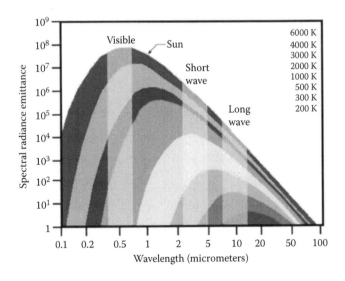

FIGURE 2.25

Emittance versus wavelength. (Redrawn from Infrared Training Center, 2014. *IR Thermography Primer*. [Online] Available at: http://www1.infraredtraining.com/view/?id=40483.)

with their solutions. The first section deals with the inherent mitigating effects, emissivity, and reflectivity. The second section explains the methods used to get the best possible information out of the imaging systems: foot powder, dye check developer, and electricians' tape.

Mitigating inherent effects: Several factors affect production and the subsequent proper interpretation of a thermal image. These include the target's emissivity, reflectance, distance from the imager, temperature, background temperature, ambient temperature, orientation, target size, and transmittance of the intervening atmosphere.

The image, as presented on the imager, is not temperature but radiosity. Imagers measure the radiant energy emitted by the target plus the radiant energy reflected from and transmitted through the target. The sum of these radiant energies is the commonly accepted definition of radiosity.

Certain practical considerations will simplify the following discussion of inherent effects. In general, the transmittance (energy transmitted through the targets) can be ignored in most, if not all, cases for targets in a power plant. Transmittance is an important factor in industries where the temperature of a thin film of plastic or other infrared semitransparent target is being observed. With the exception of absolute temperature measurements being required, transmittance through the atmosphere can be ignored as well. The major exception is a case where long distances are involved in a humid atmosphere (i.e., hydrogen igniters or spray nozzles in containment).

Shiny objects have surface thermal patterns that are hard to image, but several techniques improve the ability to establish a satisfactory image. The most common way to obtain a useful thermal image from a shiny or low-emissivity surface is to add a coating with a higher emissivity. (This is not practical and is not recommended for an energized electrical surface.) Three common nonpermanent materials have been used to improve emissivity (Zayicek, 2002):

1. Paint
2. Dye check developer
3. Electricians' tape

Paint: Paint can be sprayed on the target to obtain better emissivity. Glyptol is especially effective.

Dye check developer: Dye penetrant developer has an estimated emissivity of 0.97 and may already be formulated to conform to QA requirements for sulfur and halogen purity. Given the temperature of the target, it might take several minutes for the developer to reach thermal equilibrium, as its propellant cools the target's surface. The best way to use this in an actual survey is to apply it to all targets to be surveyed before commencing the actual survey. This will ensure all target surfaces will have reached thermal equilibrium.

Electricians' tape: Another alternative that improves the surface emissivity is the use of electricians' tape (it has an estimated emissivity of 0.95). This method is easy to use and apply, but can present problems if the glue on the tape contains chlorine or other chemicals that can attack the target surface.

Other proven inspection techniques that can be used for components difficult to image include the following:

Mirrored surfaces: A situation commonly encountered in electrical switchgear is one where there is little or no room to place the imager, and the area of interest is behind another object. One method that works well is to use a material with a high reflectivity (low emissivity), such as an infrared mirror.

Thermal transients: Another useful inspection technique for handling targets with low emissivity is to add or subtract heat from a target. Most uses of IR are in the steady-state condition. When two materials with different heat capacities are involved, a thermal transient is useful. A graphic example, shown in Figure 2.24, deals with a can of dye check developer. In the example, a thermal transient is induced on the can by spraying it. The endothermic change of state of the propellant as it evaporates causes heat to transfer from the inside of the can (warmer) to the outside. In the case of the propellant inside, the liquid has a higher heat capacity (Cp) than the vapor space above it. During the transient, the liquid causes a larger transfer and resultant temperature difference due to conduction on the can's surface. The higher emissivity of the developer on the can's surface allows it to be seen more readily.

This is an extremely useful technique. Where large masses are concerned, however, a large amount of heat transfer might be needed for observation. This technique can be used to determine relative thicknesses of material and locations of voids, delaminations, and internal structures.

When looking for voids in composite materials, a flash lamp can be used for a short pulse of energy. Hot air from a compressor can be used for containment spray ring header nozzle inspections and for locating materials near the surface of concrete. It should be noted, however, that heating is not always the most effective approach. Cooling is sometimes more effective, especially in hot areas; the evaporative cooling effects of water can be very useful (Zayicek, 2002).

2.6.4 IT Program

An IT program involves the conduction of periodic inspection surveys of critical equipment used in any production plant; for instance, in a plant for production and delivery of electric power. The critical equipment in the database will usually exhibit an abnormal thermal pattern at some point in time prior to functional or operational failure. An inspection, made with an IT camera, can detect these abnormal thermal patterns. Abnormal thermal patterns, observed on a piece of equipment, are referred to as thermal anomalies.

A thermography program uses a walk-around survey procedure, able to monitor temperatures in remote or noninstrumented areas of the plant. The intent is to verify that the surveyed components are operating within their designated temperature ranges. A typical thermography program begins with a basic approach, easily expandable to cover a variety of components as the user gains experience with the equipment and uses the techniques to monitor other areas. In addition to rotating machinery and electrical equipment, other applications can include locating condenser air in-leakage, tube leaks, roof surveys, and so on. Any system or component for which absolute or relative temperature is an indicator of abnormal performance or operating condition is a candidate for thermography.

From the periodic walk-around survey, any component seen to be operating outside normal temperature conditions is recorded with a photograph of the thermal image (thermogram) and a conventional photograph. These images are attached to a report summary with the location and temperature observed, as well as any other comments that might be appropriate.

An IT camera uses infrared sensors to make simultaneous temperature measurements of multiple points on the surface of a piece of equipment (a target) without making physical contact with the target. The measurements of the target surface are taken from a distance. IT is not an x-ray technique and will not make measurements through an object. IT inspections must be done while the equipment is in normal operating mode and after a period of run-time that is long enough to allow a component to reach normal operating temperatures.

IT data are displayed in the form of an image, commonly called a thermogram. Thermal images can be analyzed in real time, or stored electronically and analyzed at a later time. The images are analyzed to determine whether the thermal pattern is normal or abnormal.

Many plants throughout the world have systems with at least two operating components performing the same service at the same time. These similar service components allow an IR thermographer the advantage of comparing the thermal profile of one component to the other (Figure 2.26).

Throughout the industry, this process is called comparative thermography. Depending on the nature of the component(s) being observed, comparative thermography can be used to obtain qualitative or quantitative data (Zayicek, 2002).

1. Qualitative—Qualitative data are used when a component does not require numerical data to determine the severity of its condition.

 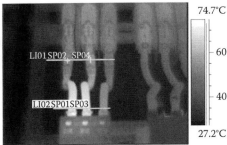

FIGURE 2.26
Thermographic analysis.

2. Quantitative—Quantitative data are used when a component requires numerical data to determine the severity of its condition.

When there are no similar service components available to perform a comparative analysis, recording baseline data is recommended. In this method, one or several thermograms are taken of the component being observed. The baseline thermogram(s) are then compared to the data taken on subsequent surveys, in the same manner that a similar service component would be analyzed. That is, any changes from the original baseline data will be recorded as abnormal.

Recording additional information, such as process data, environmental conditions, time, season, location, nameplate data, and so on, can be beneficial when performing an analysis. The type of additional data being collected will depend on the specific application. For example, a motor's thermal profile can fluctuate for many reasons, including load (amps), ambient temperature, location (inside, outside, shade, sunlight, etc.), type of coolant system, duty, casing construction, insulation type, and so on. Any one of these conditions, or a combination of them, can wreak havoc by trying to determine the condition of a motor by the external casing temperature. Even similar motors from the same manufacturer and with the same model number can have different thermal profiles due to their location. Therefore, recording additional data can assist in defining the external motor's thermal profile.

In CBM/PdM programs, IT is used to periodically inspect major plant rotating equipment, such as motors, pumps, and fans. When evaluating the operating condition of rotating equipment, IT should be used in conjunction with other diagnostic technologies, such as VA, lubricating oil analysis, motor current testing, and so on.

IT is applied to rotating equipment by comparing the thermal patterns of a component to itself over time. It can also be applied by comparing the thermal pattern of a component to the thermal pattern of a similar component operating under similar conditions. If a component is compared to itself over time, it is necessary to take a baseline thermal image of it. This image should be taken when the component is under normal operating load, is providing the performance it is expected to provide, and has had a long enough run time to reach normal operating temperature. The circuit current and ambient air temperature, at the time the baseline is taken, should be recorded.

IT is an effective tool in detecting problems with equipment used for the transmission and distribution of electric power. The equipment featured in this section can be found in locations such as power plant transformer yards, power plant switchyards, and substation distribution yards. The components include, among others, power transformers, high-voltage circuit breakers, and circuit disconnects. These components should be periodically inspected as part of an IT program.

In addition to IT, other types of diagnostics are used to evaluate the overall health of transmission and distribution components. These additional diagnostic tests include VA, oil analysis, sonic and ultrasonic evaluation, AEs, sound level, and visual inspection. They should all be used in conjunction with IT.

In general, there are four primary areas where IT is effective:

1. Electrical equipment
2. Rotating equipment
3. Performance
4. Switchyard

The primary use of IT in plants is in periodic monitoring aimed at detecting electrical deficiencies, including:

1. Overloads caused by equipment and cable failures
2. High electrical resistance due to loose, deteriorated, or corroded connections
3. Hot spots caused by inductive currents

Most mechanical applications can be classified into four areas:

1. Heating due to friction
2. Valve leakage or blockage
3. Insulation integrity
4. Building conditions that increase energy costs

Three leakage problems that plague plants are condenser tube leaks, condenser vacuum leaks, and boiler casing leaks. Utilities use IT to detect condenser leaks. As air is drawn into a condenser, the air leaks are indicated by cooler areas and are observable.

Other typical IT program components include:

1. Rotating equipment, such as pumps, fans, motors, and so on
2. Load center components, such as motor control center panels, circuit breaker compartments, transformers, and connections
3. Mechanical components, such as steam traps, relief valves, condensers, roofing, insulation, and so on.

A novice IR thermographer generally investigates electrical and rotating equipment first. One reason is that IT is the best technology (ultrasound can also be used) to identify many electrical deficiencies while energized. Another reason is that the majority of critical plant equipment is electrical or rotating.

Infrared data for rotating equipment are beneficial in identifying undesirable conditions. Some rotating equipment conditions can be defined using infrared alone, such as a motor casing being overheated due to dirty air filters. However, unlike electrical deficiencies, many rotating equipment deficiencies require support from other technologies to define the condition. For example, a hot bearing on a motor can be seen by infrared, but infrared cannot determine whether the condition stems from deteriorated oil or a deteriorated bearing. Therefore, an oil analysis is required to define the bearing condition.

2.6.5 Applications

Using IT to inspect electrical equipment, under load is probably the most popular application of the technology. The availability of this nonintrusive tool has enabled electrical maintenance personnel to monitor the condition of electrical equipment more effectively than they could previously, mainly because an IT inspection can be done without having to come into physical contact with high-voltage and current-carrying circuits. An inspection of stationed electrical equipment will yield information that identifies only the components that need to be addressed with maintenance.

This section features examples of thermal anomalies detected on electrical equipment, including:

1. Molded case circuit breakers
2. AC induction motor
3. Exciter brush compartment
4. Generator step-up transformer, fan motor, load tap changer
5. Feedwater heater

2.6.6 Image Analysis

For IT, problem severity is often categorized using temperature as the main criterion. Temperature is quite significant for most electrical system surveys, but it is not the only aspect to consider. The primary benefit of IT is its ability to locate anomalies. The analysis of severity is a complex process based on a number of factors. Most PdM programs evaluate the criticality of the equipment, normally based on replacement value and the effect on power generation, and assign one or more diagnostic technologies where it makes sense to protect the equipment.

Temperature by itself can be misleading. Variables affecting the temperature reading of an electrical problem include:

1. Load
2. Hot spot size-to-camera working distance ratio
3. Wind speed
4. Emissivity
5. Ambient and background temperatures
6. Solar effects
7. "Directness" of the reading

The last parameter considers the amount of thermal "insulation" between the actual problem spot and what the IR camera "sees." This is often categorized as a direct reading for bare connections, or connections with thin layers of insulation, and an indirect reading for heavily insulated components. However, good electrical insulation might also be thermal insulation.

All items need separate temperature severity criteria from those of direct readings. Even very small changes in temperature can indicate extremely dangerous conditions; these changes might not be detectable except under ideal inspection conditions.

Information in this section includes:

1. Reading a thermal image
2. Analyzing a thermal image
3. Software

Reading a thermal image: To analyze a thermal image, it is essential to understand how to properly read the image. There are various manufacturers of infrared equipment and many different types of systems available for use in the field. Each instrument displays its data in slightly different formats, and there is a learning curve necessary to become familiar with each one; however, there are similarities between all infrared instruments. For example, they are all designed to yield the same result, which is a thermal pattern of a particular target. An analysis of the method of one type of camera should give a basic understanding of thermal image analysis.

Analyzing a thermal image: Thermal data are analyzed comparatively. First, a reference point has to be identified. When a reference point is identified, the temperature of the reference point is compared to the temperature of the component being analyzed. A reference point can be a spot on a piece of equipment (e.g., an AC induction motor) that is similar or identical to the piece of equipment being analyzed. Because the components are similar, the reference component and the component being analyzed should exhibit similar thermal patterns.

It is assumed that when using similar service equipment as a reference, the amp load, run time, and external environmental conditions are the same for the reference component as they are for the component being analyzed. For example, if the reference component is loaded at 100 amps, the component being analyzed should be loaded at approximately 100 amps. If this is not the case, comparisons of the two components will be unreliable. Convective cooling and emissivity also result in unreliable ΔTs.

A baseline thermal image of a component could be used as a reference point. In this method, baseline data are taken for a component during normal loading conditions. Ambient air temperature and load are recorded and kept with the baseline data. If accurate baseline data of a component are available, the analyst can compare subsequent data to them.

Ambient air temperature can serve as a reference temperature, for example, for comparisons with electrical connections. Ambient air temperature is the air temperature inside the enclosure before or immediately after it is opened.

Whatever reference point is chosen, thermal characteristics of the reference point are compared to thermal characteristics of the component being analyzed. Then an assessment is made.

Software: Manufacturers of thermal imaging scanners produce software packages that digitize thermal images and convert them into black and white or color displays on video monitors.

These packages incorporate complex algorithms and image-processing features that can be categorized into four groups:

1. Quantitative thermal measurements of targets
2. Detailed processing and image diagnostics
3. Image recording, storage, and recovery
4. Image comparison

With thermal imaging software, scanning systems can provide a user with the true radiance or temperature value of any point, or all points, on the target surface. To provide true radiance values, the system calibration constants are fed into the computer on initial setup. A system of prompts assures the operator that changes in aperture settings, target distance, interchangeable lenses, and so on are fed into the keyboard each time a change in instrument operating conditions occurs. The operator must also insert an emissivity value. Some systems allow users to assign different emissivities to different areas on the target surface.

In operation, a color scale is provided along one edge of the display, with the temperature shown corresponding to each color or gray level. The operator can place one or more spots or crosshairs on the image, with the temperature value of the corresponding pixels appearing at the appropriate point on the display.

Other software components enable imaging systems to analyze each pixel of the thermal image and to present information in a wide variety of qualitative and quantitative forms for the convenience of the user. Operators can call for a profile display, draw areas on the display, and have those areas subjected to detailed analysis. Although manufacturers usually provide a standard (default) color scale, operators can create their own scales, in almost limitless variety.

By using zoom features, operators can closely examine small areas. By using three-dimensional features, they can generate isometric thermal contour maps of the target to facilitate recognition of thermal anomalies.

Thermal imaging software enables users to index, record, and retrieve images and data. Commercial thermal imaging systems can be equipped with any of a variety of digital (or analog) media storage devices, allowing stored images to be viewed on demand or transferred to a personal computer or other device for long-term storage.

The image comparison feature makes it possible to automatically compare images taken at different times. Operators can display two images side-by-side or in sequence; they can subtract one image or area from another and display a pixel-by-pixel difference thermogram. As a result, thermographers can archive thermal images of acceptable components, assemblies, and mechanisms as baselines and use them as models of comparison to subsequently produced images.

Quantitative thermal measurements of targets: This represents the temperature value of any point (or all points) on the target surface. For true radiance measurements, the system throughput attenuation must be taken into consideration, as well as losses through the measurement medium (atmosphere, in most cases).

For true temperature measurement, the target effective emissivity must be considered. To provide true radiance values, the system calibration constants are fed into the computer on initial setup. A system of prompts assures the operator that changes in aperture settings, target distance, interchangeable lenses, and so on are fed into the keyboard each time a change in operating conditions occurs. For true temperature values, it is necessary for an effective emissivity value to be inserted by the operator. The temperature readings that are then displayed assume the entire target surface effective emissivity is equal to this inserted value. An emissivity value less than 0.5 can result in unacceptable temperature measurement errors.

In operation, a color scale (or monochrome gray scale) is provided along one edge of the display with a temperature shown corresponding to each color or gray level. The operator can place one or more spots or crosshairs on the image, and the temperature value of that pixel will appear in an appropriate location on the display. Some systems allow the assignment of several different effective emissivities to different areas of the target, as selected by the operator, with the resulting temperature correction.

Detailed processing and image diagnostics: Detailed processing and image diagnostics is a phrase describing the capability of the computer to analyze each pixel of the thermal image and present information in a wide variety of qualitative and quantitative forms for the convenience of the user. The current trend by manufacturers is to offer more and more on-board image analysis capabilities. For extensive image storage and analysis, the images are more often downloaded from the cards to computers with large storage capacities and memory. The extensive image and data analysis software is resident on the computer hard drive.

Image comparison: Image comparison is a very significant capability because it allows the automatic comparison of images taken at different times. The computer allows the operator to display two images, side-by-side or in sequence, to subtract one image from another or one area from another, and to display a pixel-by-pixel difference thermogram. This provides the capability of archiving thermal images of acceptable components, assemblies, and mechanisms, and using them as models for comparison to items produced subsequently. Subtractive routines produce differential images illustrating the deviation of each pixel (picture element) from its corresponding model. Image averaging allows the computer to accumulate several scan frames and display the average of these frames. Comparison (subtraction) of images can be derived from two real-time images, two stored images, or a real-time and a stored image.

Recording, hard copy, and storage of images and data: Thermal image recording and storage has evolved dramatically from Polaroid® instant photos of the display screen, to magnetic storage and archiving of images and data (such as labels, dates, conditions of measurement, and instrument settings), to the instant digital image storage capabilities incorporated into most of today's thermal imagers. Hundreds of images can be recorded in the field and stored on removable, reusable memory cards. Thermal images are saved in any one of several digital image formats such as .bmp, .tif, and .jpeg for archiving and future analysis (Zayicek, 2002).

2.6.7 Severity Criteria

Ranking an anomaly by its severity is a goal of any diagnostic technology and can be accomplished with mature PdM teams where everyone has agreed to specific criteria and procedures. In other cases, it is extremely difficult. For the latter, bringing in other diagnostics is often helpful.

Severity criteria should address the direct/indirect temperature measurement issue to be effective. Some utilities have separate criteria for overhead and underground electrical distribution facilities.

Any published severity criterion that does not address direct/indirect issues should be revised. Industry experience indicates that a 5°C rise in a system's temperature often indicates a critical problem. Similar temperature increases in the aforementioned indirect targets can also be critical in nature. Research is needed in this area to specify severity criteria for the various types of indirect targets, especially as the failure of these devices can be extremely costly in both dollars and human life.

Attempts to correct for severity during load changes with simple equations have resulted in incorrect results. Perch-Nielsen, S. L., Bättig, M. B., and Imboden, D. (2008). Exploring the link between climate change and migration. *Climatic Change, 91*(3–4), 375–393 and Lyon Jr, B. R., Orlove, G. L., and Peters, D. L. (2000, March). Relationship between current load and temperature for quasi-steady state and transient conditions. In *AeroSense 2000* (pp. 62–70). International Society for Optics and Photonics have shown that temperature increase does not follow the simple square-of-load change rule that is widely disseminated. Similarly, for wind effects, it has been shown that the wind speed correction factors historically used by thermographers are suspect.

These variables are even more crucial for indirect targets. Small temperature increases that indicate problems can be significantly influenced by other factors, such as wind and solar insolation, because they are small.

All of these variables can appear daunting to the PdM team responsible for establishing severity criteria. But it is possible to write reasonable severity guidelines with caveats for the variables, letting the thermographer have some leeway in implementing the criteria. It is strongly recommended that written severity criteria be established with temperature increase as one of the key parameters. It is also suggested that direct criteria be differentiated from indirect criteria.

Severity criteria have two possible forms; they can be

1. Organized into general categories that identify temperature levels, or zones, versus levels of criticality

2. Applied to specific machines or components, or to like groups of machines or components. In either case, the levels are established based on a working knowledge of component operation, predicted failure modes, previous inspection histories of the component, and input of cognizant system engineers or maintenance personnel. Because the severity criteria for each machine or component group can be unique, each organization should set up a program that best applies to its equipment and operation.

Severity criteria can be established on individual machines or components, based on a number of factors, including:

1. Temperature increase versus historical data establishing rate of deterioration and time to failure

2. Criticality of the machine or component to the overall process

3. Collateral damage to other materials or equipment if a failure should result

4. Safety of personnel

Applications could include temperature increases in critical machines, mechanical components, bearing temperature increases, electrical supply or connection rises, fluid leakage losses, or the number of tubes clogged in fluid-type heat transfer equipment.

REPEATIt is important to note that these are severity guidelines. They do not imply strict adherence. The analyst has to take the temperature increase into consideration, as well as other influencing factors such as operational load, environmental factors, and the importance of the component to production or operation. A judgment is then made by the analyst as to what level of severity should be assigned. Temperature increase is only used as a benchmark reference. The user must also understand that these severity guidelines pertain to electrical connections.

When evaluating equipment other than electrical connections, severity classifications should be based on a working knowledge of component operation, predicted failure modes, operational load, environmental factors, the importance of the component to production or operation, and the experience and interpretations of the analyst. The interpretations are made in light of the function of the component being analyzed on a case-by-case basis. In these cases, comparisons to similar service equipment and baseline data are heavily relied upon.

2.7 Acoustic Technology: Sonic and Ultrasonic Monitoring

AE is defined as the science that deals with the generation, transmission, reception, and effects of sound. The sound referred to is the detectable structural or airborne sound that can manifest itself as a signal on mechanical objects, the pressure waves associated with leaking vapors or gases, or the humming of electrical equipment. Acoustic technology considers frequencies as low as 2 Hz and as high as the mega-Hertz range (Bauernfeind, 2001). Through a process of filtering, frequency bandpassing, and sensor selection, the potential uses for acoustic testing to diagnose equipment condition and operability are virtually unlimited (Figure 2.27).

Acoustic work can be performed in either the noncontact or contact mode. In either case, it involves the analysis of wave shapes and signal patterns, and the intensity of the signals that can indicate severity (Madanhire et al., 2013). Because acoustic monitors can filter background noise, they are more

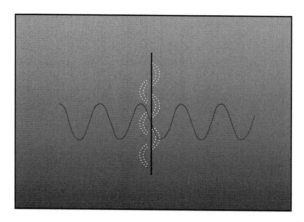

FIGURE 2.27
Low-frequency waves pass the borders as transparent frontiers.

sensitive to small leaks than the human ear and can detect low-level abnormal noises earlier than conventional techniques. They can also be used to identify the exact location of an anomaly.

Acoustic systems can be simple portable devices that can detect anomalies, either structural or airborne. They provide a digital indication of the sound intensity level and can locate the source of the sound. If it is necessary to know the wave shape and the frequency content of the signal, then a more sophisticated portable waveform analyzer type is needed. When it is necessary to monitor critical equipment on a continuous basis, the sensors are permanently attached to the equipment and the signals are transmitted to an online acoustic monitoring system.

Because of its broad frequency spectrum, acoustics are further delineated into two ranges, the sonic range and the ultrasonic range, and their applications are as follows (International Atomic Energy Agency, 2007):

1. Sonic range (0 Hz–20 kHz): The sonic range includes all frequencies in the hearing range of humans and all frequencies used in mechanical VA and low-frequency leak detection (2 Hz–20 kHz)

2. Ultrasonic range (20 kHz–1 MHz): Ultrasonic frequencies are used in cavitation detection, AE, high-frequency leak detection, and corona and partial discharge detection.

Each of these frequency ranges makes use of contact and noncontact transducers, such as microphones, accelerometers, and high-frequency resonant transducers.

The rest of the section is divided into acoustic leak detection and acoustic crack detection.

2.7.1 Acoustic Leak Detection

When applying acoustics for CM and fault detection, it is advisable to select the anticipated frequency band and filter out all other unwanted background noises.

The following provides some guidance for selecting the sonic bands:

1. 100 Hz to 20 kHz—General fault detection
2. 1.0 Hz to 20 Hz—Leak detection that eliminates low-frequency background noises
3. 3.0 Hz to 20 kHz—Leak detection that eliminates additional background noise

A tube leak acoustic trace that is typical of any leak is shown in Figure 2.28. Note that the filtered leak signal amplitude can be significantly higher than that of the noise level, and can, therefore, be readily detectable with acoustics. The ordinate scale in Figure 2.28 is in volts RMS, but these units are directly related to gravity acceleration units (g) (IAEA, 2007).

The following intensities are provided for guidance:

1. Less than 0.1 V = small or no leak
2. Greater than 0.1 V but less than 0.3 V = medium leak (schedule repair)
3. Greater than 0.3 V = large leak (immediate repair needed)

Acoustic monitoring devices have been applied in several different areas, including leak detection for feed water heaters, valve internals, valve externals, and boiler tubes. Feedwater heater tube leaks are commonly detected

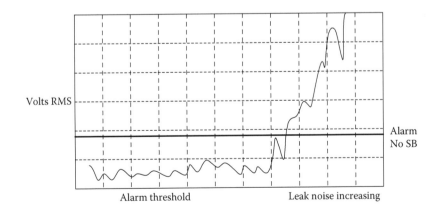

FIGURE 2.28
Acoustic leak trace.

FIGURE 2.29
Acoustic monitoring in several areas.

by changes in heater water level, flow rate, or water chemistry (Figure 2.29). Because of the large total fluid volume, all methods require large leaks for the change to be noticeable, and it might be hours or days before the leak is discovered. Such a delay in the detection of the tube leak can reduce plant efficiency through thermal efficiency losses. If the leak remains undetected, even for a short period of time, impingement of the escaping feedwater can damage adjacent tubes, further increasing repair time and expense.

Determining which heater among many is leaking can be difficult. Not only is there a delay in detecting the leak but, in some cases, the wrong feedwater is taken off-line, adding to an already inefficient situation. When a feedwater leak is detected early, it is possible that the leaking heater can be taken off-line for repair while the others remain in service. This avoids a complete shutdown and minimizes performance efficiency degradation.

In a feedwater leak detection system, sensors are mounted at various locations on the heaters. When a hole develops in a heater tube, the fluid flowing through the orifice generates acoustic pressure waves. These waves are detected by the sensors and converted to electrical signals.

Leaks are annunciated by alarm lights and contacts for remote alarm indication. An analog output from the signal processing unit is available to provide an indication of the acoustic level on any auxiliary peripheral device, such as a data analysis computer, control room recorder, or distributed control system. The system can provide trending, alarm, and graphics capabilities that are used by the technician to predict feedwater heater tube failure.

In a boiler tube leak detection system, steam escaping through a hole in a boiler tube generates a broadband noise that peaks in the range of 1–5 kHz. High-temperature sensors, mounted onto waveguides that couple the acoustic pressure waves in the boiler to the sensor, are installed in locations of the boiler with a history of failures.

Acoustic pressure waves generated in the boiler from a tube leak are detected by the sensors and converted to an electrical signal. The signal from the sensors near the leak increases proportionately with the size of the leak. The increase in signal level is monitored on a trend plot until the level exceeds a preset alarm point. The relative magnitude of the various sensor readings allows approximation of the leak location.

TABLE 2.4

Correlation of Leak Detection with Other Technologies

Technology	Correlation Method	Indication	When Used
Thermal analysis	Time coincident	Abnormal temperature coincident with acoustic signals indicating internal leak of fluid or gas	On condition of suspected leak especially in systems with many potential leak points
Nonintrusive flow	Time coincident	Flow downstream of shut valve, giving acoustic indication of internal leakage	On condition of suspected leak and with many choices of valves to open for repair
Visual inspection	Time sequence	Visual indication of valve disks or seal damage sufficient to cause internal leakage	Use for confirmation before valve disassembly. Use after removal for correlation between acoustic signal and visually observed degree of leak-causing damage

Acoustic detection of valve leaks can be performed with either permanently installed or portable devices (Table 2.4).

Fluid flow through the valve will provide a signature of acoustic activity. When the valve is closed, the acoustic level should drop to zero, taking background noise levels into account. This indicates the valve has sealed and no leak is present. Should a leak be detected, the valve is scheduled for maintenance action (Figure 2.30).

Acoustic leak detection applications include:

1. Feedwater heater tube leaks
2. Valve internal leaks in liquid systems
3. External leaks in the valves and fittings of high-pressure air systems
4. Boiler tube leaks

Areas

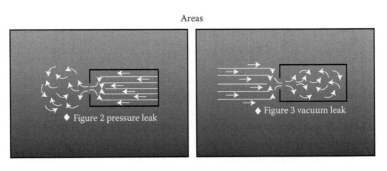

FIGURE 2.30
Pressure and vacuum leaks.

Another effective option for leak detection using AE is the additional use of tone generator as an emitter inside the device to be tested. This is done to transmit signals and produce the AE in the undesired holes in ultrasound waves detectable by an external device. In summary, the tone generator feeds the system and forces it to produce AE suitable to be detected by the external device.

2.7.2 AE Crack Detection

AE is defined as the transient elastic waves generated by the rapid release of energy. The sound results from a crack developing in solid material. As this extremely small, low-level sound propagates through the material, it can be picked up by highly sensitive piezoelectric or strain gauge sensors. The advantage of AE is that very early crack growth can be detected, well before a highly stressed component might fail (IAEA, 2007).

AE technology is intended to provide analysts with an early indication of the onset of degraded strength in metal components, such as pressure vessels. By trending the AE event occurrence, analysts can track and project the progression of grain structure breakdown. With this information, a plant can remove the metal component from service before total loss of function occurs (Figure 2.31).

AE technology has been successively used on the following equipment:

1. Reactor vessels and related piping
2. Control rod housings
3. Main steam lines

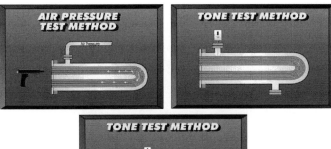

FIGURE 2.31
AE methods.

FIGURE 2.32
Electric inspections like crown effect or arc discharge not visible using thermography.

4. Transformers

5. Fossil high-energy piping (Figure 2.32)

Microgranular material, such as steel, when put under tension or compression beyond its yield point, breaks and tears along grain boundaries (intergranular cracking). This can happen in pressure vessels, structural supports, and high-energy piping. AE monitors for microcracks in heavy metal components and, specifically, for the growth of cracks. The breakdown in a plant component's metal crystalline structures can lead to equipment failure in extreme cases.

AE sensors have been very effective when applied to reactor vessels and related piping for the early determination of developing cracks. In these applications, the sensors are monitored continuously during all levels of plant operations. Other applications include monitoring the integrity of control rod housings and main steam lines.

AE monitors use many acoustic pickup transducers to detect the energy bursts caused by intergranular cracking and other sources. The equipment sends these signals to a computer console for signal conditioning, measuring, and comparison of arrival time. Arrival time comparison locates the signal source.

Different AE equipment versions present the data in several ways. Most show the acoustic events per second, cumulative total events over time, change in count rate, or some combination of these three. With this information, analysts can trend the data.

AE sensing techniques can be particularly useful to monitor in-service power transformers (Table 2.5). A major concern of transformer failure is the partial discharge associated with the degradation of insulation inside the transformer. This insulation breakdown causes electrical arcing, which deteriorates the oil insulation factor and, if continued, produces highly explosive gases.

Each partial discharge propagates to the tank wall. These stress waves are similar in character to stress waves propagated in solids during crack

TABLE 2.5

Correlation of AE with Other Technologies

Technology	Correlation Method	Indication	Usage
Ultrasonic imaging	Time sequence	Cracking in heavy metal weld joints	In conjunction with code requirements for periodic (10-year) inspection, or after a rise in AE events in a specific region indicates cracks might be developing
Dynamic radiography	Time sequence	Cracking in heavy metal weld joints	In conjunction with code requirements for periodic (10-year) inspection, or after a rise in AE events in a specific region indicates cracks might be developing
Stress/strain measurement	Time coincident	High levels on strain gauges or distortion indicated by other methods	When monitoring for pressure vessel or heavy-section weld deterioration during hydro testing or some other high-stress event

formation, and they generate AE signals that contain an appreciable amount of energy (in the 150 kHz frequency range). These signals can readily be differentiated from other signals emanating from the transformer.

By taking into account the intensity of the AE signal, the approximate location of the emitting source, and the estimated level of the activity involved, it is often possible to estimate the severity of the problem and make a reasonable assessment of its cause.

Detecting AEs from partial discharge events in transformers is valid, and instrumentation is available for this detection; however, AEs from transformers have been detected even in the absence of partial discharge. It has been shown that these signals are produced as a result of the inception of bubbles.

The operating transformers can generate AEs for a variety of reasons. These can be categorized into heating sources, electrical sources, and background noises.

Partial discharge inside a transformer produces AEs. Other sources that can produce AEs prior to the actual occurrence of partial discharge in the transformer include:

1. Localized heating in oil or paper, which can sometimes produce AEs
2. Paper tracking or carbonization, which produces AE signals
3. Energy released during hydrogen gas evolution by partial discharge from either heating or implosion
4. Cavitation, which occurs when nitrogen is released from a solution and generates AEs

These mechanisms produce AE that is directly related to a breakdown in the transformer. Some other mechanisms which will generate AE activity not directly related to a problem in the unit are (Figure 2.33):

1. Environmental sources, due to the impact of rain, snow, ice, or dust against the transformer
2. Areas or pockets of turbulent oil flow within the unit
3. During pump flow with colder oil (note: these observed AEs may be caused by static discharge and are not associated with gassing)

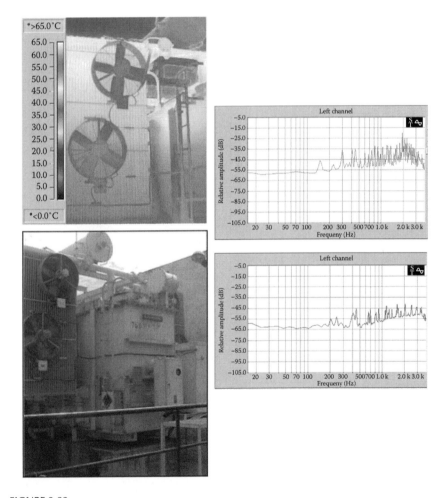

FIGURE 2.33
Problems not detected as hot spots with thermography in transformers may be detected using ultrasonic emission.

4. Oil pyrolysis (chemical change as a result of heat) and core winding
   irregularities

Data filtering techniques allow separation of the relevant and nonrelevant
AE data. There are several different filtering schemes, depending on the situation. In addition, the use of the newer digital signal processing techniques
has improved the accuracy and speed of analysis. Testing transformers *in
situ* can produce a large amount of data if the transformer is acoustically
active, but the speed and accuracy of data analysis are critical.

Several different sources of AE can be detected within a transformer. They
can be classified as either burst emissions or continuous emissions. Figure
2.34 shows some samples of these emissions.

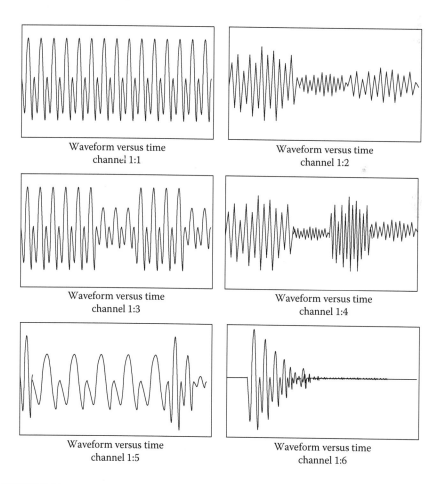

Waveform versus time
channel 1:1

Waveform versus time
channel 1:2

Waveform versus time
channel 1:3

Waveform versus time
channel 1:4

Waveform versus time
channel 1:5

Waveform versus time
channel 1:6

FIGURE 2.34
Continuous (left) and burst (right) emission waveforms from transformers.

2.8 Performance Monitoring Using Automation Data, Process Data, and Other Information Sources

Process control systems are used in the oil and gas industry, pulp and paper industry, or other process industries. These systems host or serve as an operator workstation for one or more process control and instrumentation devices, such as valve positioners, switches, transmitters, and sensors (e.g., temperature, pressure, and flow rate sensors), whose functions include opening or closing valves and measuring process parameters (Galar et al., 2012).

The information from the field devices and the controller is made available for one or more applications run by an operator workstation. This allows an operator to view the actual status of the process, change the operation of the process, and so on. A common process control system has many process control devices and instrumentation, such as valves, transmitters, sensors, connected to one or more process controllers which, in turn, run the software controlling the devices.

Many other supporting devices are related to the process operation; these include, for example, power supply equipment, power generation and distribution of energy, and rotating equipment, as found in a typical plant. This additional equipment does not necessarily create or use process variables and, in many instances, is not controlled by or even attached to a process controller. Nevertheless this type of equipment is necessary for a process to work correctly. In the past, process controllers were not necessarily aware of these other devices, and simply assumed they worked properly when process control was performed. Figure 2.35 shows a fan monitored by two accelerometers in such a way that the information generated by vibration can close the loop and couple the device into the control system.

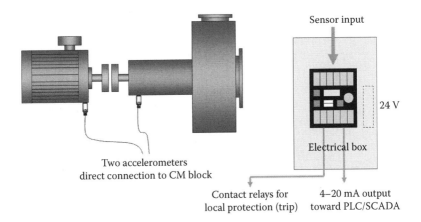

FIGURE 2.35
Monitoring of a fan and its integration to the control loop.

Many plants have other associated software systems that execute applications related to business functions (enterprise resource planning [ERP]) or maintenance functions (CMMS). In fact, many processing plants, especially those with smart field devices, use equipment monitoring applications to aid, monitor, and maintain the devices within the plant, regardless of whether they are process control, instrumentation, or other types of devices.

The integration of maintenance information, management, and monitoring is essential to close the loop of the process; accordingly, CMMS systems have evolved. For example, enterprise asset management (EAM) is a more sophisticated software than CMMS. It allows normal communication and data storage related to field devices to monitor the operational status of field devices. In some cases, the EAM application can be utilized to communicate with devices to modify parameters within the device, or to make the device execute applications on itself such as self-calibration or self-diagnostic routines, and, thus, obtain information about the status or health of the device.

This information can be stored and used by a maintenance person who monitors and maintains the devices. In the same manner, there are other types of applications used to monitor other types of devices, such as power generation equipment, rotating equipment, and supply devices. These other applications are occasionally available to the maintenance personnel and can monitor and maintain the devices within a processing plant. In many instances, outside service organizations offer services related to monitoring process and equipment. In these instances, external service organizations obtain the data required and run applications to analyze the data, but are limited in providing results and recommendations to the process plant staff. At the same time, however, the plant personnel have little or no ability to view the raw data measured or to utilize the analyzed data in any other way (Figure 2.36). Thus, in a typical plant, functions associated with the activities

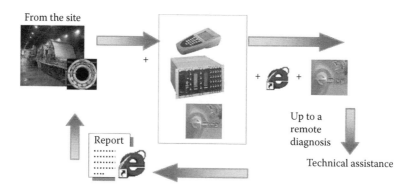

FIGURE 2.36
Typical process of outsourcing in CM.

of process control, equipment and device maintenance, monitoring activities, and business activities such as performance monitoring processes, are separated, both in the location in which they are carried out and the staff who usually perform these activities. Of course, the people involved in these functions often use different tools, for example, various applications running on different computers to perform different functions. Moreover, these various tools often collect or use other types of data and, thus, are configured differently. However, there should be cooperation among departments in an enterprise and between experts in their respective knowledge domains if the maintenance policy is to succeed.

Process control operators are primarily responsible for ensuring the quality and continuity of the process, supervising the daily operation of the process; they also affect the process, usually by setting and changing set points in the process, tuning loops of the process, scheduling process operations, and so on. They use the tools available to diagnose and correct problems within a process control system, and they receive variable operation information through one or more process controllers, including alarms generated in the process.

Maintenance personnel are primarily responsible for ensuring the efficient operation of the actual equipment in the process, along with the repair and replacement of malfunctioning equipment, and the use of tools such as maintenance interfaces maintenance, the EAM application discussed above, as well as many other diagnostic tools that provide information about the operating status of the various devices. In addition, they are responsible for scheduling maintenance activities that may require off parts or points of the plant.

Many new types of devices and process equipment, usually called intelligent field devices, include screening and diagnostic tools that automatically detect problems and report these problems to a maintenance person through a standard maintenance interface. For example, EAM software reports the device status and diagnostic information to the maintenance person and provides communication and other tools that allow him/her to know what is occurring in the device and access the information provided by it.

Although maintenance interfaces and maintenance staff are a huge part of the data network, they are located apart from process control operators, as shown in Figure 2.37. This is not always the case; in some plants, process control operators can perform the duties of maintenance people, or vice versa, or other people responsible for these functions can use the same interface.

The general lack of connectivity seriously affects the performance of maintenance functions. Many applications with different functions in a plant, such as process control operations, maintenance operations, and business operations, are not integrated and, therefore, do not share data or information. Some activities, such as monitoring equipment or operational testing of devices to determine if the plant is working in an optimal manner, and so on, are performed by external consultants or service companies. As noted above,

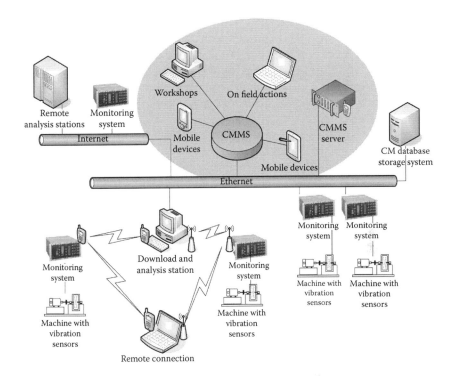

FIGURE 2.37
Typical architecture of maintenance information system.

these companies measure the data needed, perform an analysis, and supply only the results of the analysis to the plant staff. In these instances, as also mentioned, the data are normally collected and stored in a unique manner and are rarely made available to the plant personnel for other reasons.

Furthermore, even if all applications are within a single plant, because different staff use different applications and analysis tools, if these are normally located at different hardware locations within the plant, there is usually little or no information flow from a functional area of a plant to other areas, even when this information may be useful to other functions within the plant.

For example, a tool, such as a CM data analysis, can be used by a maintenance person to detect a malfunction. This tool can detect a problem, and alert the maintenance personnel that the device needs to be calibrated, repaired, or replaced. Nevertheless, the process control operator (either a human or software) does not have access to this information, although the device malfunction may be causing a problem that is affecting a loop or some other component being monitored by the process control operation. Similarly, the business person is not aware of this fact, even though the device may be a critical factor in plant optimization. Because the process control expert is not aware of a problem with the device, ultimately causing poor performance of a loop or the drive in the process control system and because he/she has assumed this computer is running smoothly, he/she may misdiagnose the problem detected within the loop or may try to apply a tool, such as a tuner loop, that will not correct the problem. Similarly, a business person can make an erroneous business decision to operate the plant in a manner that will not achieve the desired business effects (such as profit maximization) simply because the device does not work properly and he/she does not know.

Given the plethora of data analysis tools and other tools for the detection and diagnosis of process control, either in the plant or through external company services or consultants, there is a wealth of information on the health and performance of the devices available to the maintenance person that could be helpful to the process operator and business person. Similarly, there is a wealth of information available to the process operator on the current operating status of process control loops and other routines that may be useful to the maintenance person. In addition, information is generated by or used in the course of performing the business functions of the company, and this could be useful for the maintenance staff or the process control operator in optimizing process operation. Nevertheless, in the past, because these functions were separated, the information generated or collected in one functional area was not used, or not used well, in other functional areas, leading to a suboptimal use of assets within a plant.

2.8.1 Data Fusion: A Requirement in the Maintenance of Processes

A process control system consists of a system of collection and distribution of data from different sources, each of which can use its own unique way to acquire or generate data in the first place. The system of collection and distribution of data makes the stored data available to other applications for use in any manner desired. This way, applications can use the data from many different data sources to provide a better overview of information on the current operating status of a plant, to make better and more complete diagnostic or financial decisions about the plant, and so on.

Combined applications can provide or use data from collection systems previously disparate, such as process control monitoring systems, CM and performance models of processes, to arrive at a better overview of the state of a plant's process control, to facilitate the diagnosis of any problems and to recommend or take actions in production planning and maintenance, as seen in Figure 2.38. For example, information or data may be collected by maintenance functions related to health, variability, performance, or use of a device, loop, unit, and so on. This information can be sent and displayed to a process operator or maintenance personnel to inform those persons of a current or future problem. This same information can be used by a process operator to correct a current problem within a loop; for example, taking into account and correcting a device that is working suboptimally. Diagnostic

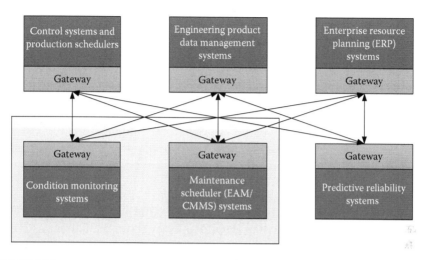

FIGURE 2.38
Sources of information potentially suitable for integration and fusion.

applications may generate information about devices using nonprocess variables, such as measurement, control, or rates, thereby determining, for example, the health of a device. Or team performance can be determined from models calculating key performance variables, such as the efficiency and cost of production.

A process control expert can use these measurement, control, and device rates with process variables to optimize the operation. Process variable data, and variable data of nonprocesses can be combined, for example, to generate models of processes.

Similarly, the detection of a problem in a device, such as one that requires the shutdown of process, can cause business software to automatically order replacement parts or alert the relevant business person that the selected strategic actions will not produce the desired results. The change of strategy control within the process control function may allow business software to automatically order new or different raw materials.

Of course, there are many other types of applications for which merged data relating to the process control, equipment monitoring, and performance monitoring data can be helpful, as these provide different and more complete information about the status of activity in all areas of a process control plant (Hall and Llinas, 1997).

Figure 2.34 shows a typical process control plant integrated into a number of businesses and other information systems, and interconnected with control and maintenance systems in one or more communication networks. Operator interfaces can store and run the tools available for process controls, including, for example optimizing control, diagnosticians, neural networks, and tuners.

Maintenance systems, such as computers running the CMMS/EAM application or any other device or equipment control and communication applications can be connected to the process control systems or to individual devices for monitoring and maintenance activities. For example, a maintenance team can be connected to the controllers and/or devices through the desired communication lines or networks (including wireless networks and hand-held devices) to communicate and, in some cases, reconfigure or perform other maintenance activities on the devices. Similarly, maintenance applications, such as the implementation of EAM/ CMMS, can be installed and run by one or more of the user interfaces associated with the processing control system to perform maintenance functions, including data collection related to the state of device operation.

Some plants include assets, such as motors, turbines, and so on, which are connected to a maintenance computer through a permanent or temporary communication link (i.e., a wireless communication system or a hand-held device connected to the equipment to take readings and then removed). The maintenance computer can store and run applications for diagnostic and known monitoring applications to monitor, diagnose, and optimize the operating state of the assets. A plant's maintenance staff generally uses the applications to maintain and oversee the performance of rotating equipment, determine problems with it, and decide when it should be repaired or replaced. In some instances, systems for the generation and distribution of energy may use external consultants or services to use the data to detect equipment problems, poor performance, or other issues. In these cases, the computers running the analyses cannot be connected with the rest of the system through any communication line; they can only be connected temporarily.

In the past, process control, energy generation, and maintenance systems were not attached to each other in a manner allowing them to share data generated or collected by any of these systems in a useful manner. As a result each of the different functions, such as the process control functions, energy generation functions, and rotating equipment functions, have operated under the assumption that other equipment in the plant may be affected by, or have an effect on, that particular function which is operating perfectly (this is seldom the case). Because the functions are so different and the equipment and personnel used to supervise these functions are different, there has been little or no meaningful data sharing between the various plant systems.

To solve this problem, we propose a data collection and distribution system, henceforth referred to as the asset cloud (see Section 2.8.3). This system acquires data from the disparate sources, formatting these data to a common data format or structure and then providing them, as needed, to any set of applications which are run by a computer system or dispersed among workstations throughout the process control network. The proposed application is able to integrate the use of data from previously separate sources to provide better measurement, monitoring, control, and understanding of all plant systems.

2.8.2 XML: Protocol for Understanding Each Other

2.8.2.1 Common Standards for Maintenance Information Exchange

The complexity of connectivity between applications is enormous, as control systems, maintenance management, CM, and enterprise applications are all involved in the management of complex, asset-intensive operations. Unfortunately, standards for information exchange have evolved independently in each area. OLE for Process Control (OPC) has become a popular standard for sharing information between control systems and associated manufacturing applications, as has the Machinery Information Management Open Systems Alliance (MIMOSA) OSA-EAI standard for sharing CM and asset health information with maintenance, operations, and enterprise systems. Meanwhile, the Instrumentation, Systems & Automation Society ISA-95 standard for integrating enterprise and production management systems is being adopted by a wide range of the relevant suppliers and users. All address an important issue and each has made significant progress.

OpenO&M recognizes that combining standards provides a good way to address many asset management challenges. Accordingly, it is being developed by a joint working group of professionals and is based on MIMOSA, OPC, and ISA-95 standards. Simply stated, it seeks to enable optimal asset performance through collaborative decision making across operating and maintenance organizations. While the standards used to develop OpenO&M originate in process manufacturing, the joint working group is also addressing the needs of the broader asset management community, including facilities and fleets in both public and private sectors.

OpenO&M is concerned with the integration of information among four areas: asset status assessment, CM, specialized sensors, and recent analysis tools. At this point, CBM and condition-based operations (CBO) strategies, that is, the performance of maintenance actions based on information collected by CM, are realizable. CBM and CBO attempt to avoid unnecessary maintenance tasks by taking action only when a physical asset shows evidence of abnormal behavior. In many organizations, however, this information is used by local technicians who maintain the equipment and is not yet accessible to other personnel.

Integration of asset CM information with control systems and operations (OPS), EAM, and other decision support systems (DSS) is imperative today, and this is where OpenO&M can play a key role. OpenO&M exploits the benefits of MIMOSA's common Asset Registry model to eliminate asset identification issues across multivendor systems and across organizational solutions. Integrating this with the standard models of OPC creates a recognized interface that includes automation and control systems and all supporting solutions; it often includes EAM as well.

2.8.3 XML: Protocol to Destroy Communication Barriers

To pick up on the final point above, working within the context of ISA-95 ensures that derived information can be used by higher-level enterprise applications such as ERP or EAM. This emerging standard is specifically focused on providing value to end users by creating plug-and-play capabilities for faster implementation by allowing them to select the best solution from complying suppliers. Implicit in OpenO&M is an extensible, open architecture based on extensible markup language (XML) and service-oriented interfaces, one that leverages leading-edge technology and supports practical interoperability and compliance.

XML may now be the most popular protocol for the communication exchange of maintenance information. While HTML is focused on document format, XML is focused on information content and relationships. A class of software solutions is evolving, which enables tighter coupling of distributed applications and hides some of the inherent complexities of distributed software systems. The general term for these software solutions is middleware. Fundamentally, middleware allows application programs to communicate with remote application programs as if the two were located on the same computer.

The process to transfer information between disparate sources in the XML environment is as follows. An XML wrapper wraps data from each of the computers involved in asset data exchange and sends them to an XML data server. Because XML is a descriptive language, the server can process any type of data. At the server, the data are encapsulated and mapped to a new XML wrapper, if necessary. Put otherwise, data are mapped from one XML schema to one or more other schemas created for each receiving application.

Essentially, XML is a model for describing the structure of information. For example, XML schemas can be used to test document validity; this is especially important when web-based applications are receiving and sending information to and from many sources.

When we are mapping the process model to an XML schema, we must establish rules, as there are many ways to accomplish the same output data structure, and we need a certain degree of regularity to simplify the data conversion. Once these rules are devised, creating the schemas is relatively straightforward. Figure 2.39 shows XML schema syntax.

XML can model all existing data (e.g., assets, events, failures, alarms). The most difficult data layer to represent is also the most critical; this is the layer containing information on sensory inputs and outputs, whether a single scalar value or an array of complex data points. Sensory data are especially relevant in CM and process control; they may be as simple as a single value or as complex as several synchronous sampled waveforms. The standards suggest a number of data formats that may represent sensory information.

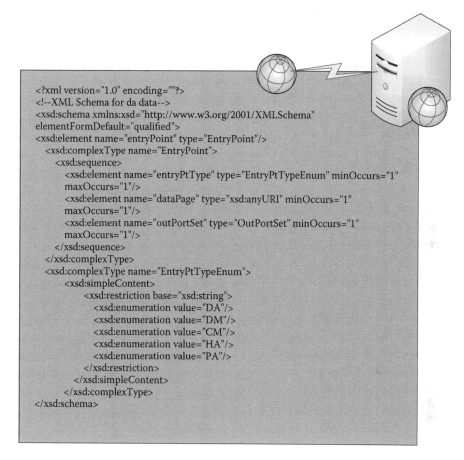

```
<?xml version="1.0" encoding=""?>
<!--XML Schema for da data-->
<xsd:schema xmlns:xsd="http://www.w3.org/2001/XMLSchema"
elementFormDefault="qualified">
<xsd:element name="entryPoint" type="EntryPoint"/>
  <xsd:complexType name="EntryPoint">
    <xsd:sequence>
      <xsd:element name="entryPtType" type="EntryPtTypeEnum" minOccurs="1"
      maxOccurs="1"/>
      <xsd:element name="dataPage" type="xsd:anyURI" minOccurs="1"
      maxOccurs="1"/>
      <xsd:element name="outPortSet" type="OutPortSet" minOccurs="1"
      maxOccurs="1"/>
    </xsd:sequence>
  </xsd:complexType>
  <xsd:complexType name="EntryPtTypeEnum">
    <xsd:simpleContent>
        <xsd:restriction base="xsd:string">
          <xsd:enumeration value="DA"/>
          <xsd:enumeration value="DM"/>
          <xsd:enumeration value="CM"/>
          <xsd:enumeration value="HA"/>
          <xsd:enumeration value="PA"/>
        </xsd:restriction>
      </xsd:simpleContent>
    </xsd:complexType>
  </xsd:schema>
```

FIGURE 2.39
XML schema of transformed and transferred data.

With XML, each data originator can wrap its own data using a schema understood or convenient for that particular device or application, and each receiving application can receive the data in a schema it recognizes and/ or understands. The XML server may be configured to map one schema to another depending on the data source and destination(s); it also may perform data processing functions or other functions based on data receipt. The rules for mapping and processing functions are set up and stored in the server before a series of data integration applications begins. This allows data to be sent from any one application to one or more other applications.

2.8.4 Example of Asset Data Integration Using XML

Web-based technologies are widely accepted for eMaintenance purposes.

For data from different data sources to be collected and used in a single system, a configuration database or another integrated configuration system, such as XML, is required. An explorer-type display or hierarchy should also be provided to allow the data to be manipulated, organized, and ultimately used by other applications.

Figure 2.40 shows an architectural overview of such a system, in this example, a process control system. Generally, the system can include a

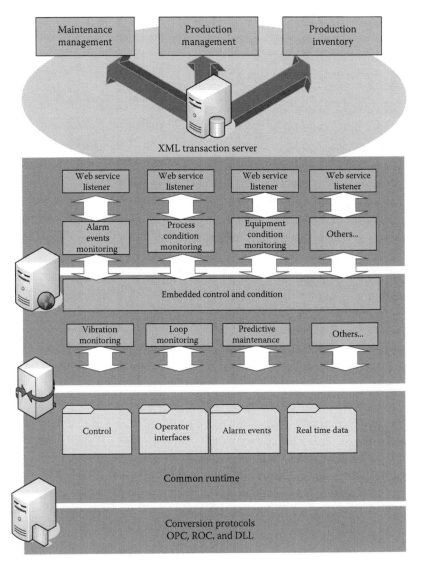

FIGURE 2.40
Integration of disparate data sources.

maintenance management system, a product inventory control system, a production scheduling system, along with other systems connected by LAN, internet, and so on. In this case, XML is used as a transaction server; the server sends XML-wrapped data to the web services.

The web services include a series of web service listeners; they listen for or subscribe to certain data from other data sources and supply these data to the various subscribing applications. The web listening services may be part of the data collection and distribution system; their functions include listening for and redistributing alarms and events data, processing CM data, and equipment CM data. Interfaces convert the data to a standard format or protocol, such as Fieldbus or ML6, as required.

The web services are in contact with and receive data from other external data sources via web servers, including, for example, vibration monitoring, real-time optimization, expert system analysis, PdM, and loop monitoring data sources, and so on.

Finally, a configuration database stores and organizes the data from the process control runtime system; this includes data from remote sources, such as external web servers.

2.8.4.1 Cloud Computing in Asset Management: Natural Data Repository

2.8.4.1.1 Introduction to Asset Cloud

Cloud computing, the next stage in internet evolution, provides the means through which everything from computing power to computing infrastructure, applications, business processes, and personal collaboration can be delivered as a service wherever and whenever we may want it. The cloud comprises a set of hardware, networks, storage, services, and interfaces that allow computing to be delivered as a service. Cloud services include the delivery of software, infrastructure, and storage over internet (either as separate components or a complete platform) based on user demand. Cloud computing, in all of its forms, is transforming the computing landscape. It will change the way technology is deployed and how we think about the economics of computing. Cloud computing is more than a service sitting in some remote data center. It is a set of approaches that can help organizations quickly and effectively add and subtract resources in almost real time. Unlike other approaches, the cloud is as much about the business model as it is about technology. Companies clearly understand technology is at the heart of how they operate their businesses. Business executives have long been frustrated with the complexities of getting their computing needs met quickly and cost effectively.

In the case of asset management, the cloud may solve the problem of dispersed data in many different repositories. The end user (maintenance or operators) does not need to understand the underlying technology. The data collection and distribution applications may be dispersed throughout the network, with data being collected at various locations. The collected data can be converted to a common format at these locations and sent to one or more central databases for distribution. These distributed databases constitute the asset cloud.

Thus, generally speaking, we need to establish routines to collect data from disparate sources and then provide these data to the cloud in a common or consistent format. The applications within the cloud may use the collected data and other information generated by the process control systems, the maintenance systems, and the business and process modeling systems, as well as the information generated by data analysis tools executed in each of these systems. However, the cloud may use any other desired type of expert system, including, for example, any type of data mining system.

It may also include other applications that integrate data from various functional systems for other purposes, such as user information, diagnostics, or for taking actions within a plant, such as process control actions, equipment replacement, or repair actions, altering the type or amount of product based on financial factors, process performance factors, and so on.

2.8.4.2 Services Provided by the Asset Cloud

In a sense, the cloud operates as a data and information clearing house in a processing plant; it coordinates the distribution of data or information from one area, that is, maintenance, to other areas, that is, the process control or business areas. The cloud may use the collected data to generate new information or data; these data can then be distributed to other computer systems associated with different functions in the plant. Finally, the cloud may execute other applications (or oversee their execution) to use the collected data to generate new types of data to be used within the process control plant.

A cloud-associated application may also diagnose conditions or problems within a process control plant based on data from two or more process control monitoring applications, process performance monitoring applications, or equipment monitoring applications. The applications may respond to a diagnosed or detected problem in the plant or may recommend actions to be taken by a user, that is, an operator, maintenance technician, or business executive responsible for the overall plant operation.

The cloud either includes or executes index generating software that collects or creates indexes associated with devices (process control and instrumentation, power generation, rotating equipment, units, areas etc.) or process control entities (loops etc.) within the plant. These indexes can be used to optimize process control or to provide business managers with more complete or understandable information about the operation.

The asset cloud must provide maintenance data (e.g., device status information) and business data (e.g., data associated with scheduled orders, etc.) to a control expert associated with the process control system to facilitate such activities as optimizing control. The control experts may also incorporate and use data related to the status of devices or other hardware within the process control plant. In addition, they may generate performance data using process performance models of the decision making. In the past, software control experts could only use process variable data and some limited device status data to make recommendations to the process operator.

With the cloud's ability to collect communication data, especially device status information and data analysis tools, the control expert can incorporate device status information, including health, performance, utilization, and variability information, into its decision making. In addition, the asset cloud can provide information on the state of the devices and the operation of the control activities within the plant to the business systems; a work order application or program can automatically generate work orders based on problems detected in the plant, or an order for parts based on work performed.

Figure 2.41 is a simplified block diagram of data flow and communication associated with or used by the asset cloud. The diagram includes the data collection and distribution system, which receives data from many sources. A process control data source may include traditional process control activities and applications, such as process control and monitoring applications, process control diagnostic applications, process control alarming applications, and so on. Any or all of these can send data to the cloud (Figure 2.41).

Equipment or process health data sources also send information to the cloud. These include traditional equipment monitoring applications,

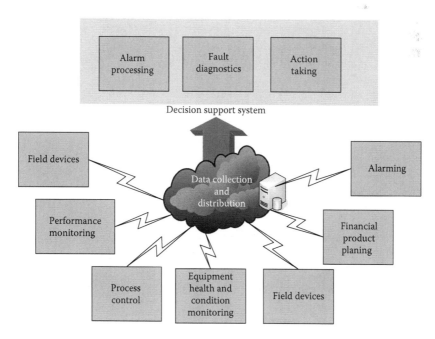

FIGURE 2.41
Services provided by the asset cloud.

equipment diagnostic applications, equipment alarming applications, abnormal situation analysis applications, environmental monitoring applications, and so on. In other words, the source may send data generated by any of a number of diagnostic applications or traditional equipment monitoring.

A performance monitoring data source is another source of data. These include performance monitoring applications such as process models used to monitor or model process operation, process or equipment health, and so on. Data may also be acquired by or generated by any type of performance monitoring equipment or application.

Finally, a financial or production planning data source may be connected to the cloud. These applications perform financial or cost analysis functions within the process control system, including deciding how to run the plant to maximize profits, how to avoid environmental fines, what or how much of a product to make, and so on. Field devices, such as smart field devices, may provide further data; this includes any data measured or generated by these field devices, such as alarms, alerts, measurement data, calibration data, and so on.

2.8.4.3 Asset Cloud as a Decision Support System

The data collection and distribution system cloud collect data from the various data sources in a common format and/or convert the received data to a common format for storage and later use by the other elements, devices, or applications in the process control system. Once received and converted, data are stored in a database; they must be both accessible and available to applications or users within the asset cloud. Applications related to process control, alarms, device maintenance, fault diagnostics, PdM, financial planning, optimization, and so on can use, combine, and integrate the data from one or more of the many data sources, thus allowing them to operate better than in the past when data were both disparate and inaccessible.

Figure 2.42 provides a detailed diagram of data flow in a process control plant. At the left side of the diagram, data associated with the process plant are collected through various functional areas or data sources:

1. Control data are collected by typical process control devices: field devices, input/output devices, hand-held or remote transmitters, or any other devices, that is, communicatively connected to process controllers.

2. Equipment monitoring data associated with traditional equipment monitoring activities are collected by sensors, devices, transmitters, and so on. Process performance data are collected by the same or other devices.

3. Financial data are collected by other applications as part of the performance monitoring data.

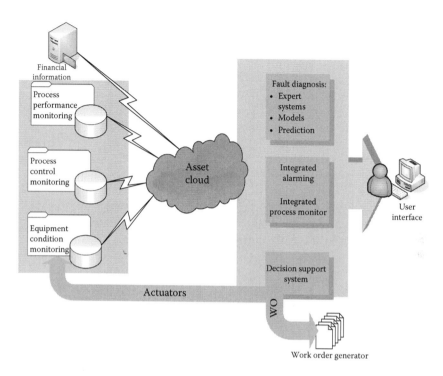

FIGURE 2.42
Data flow diagram of information sources in process control.

4. Collected data from applications or sources outside the traditional process control network, for example, applications owned and operated by service organizations or vendors.

These data may be collected automatically or manually, as many diverse sources can compose equipment monitoring. Accordingly, data collectors include the following: hand-held collection devices, a laboratory's chemical and physical measurements, fixed or temporary online devices, which periodically (e.g., RF) telemeter data from remote process and equipment measurement devices, online device inputs or remote multiplexers, and/or concentrators or other data collection devices.

The process control data, equipment monitoring data, and process performance data can be reconciled, verified, validated, and/or formatted by data collection and reconciliation applications. Note: these may be part of the cloud. These applications run within the data collection device or within any other device; this includes central data historians, process controllers, equipment monitoring applications, and the like, which receive or process this data.

After being reconciled or, in some cases, not reconciled at all, the collected data may be provided to applications associated with the various functional areas of the process control system. One or more diagnostic applications may use the collected data to perform process control diagnostics, including applications that help an operator pinpoint problems within process control loops, instruments, actuators, and so on. The diagnostic applications can include expert diagnostic engines as well. *Note:* The process diagnostic applications can take the form of any typical or standard process diagnostic application and are not limited to those mentioned here.

The outputs of these diagnostic applications can take any form. They may indicate faulty or poorly performing loops, functions blocks, areas, and units within the process control system or they may indicate where the loops need to be tuned.

The equipment monitoring functional block will receive the reconciled equipment condition data version. This block can include equipment or CM applications; these may, for example, accept or generate alarms indicating problems with equipment, detect poorly performing or faulty equipment, or detect other equipment problems or conditions of interest to a maintenance person. Well-known equipment monitoring applications include utilities adapted to the different types of equipment within a plant. Equipment diagnostic applications can be used to detect and diagnose equipment problems based on measured raw data on the equipment. Examples of equipment diagnostic applications include vibration sensor applications, rotating equipment applications, power measurement applications, and so on.

Of course, there are many types of equipment CM and diagnostic applications that produce data associated with the state or operating condition of equipment within a process control plant. In addition, a historian may store raw data detected by equipment monitoring devices, store data generated by the equipment CM and diagnostic applications, and provide data to those applications as needed. Finally, equipment models may be provided and used by the equipment CM and diagnostic applications in any manner deemed appropriate.

2.8.4.3.1 *Context as a Result of Fused Data: The Future*

In the past, process monitoring, equipment monitoring, and performance monitoring were performed independently. Each attempted to "optimize" its own functional area, ignoring the effect its actions might have on the other functional areas. As a result, a low-priority equipment problem causing a larger problem in the failure to achieve a desired process control performance level may not have been corrected because it was not considered important in the context of equipment maintenance. With the asset cloud providing data, however, end users have access to a broader view of the plant based on two or more of three possible data sources: equipment monitoring data, process performance data, and process control monitoring data. Similarly, diagnostics performed for the plant can now consider data associated with process operation and equipment operation, providing a better overall diagnostic analysis.

Simply stated, the collected data (process control data, process monitoring data, equipment monitoring data) can be provided to different people, collected and used in different formats, and used by completely different applications for different purposes. As a result, some of these data may be measured or developed by service organizations that use proprietary applications incompatible with the rest of the process control system. By the same token, if data are collected by or generated by financial applications typically used in a process control environment, they may not be in a format or protocol recognizable or useable by process control or alarm applications. A maintenance person and the equipment monitoring and diagnostic applications he/she typically uses may not have access to data collected by or generated by any of the process control applications, process models, or financial applications; they may not be constructed in such a fashion so as to be able to use them either. Finally, the process control operator and the process control monitoring and diagnostic applications used by him/her do not usually have access to data collected by or generated by the equipment monitoring applications and performance modeling or financial applications; again, they may not be constructed to use them, in any event.

The asset cloud overcomes the problem of restricted or no access to data from various external sources. It collects data and converts those data, if needed, into a common format or protocol that can be accessed and used by applications. The integration of the various types of functional data promises improved personnel safety, higher process and equipment uptime, avoidance of catastrophic process and/or equipment failures, greater operating availability (uptime) and plant productivity, higher product throughput stemming from higher availability, and the ability to safely and securely run faster and closer to design and manufacturing warranty limits. There will also be a higher throughput because of the ability to operate the process at the environmental limits, and improved quality with the elimination or minimization of equipment-related process and product variations.

References

Abbott, P., 2001. *Guideline for Developing and Managing and Infrared Thermography (IRT) Program*. Palo Alto, CA: EPRI.

Bauernfeind, J., 2001. *Developments of Diagnostic Methods for Online Condition Monitoring of Primary System Components*. u.o.: Kerntechnik 58.

Bethel, N., 1998. *Identifying Motor Defects Through Fault Zone Analysis*. [Online] Available at: http://www.reliableplant.com/Read/4288/motor-defects-fault-zone.

BINDT, 2014. *About CM*. [Online] Available at: http://www.bindt.org/What-is-CM/

BINDT, 2014. *About NDT*. [Online] Available at: http://www.bindt.org/What-is-NDT/ Bob is the Oil Guy, 2014. *What is Oil Analysis?* [Online] Available at: http://www.bobistheoilguy.com/what-is-oil-analysis/.

Commtest Instruments, 2014a. *How is Machine Vibration Described?* [Online] Available at: http://reliabilityweb.com/index.php/articles/how_is_machine_ vibration_ described/.

Commtest Instruments, 2014b. *How is Vibration Measured?* [Online] Available at: http://reliabilityweb.com/index.php/articles/how_is_vibration_measured/.

Dennis, H. S., 1994. Introduction to vibration technology. [Online] Available at: http://www.irdbalancing.com/downloads/vt1_2.pdf.

Emerson. 2014. *Measurement Types in Machinery Monitoring.* White Paper. www.assetweb.com.

Galar, D., Kumar, U., Juuso, E. and Lahdelma, S., 2012. *Fusion of Maintenance and Control Data: A Need for the Process.* Durban, South Africa: UN.

GE Measurement & Control, 2013. *Beginner's Guide: Machine Vibration.* [Online] Available at: http://www.commtest.com/training_services/training_materials/beginners_guide_to_vibration_analysis?download=1.

Gorgulu, M. and Yilmaz, M., 1994. *Three-Dimensional Computer Graphics Visualization of Target Detection.* Monterey, CA: Naval Postgraduate School.

Grisso, R. D. and Melvin, S. R., 1995. *NF95-225 Oil Analysis.* u.o.: Historical Materials from University of Nebraska-Lincoln.

Hall, D. L. and Llinas, J., 1997. An introduction to multisensor data fusion. *Proceedings of the IEEE*, 85(1), 6–23.

Hellier, J. C., 2003. *Handbook of Nondestructive Evaluation.* New York: McGraw-Hill.

IAEA, 2007. *Implementation Strategies and Tools for Condition Based Maintenance at Nuclear Power Plants.* Vienna, Austria: International Atomic Energy Agency.

Infrared Training Center, 2014. *IR Thermography Primer.* [Online] Available at: http://www1.infraredtraining.com/view/?id=40483.

ISO 18434-1:2008, 2008. *Condition Monitoring and Diagnostics of Machines: Thermography—Part 1: General Procedures.* u.o.: UN.

Kaplan, H., 2007. *Practical Applications of Infrared Thermal Sensing and Imaging Equipment.* 3rd edition. Washington, DC: SPIE.

Kastberger, G. and Stachl, R., 2003. Infrared imaging technology and biological applications. *Behavior Research*, 35(3), 429–439.

Lindley, H. R., Mobley, R. K. and Wikoff, D., 2008. Vibration: Its analysis and correction. i: *Maintenance Engineering Handbook.* New York: McGraw-Hill Companies.

Madanhire, I., Mugwindiri, K. and Mayahle, L., 2013. Application of online intelligent remote condition monitoring management in thermal power plant maintenance: Study of ThermPower plant in Zimbabwe. *International Journal of Science and Research (IJSR)*, 2(1), 12–25.

Marwan, F. B., 2004. Nondestructive testing technologies for local industries. *NDT. NET*, 9(4), 1, 133–141.

Mobley, R. K., 2001. *Plant Engineer's Handbook.* Woburn, MA: Butterworth-Heinemann.

Monacelli, B., 2005. *Spectral Signature Modification by Application of Infrared Frequency Selective Surfaces.* Orlando, FL: University of Central Florida.

Penrose, H. W., 2004. *Motor Circuit Analysis Concept and Principle.* [Online] Available at: http://www.reliabilityweb.com/art04/mca_concept.htm.

Randall, R. B., 2011. *Vibration Based Condition Monitoring.* u.o.: John Wiley and Sons.

Stephen, R., 2006. *Certification of Condition Monitoring Personnel and How It Relates to NDT.* Northampton, UK: UN.

Symphony Teleca, 2013. *M2M Telematics & Predictive Analytics.* [Online] Available at: http://www.symphony-analytics.com/media/159393/analytics_wp_2013_v1_m2m_telematics_predictive_analytics.pdf.

Vibratec, 2008. *Análisis de Circuito Eléctrico en Motores.* [Online] Available at: http://www.vibratec.net/pages/tecnico4_anacircelec.html.

Vibration Training Course Book Category II. 2013. Vibration school website: http://www.vibrationschool.com.

Vyas, M., 2013. *Vibration Monitoring System Basics.* [Online] Available at: http://www.for-besmarshall.com/fm_micro/downloads/shinkawa/FUNDAMENTALS%20OF%20VIBRATION%20By%20FM%20-%20Shinkawa.pdf.

Wenzel, R., 2011. *Condition-Based Maintenance: Tools to Prevent Equipment Failures.* [Online] Available at: http://www.machinerylubrication.com/Read/28522/condition-based-maintenance.

Zayicek, P., 2002. *Infrared Thermography Guide (Revision 3).* Palo Alto, CA: EPROI.

3

Challenges of Condition Monitoring Using AI Techniques

3.1 Anomaly Detection

The anomaly detection task is to recognize the presence of an unusual (and potentially hazardous) state within the behaviors or activities of a system, with respect to some model of "normal" behavior that may be either hard coded or learned from observation. We focus here on learning models of normalcy at the user behavioral level. An anomaly detection agent faces many learning problems including learning from streams of temporal data, learning from instances of a single class, and adapting to a dynamically changing concept. In addition, the domain is complicated by considerations of the trusted insider problem (recognizing the difference between innocuous and malicious behavior changes on the part of a trusted user) and the hostile training problem (avoiding learning the behavioral patterns of a hostile user who is attempting to deceive the agent).

We propose an architecture for a learning anomaly detection agent based on a hierarchical model of user behaviors. The leaf level of the hierarchy models the temporal structure of user observations, while higher levels express interrelations between descendant structures. We describe approaches to the fabrication of such models, employing instance-based learning models and hidden Markov models (HMMs) as the fundamental behavioral modeling units. Approaches to the trusted insider and hostile training problems are described in terms of the hierarchical behavior model (Lane, 1998).

In this chapter, we will discuss the goals of the anomaly detection domain, related background work, and the issues raised by the proposed research. Although we make an effort to divide the issues into those that are most closely related to learning and those most closely related to security, we note the nature of the domain is such that it can be difficult to completely separate the two.

Before examining the anomaly detection problem in detail, we discuss the overall performance goals of an adaptive anomaly detection system. When examining machine-learning algorithms for anomaly detection, we must keep in mind several practical requirements imposed by the domain and the intended use. Specifically, the purpose of any system is to enhance the users' ability to accomplish their desired tasks. In the context of anomaly detection, this enhancement increases the individual's confidence in the privacy and confidentiality of personal systems. From a global perspective, strong anomaly detection systems increase confidence in authenticity (the belief that actions originating with a particular account are actually associated with the owner of that account) and increase assurance that shared system resources and data are being used properly (Lane, 1998). In this section, we review personal versus global definitions of "anomaly" requirements for system accuracy, and space and time resource issues.

Anomaly detection refers to finding patterns in data that do not conform to expected behavior. Such nonconforming patterns may be called anomalies, outliers, discordant observations, exceptions, aberrations, surprises, peculiarities, or contaminants in different application domains. Anomalies and outliers are two most commonly used terms in the context of anomaly detection and are sometimes used interchangeably. Anomaly detection is used in a wide variety of applications, including credit card fraud detection, insurance or health care, intrusion detection for cyber security, fault detection in safety-critical systems, and military surveillance of enemy activities. Anomaly detection is important because anomalies in data translate into significant (often critical) actionable information in many application domains; for example, a machine's anomalous behavior (Chandola et al., 2009).

Th detection of outliers or anomalies in data has been studied by statistics researchers since the nineteenth century. Over time, other research communities have also developed anomaly detection techniques. Many of these techniques have been developed for specific domains, but others are more generic.

This chapter provides a structured, comprehensive overview of the research on anomaly detection to explain the many different research directions and show how techniques developed in one area can be applied in other domains.

3.1.1 What Are Anomalies?

Data patterns that do not conform to a well-defined notion of normal behavior are anomalies. Figure 3.1 uses a simple two-dimensional data set to illustrate anomalies. In the figure, the data have two normal regions, N_1 and N_2; most observations lie in these two regions. Points sufficiently far away from the two regions are anomalies, for example, points O_1 and O_2, and points in region O_3. Anomalies are induced in data for a variety of reasons, such as malicious activity, including credit card fraud, cyber intrusion, terrorist activity, or a system breakdown. However, all these reasons have a common characteristic that an analyst finds interesting: they are all related to real life. In fact, "interestingness" or reallife relevance is a key feature of anomaly detection.

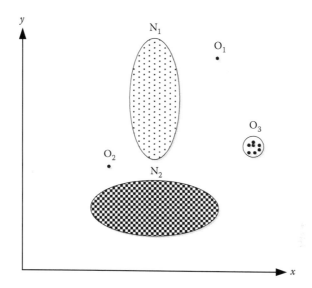

FIGURE 3.1
A simple example of anomalies in a two-dimensional data set. (Redrawn from Chandola, V. et al., 2009. Anomaly detection: A survey. *ACM Computing Surveys (CSUR)*, 41(3), 15.)

Noise removal and accommodation are related to anomaly detection as they deal with unwanted noise in the data. Noise is not of interest to the analyst; it hinders data analysis, so it must be removed before data are analyzed. Noise accommodation refers to immunizing models against anomalous observations or noise. Novelty detection, also related to anomaly detection, finds previously unobserved (emergent, novel) patterns in the data, for example, a new topic of discussion appearing in a news group. Unlike anomalies, however novel patterns are typically incorporated into the normal model after being detected.

Note: because the problems and solutions mentioned above often appear in anomaly detection and vice versa, they are discussed in this chapter (Chandola et al., 2009).

3.1.2 Challenges

In an abstract sense, an anomaly is a pattern that does not conform to expected (i.e., normal) behavior. Therefore, a straightforward anomaly detection approach will define normal behavior; any observations that do not belong to this normal region are anomalies. It is not always so simple, however (Chandola et al., 2009):

- It is difficult to find a normal region that encompasses every possible normal behavior. Moreover, the borders between normal and anomalous behavior may be blurred: an anomalous observation close to the border can actually be normal, and vice versa.

- When anomalies result from malicious actions, the instigators of these actions often adapt themselves to make the anomalous observations appear normal; obviously, this complicates the task of defining normal behavior.

- Given the tendency of normal behavior to evolve, a current definition of normal behavior might not be accurate in the future.

- The precise notion of an anomaly differs in different application domains. In the condition monitoring domain, for example, a slight deviation from normal (e.g., fluctuations in bearing temperature) might be an anomaly, while a similar deviation in the stock market domain (e.g., fluctuations in the value of a stock) might be acceptable. In other words, applying techniques across domains is not likely to work.

- The availability of labeled data to train and/or validate the models used by anomaly detection techniques is generally problematic.

- The data frequently contain noise that resembles the anomalies; it is hard to distinguish and remove this noise.

Given these challenges, it is not easy to solve the anomaly detection problem, at least in its most general form. In fact, most anomaly detection techniques solve a specific formulation of the problem induced by factors including the nature of the data, availability of labeled data, type of anomalies to be detected, and so on. These factors are often determined by the nature of the application domain where anomalies must be detected. Researchers regularly adopt concepts from a range of disciplines, including statistics, machine learning, data mining, information theory, and spectral theory, and apply them to their specific problem formulations.

Figure 3.2 shows the key components of anomaly detection techniques.

3.1.2.1 Different Aspects of an Anomaly Detection Problem

As mentioned above, a specific formulation of the anomaly detection problem is determined by such factors as the nature of the input data, the availability (or unavailability) of labels, and the constraints and requirements of the application domain. Given the diversity, we obviously need a broad spectrum of anomaly detection techniques (Chandola et al., 2009).

3.1.2.1.1 Nature of Input Data

Input data are usually a collection of data instances (also called object, record, point, vector, pattern, event, case, sample, observation, entity). Each data instance has a set of attributes (also called variable, characteristic, feature, field, dimension). The attributes can be binary, categorical, or continuous. A data instance might have one attribute (univariate) or multiple attributes

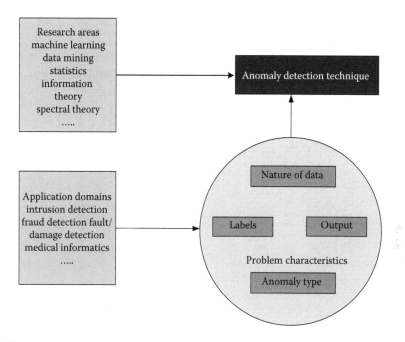

FIGURE 3.2
Key components associated with an anomaly detection technique. (Redrawn from Chandola, V. et al., 2009. Anomaly detection: A survey. *ACM Computing Surveys (CSUR)*, 41(3), 15.)

(multivariate). In the case of the latter, all attributes might be the same type or a mixture of types.

The applicability of an anomaly detection technique depends on the nature of the attributes. If we are using statistical techniques, our choice of models will depend on whether we have continuous or categorical data. Similarly, if we are using nearest-neighbor-based techniques, the nature of attributes would determine which distance measure we select. Instead of the actual data, the pairwise distance between instances may be provided as a distance (or similarity) matrix. Techniques requiring original data instances are not applicable in such cases, including numerous statistical and classification-based techniques.

The categories of input data can also be based on the relationship of data instances. Most anomaly detection techniques deal with record data (or point data), with no relationship assumed among data instances.

Data instances that can be related to each other include sequence data, spatial data, and graph data. In sequence data, the instances are ordered linearly, as in time-series data, genome sequences, protein sequences. In spatial data, each instance is related to neighboring instances, as in vehicular traffic data, ecological data. If spatial data have a temporal (sequential) component, we call them spatiotemporal data, as in climate data. In graph data, data instances are represented as vertices on a graph and connected to other vertices by edges. In certain situations, such relationships among data instances become relevant for anomaly detection (Chandola et al., 2009).

3.2 Types of Anomaly

The nature of the anomaly is an important aspect of an anomaly detection technique. Anomalies can be classified into three categories: point, collective, and contextual anomalies (Chandola et al., 2009).

3.2.1 Point Anomalies

If an individual data instance can be considered anomalous to the rest of the data, it is called a point anomaly. This is the simplest type; the majority of research on anomaly detection focuses on it.

In Figure 3.1, points O_1 and O_2 and points in region O_3 lie outside the boundary of the normal regions; they are point anomalies because they differ from normal data points. Consider credit card fraud detection as a case in point. Let the data set correspond to a person's credit card transactions. For simplicity, assume the data are defined using a single feature, namely, amount spent. A transaction for which the amount spent is much higher than the normal range of expenditure is a point anomaly.

3.2.2 Contextual Anomalies

If a data instance is anomalous in a specific context (but not otherwise), we call it a contextual anomaly (also a conditional anomaly). The context is induced by the data set's structure and must be specified in problem formulation. Each data instance is defined using two sets of attributes: contextual and behavioral.

1. Contextual attributes are used to determine the context (or neighborhood) for that instance. In spatial data sets, for example, the contextual attributes of a location are its longitude and latitude. Meanwhile, in time-series data, time is a contextual attribute that determines the position of an instance in the whole sequence.

2. Behavioral attributes define the noncontextual characteristics of an instance. In a spatial data set describing the average rainfall of the entire world, for example, the amount of rainfall at any one location is a behavioral attribute.

An instance of data could be a contextual anomaly in a given context but considered normal in another (in terms of behavioral attributes). In contextual anomaly detection, this property is key.

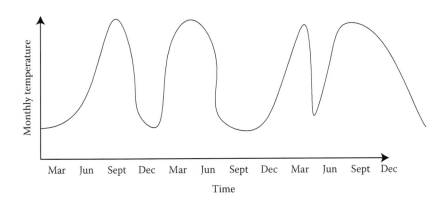

FIGURE 3.3
Contextual anomaly *t*2 in a temperature–time series. The temperature at time *t*1 is same as at time *t*2 but occurs in a different context and is not considered as an anomaly. (Redrawn from Chandola, V. et al., 2009. Anomaly detection: A survey. *ACM Computing Surveys (CSUR)*, 41(3), 15.)

Contextual anomalies are often explored in time-series and spatial data. Figure 3.3 gives one such example for a temperature–time series, showing monthly temperatures for an area over the past few years.

A temperature of 35°F might be normal in winter (time *t*1) in that particular place, but in summer (time *t*2), it would be an anomaly. Selecting a contextual anomaly detection technique is based on the meaningfulness of the contextual anomalies in the target application domain. Another factor to consider is the availability of contextual attributes. When defining a context is straightforward, applying a contextual anomaly detection technique makes sense. When it is difficult, such techniques are hard to apply.

3.2.3 Collective Anomalies

If a collection of related data instances is anomalous to the entire data set, we call this a collective anomaly. While, the individual data instances may not be anomalies, their occurrence together as a collection is anomalous. In Figure 3.4, an illustration of a human electrocardiogram output, the highlighted region denotes an anomaly; the same low value exists for an abnormally long time (corresponding in this instance to an atrial premature contraction). By itself, however, the low value is not an anomaly.

Collective anomalies have been explored in sequence data, graph data, and spatial data. While point anomalies can occur in any data set, collective anomalies can occur only in sets where data instances are related. The appearance of contextual anomalies, meanwhile, depends on the availability of context attributes in the data. *Note*: a point anomaly or a collective anomaly can be transformed into a contextual anomaly detection problem by incorporating the context information (Chandola et al., 2009).

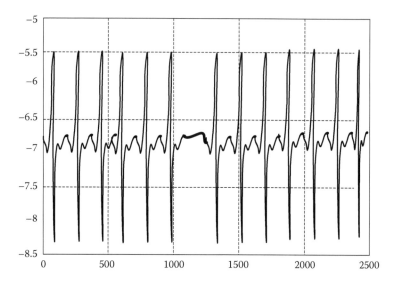

FIGURE 3.4
Collective anomaly (atrial premature contraction in a human electrocardiogram output).
(Redrawn from Chandola, V. et al., 2009. Anomaly detection: A survey. *ACM Computing Surveys (CSUR)*, 41(3), 15.)

3.2.4 Data Labels

The labels for a data instance indicate if it is normal or anomalous. Obtaining labeled data that are both accurate and representative of all types of behaviors may be extremely expensive, as labeling is often done manually by a human expert, thus requiring a great deal of effort. By and large, deriving a labeled set of anomalous data instances covering all possible types of anomalous behavior is more difficult than getting labels for normal behavior. Added to this, anomalous behavior is often dynamic; new types of anomalies might arise, for example, for which there are no labeled training data. In air traffic safety, for example, anomalous instances may include catastrophic events and, thus, be extremely rare.

Depending on the availability of labels, anomaly detection techniques operate in one of three modes: supervised, semisupervised, or unsupervised anomaly detection (Chandola et al., 2009).

3.2.4.1 Supervised Anomaly Detection

The supervised mode assumes the availability of a training data set with labeled instances for normal and anomaly classes. A typical approach is building a predictive model for normal versus anomaly classes. In this approach, any unseen data instance is compared against the model to determine its class. Supervised anomaly detection has two problems. First, there are fewer anomalous instances than normal instances in the training data, creating imbalanced class distributions. Issues arising from this are addressed in the literature on data mining and machine-learning. Second, deriving accurate and representative labels for the anomaly class is generally challenging. A number of techniques propose injecting artificial anomalies into a normal data set to obtain a labeled training data set.

Otherwise, supervised anomaly detection is similar to building predictive models, and we will, therefore, not continue the discussion.

3.2.4.2 Semisupervised Anomaly Detection

Techniques using a semisupervised mode assume training data only need labeled instances for the normal class, making them more widely applicable than supervised techniques. In spacecraft fault detection, for example, an anomaly would signify an accident, and this is difficult to model. Typically, such techniques build a model for the class of normal behavior and use the model to identify anomalies in the data.

A limited set of anomaly detection techniques assumes the availability of only anomaly instances for training. These are seldom used because it is difficult to obtain a training data set covering all possible anomalous behaviors in the data (Chandola et al., 2009).

3.2.4.3 Unsupervised Anomaly Detection

Techniques using the unsupervised mode do not require training data, making them widely applicable. These techniques implicitly assume normal instances occur far more frequently than anomalies in the test data. If this assumption turns out to be false, they have a high false alarm rate.

We can adapt many semisupervised techniques to operate in an unsupervised mode if we use a sample of the unlabeled data set as training data. This assumes the test data contain few anomalies; it also assumes the model learned during training is robust to these anomalies (Chandola et al., 2009).

3.2.5 Anomaly Detection Output

Any anomaly detection technique must consider how anomalies are reported. The outputs of anomaly detection techniques are usually either scores or labels (Chandola et al., 2009).

3.2.5.1 Scores

Scoring techniques assign an anomaly score to each instance in the test data; the score depends on the extent to which that instance is considered an anomaly. Thus, the output is a ranked list of anomalies. An analyst may opt to analyze the top few anomalies or to make use of a cut-o® threshold.

3.2.5.2 *Labels*

Labeling techniques assign a label (normal or anomalous) to each test instance.

The scoring-based anomaly detection techniques discussed above permit an analyst to use a domain-specific threshold to select the most relevant anomalies. In contrast, techniques providing binary labels to the test instances do not directly permit him/her to make such a choice. However, this can be indirectly controlled by the parameter choices of each technique.

3.2.6 Industrial Damage Detection

Industrial units suffer damage from continuous use and normal wear and tear; early detection is crucial to prevent escalation and further losses. Data in this domain are usually called sensor data because they are recorded by various types of sensors and collected for analysis. Anomaly detection techniques are widely applied to these data to detect industrial damage.

Industrial damage detection can be subdivided into a domain dealing with defects in mechanical components, such as motors, engines, and so on, and one dealing with defects in physical structures. The former is also called system health management.

3.2.6.1 *Fault Detection in Mechanical Units*

As the name suggests, anomaly detection techniques in mechanical units monitor the performance of industrial components, including motors, turbines, oil flow in pipelines, and other mechanical components; they detect possible defects caused by anticipated wear and tear or unforeseen circumstances.

Such anomalies appear as an observation in a specific context (contextual anomalies) or as a sequence of anomalous observations (collective anomalies). For the most part, normal data (i.e., in components without defects) are readily available, and semisupervised techniques are applicable. As preventative measures must be taken as soon as possible, these anomalies are generally detected online (Chandola et al., 2009).

3.2.6.2 *Structural Defect Detection*

Structural defect and damage detection techniques detect structural anomalies, such as cracks in beams or strains in airframes. These data have a temporal aspect and may also have spatial correlations. The anomaly detection techniques are like novelty detection or change point detection techniques in that they detect changes in data collected from a structure. Normal data and the models learned from those data tend to be static over time (Chandola et al., 2009).

3.2.6.3 Image Processing

Anomaly detection techniques for images are interested in changes in an image over time (motion detection) or in regions appearing abnormal on a static image (thermography in maintenance).

These anomalies are caused by motion, by the insertion of a foreign object, or by instrumentation errors. Data have both spatial and temporal attributes. Each data point has some continuous characteristics such as color, lightness, texture, and so on. The interesting anomalies are either anomalous points or regions in the images (point and contextual anomalies, respectively).

A major challenge is the large size of the input, for example, video data. Online anomaly detection techniques are required for these types of data (Chandola et al., 2009).

3.2.6.4 Anomaly Detection in Text Data, WOs, and Other Maintenance Documents

In these domains, anomaly detection techniques detect novel topics, events, or news stories in a collection of documents or news articles, and so on. The anomalies are caused when new events or anomalous topics are introduced. Data are typically high dimensional and extremely sparse; they also have a temporal aspect, as documents are collected over time. Handling the large variations in documents belonging to one category or topic presents a challenge in this domain.

3.2.6.5 Sensor Networks

Sensor networks have lately become an important topic of research due to the large number of sensors deployed in equipment for health assessment. The interest tends to take a data-analysis perspective, as the data have some unique characteristics. Anomalies in data collected from a sensor network can mean one or more sensors are faulty, or sensors are detecting events (e.g., intrusions) that are interesting for analysts. Thus, anomaly detection can capture sensor fault detection or intrusion detection or both.

A single sensor network may consist of sensors collecting different types of data, including binary, discrete, continuous, audio, video, and so on. Data are generated in a streaming mode. The environment in which the various sensors are deployed and the communication channel frequently induce both noise and missing values in the data.

Anomaly detection in sensor networks poses unique challenges. For one thing, the techniques must operate online. For another, because of severe resource constraints, they must be lightweight. Moreover, data are collected in a distributed manner, thus requiring the use of a distributed data-mining approach to their analysis. Finally, the presence of noise in the data collected from the sensor requires the detection technique to distinguish between interesting anomalies and unwanted noise/missing values (Chandola et al., 2009).

3.3 Rare Class Mining

Rare class classification is the data-mining task aimed at building a model to correctly classify both majority and minority classes. Classifying these is difficult because the rare class is very small, but many researchers have tried to solve this problem (Chomboon et al., 2013).

Rare class mining is found in many real-world data-mining applications, including network intrusion detection, *video surveillance* (Medioni 2001), oil spills detection *in satellite radar images,* diagnoses of rare medical conditions, *text categorization,* and so on. In all these applications, samples from one class are extremely rare, but the number of samples from other classes is sufficiently large. Problematically, the correct detection of the rare samples is significantly more important than the correct classification of the majority samples (Han et al., 2009).

The network intrusion detection domain receives hundreds of thousands of access requests every day. Among these, the number of malicious connections is generally very small compared to the number of normal connections. However, building a model that can detect attacks is crucial; the system must be able to respond promptly to any network intrusions. Samples from a rare class are sometimes called rare events or rare objects. Because the rarely occurring samples are usually overwhelmed by the majority class samples, they are much harder to identify. This represents a major problem.

First, traditional machine-learning algorithms try to achieve the lowest overall misclassification rate, creating an inherent bias in *favor* of the majority classes because the rare class has a smaller impact on accuracy. Second, if noisy data resemble the rare objects, they may be difficult to distinguish. Given these issues, the rare class mining problem is attracting considerable attention from the research community.

The following sections note various issues associated with rare class mining and give a systematic review of techniques proposed to mine rare events. *Note*: the domain of rare class mining is not clearly defined in the literature, and is often combined with the imbalanced dataset problem.

3.3.1 Why Rare Cases Are Problematic

Rare cases can be problematic for data-mining systems for many reasons. A basic problem is the lack of data, as rare cases generally cover only a few training examples (i.e., absolute rarity). The lack of data makes the detection of rare cases extremely difficult. Then, even if we manage to detect the rare

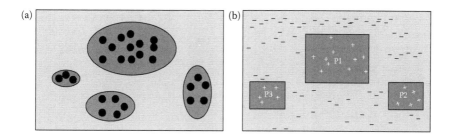

FIGURE 3.5
Rare and common cases in unlabeled (a) and labeled (b) data. (Redrawn from Weiss, G. M., 2005. Mining with rare cases. In: Maimon, O. and Rokach, L. (Eds.), *Data Mining and Knowledge Discovery Handbook*. USA: Springer, pp. 765–776.)

case, generalization will cause problems as it is hard to identify regularities from only a few data points. Consider the classification task shown in Figure 3.6, which focuses on the rare case, P3, from Figure 3.5b. Figure 3.6a reproduces the region from Figure 3.5b surrounding P3. Figure 3.6b shows what happens when the training data are augmented with only positive examples while Figure 3.6c shows the result of adding examples from the underlying distribution (Weiss, 2005).

Figures 3.6a and b show the learned decision boundaries using dashed lines. The learned boundary of Figure 3.6a is far away from the "true" boundary and excludes a substantial part of P3. Figure 3.6b's inclusion of additional positive examples addresses the problem of absolute rarity and allows all of P3 to be covered/learned, even though some examples not belonging to P3 will be erroneously assigned a positive label. By including additional positive and negative examples, Figure 3.6c corrects this last problem (*note*: the learned decision boundary almost overlaps the true boundary and is therefore not shown). As Figures 3.6b and c demonstrate, the problem of absolute rarity can be addressed by providing additional data. In practice, however, it is not always possible to obtain additional training data. See Figures 3.5 and 3.6.

FIGURE 3.6
The problem with absolute rarity. (Redrawn from Weiss, G. M., 2005. Mining with rare cases. In: Maimon, O. and Rokach, L. (Eds.), *Data Mining and Knowledge Discovery Handbook*. USA: Springer, 765pp.)

Another problem associated with mining rare cases is reflected in the aphorism: "it's like finding a needle in a haystack." The difficulty is not so much trying to find a small object or even the fact of having only one needle. Rather, the needle is hidden by the hay. If we extend this analogy to data mining, we see that rare cases may be obscured by common cases (relative rarity). This is especially problematic when data-mining algorithms use greedy search heuristics that examine one variable at a time; as rare cases may depend on the conjunction of many conditions, looking at any single condition in isolation may provide little guidance. As an example of relative rarity, consider the association rule mining problem described earlier, where we want to detect the association between a mop and a broom. Since this association rarely occurs, it can only be found if the minimum support (minsup) threshold (number of times the association is found in the data) is set very low. However, setting the threshold this low will cause a combinatorial explosion, as frequently occurring items will be associated with one another in an enormous number of ways, and most will be random and/or meaningless (Liu et al., 1999).

The metrics used in data mining and to evaluate the results of data mining can make it difficult to mine rare cases as well. Because a common case covers more examples than a rare case, classification accuracy will cause classifier induction programs to focus attention on common cases, not on rare cases. Rare cases may be totally ignored as a consequence. Or consider how decision trees are induced. Most are grown in a top-down manner; test conditions are repeatedly evaluated and the best one selected. The metrics (e.g., information gain) used to select the best test usually opt for those likely to result in a balanced tree; however, these tests yield high purity for a relatively small subset of the data but low purity *for the rest* (Riddle et al., 1994). Rare cases correspond to high-purity branches covering a few examples, making them likely to be excluded from the decision tree. Finally, consider association rule mining; in this case, rules that do not cover at least minsup examples will never be considered.

A data-mining system's bias is critical to its performance. Its extra, evidentiary bias allows it to generalize from specific examples. Unfortunately, the bias in most data-mining systems has an impact on their ability to mine rare cases. Many data-mining systems, especially those used to induce classifiers, have a maximum-generality bias (Holte et al., 1989). Therefore, when a disjunct covering some set of training examples is formed, only the most general set of conditions satisfying these examples are chosen. In contrast, a maximum specificity bias would add all possible, shared conditions. While the maximum-generality bias works well for common cases/large disjuncts, it does not work well for rare cases/small disjuncts, creating the problem with small disjuncts described previously.

Noisy data may make it difficult to mine rare cases, as well. Given a sufficiently high background noise level, a learner may not be able to distinguish between the true exceptions (i.e., rare cases) and the noise-induced exceptions (Weiss, 1995). Consider the rare case, P3, shown in Figure 3.5b. Since P3 contains so few training examples, if attribute noise causes even a few negative examples to appear, this will prevent P3 from being learned correctly. Common cases like P1 are not nearly as susceptible to noise, but little can be done to minimize the impact of noise on rare cases. Pruning and other *overfitting avoidance techniques* as along with inductive *biases that foster generalization* can minimize the impact of noise, but because these methods have a tendency to remove both the rare and the "noise-generated" cases, they do so at the expense of the former.

3.3.2 Methods for Rare Class Problems

3.3.2.1 Evaluation Metrics

Evaluation metric is used to evaluate the performance of a classifier, and in many algorithms, it guides the learning process. Although accuracy is a widely used evaluation metric, it is not a good option for the rare class mining issue because of its strong bias against that class. Consider a classification problem where the rare class accounts for only 0.1% of the training set. A classifier that predicts every sample as the majority class can still achieve an apparently satisfactory overall accuracy of 99.9%. But it has no significance for the rare class mining application as the classifier has learned nothing about the rare class. In other words, overall accuracy is not a meaningful metric for rare class mining. Precision and Recall are the two preferred metrics. Usually, the rare class is denoted as containing positive samples. Following this denotation, Precision is the percentage of true positive samples among all samples identified as positive, and Recall is the percentage of the correctly predicted true positive samples. Precision measures exactness, while Recall measures completeness; both are directly related to the objective of the rare class mining problem (Han et al., 2009).

Since Recall and Precision are conflicting metrics, an acceptable trade-off is often sought, depending on the application. Two popular metrics seeking to balance Recall and Precision are Geometric Mean of Recall and Precision (GMPR) (Joshi, 2002) and F-measure. GMPR is defined as the square root of the product of Precision and Recall, while F-measure, a metric widely used in the information retrieval community, is defined as

$$F_\beta = \frac{(1 + \beta^2)\,(\text{Precision} \cdot \text{Recall})}{\beta^2\,\text{Precision} + \text{Recall}} \tag{3.1}$$

where β measures the importance of Precision against Recall. The metrics achieve high values only if the values of both Precision and Recall are high.

Other proposed metrics include sum of recalls, geometric mean of recalls, information score, and so on.

3.3.2.2 Sampling-Based Methods

Sampling is a common data *preprocessing* technique. It purposely manipulates the distribution of samples so the rare class is well represented in the training set. Sampling was originally used to handle the class imbalance problem, but more recent studies discuss the relevance of sampling for rare class mining (Seiffert et al., 2007).

3.3.2.2.1 Sampling Methods

The basic sampling methods are undersampling and oversampling. On the one hand, undersampling randomly discards majority class samples; on the other hand, oversampling randomly duplicates minority class samples to modify the class distributions. Although these methods alleviate the rare class problem, they introduce some new issues. In random undersampling, some possibly helpful majority samples may be left out, leading to information loss and creating a less-than optimal model. Meanwhile, in random oversampling, the size of the training set is significantly increased, thus increasing the computational complexity. Finally, since oversampling makes exact copies of rare class samples and adds no new information to the dataset, the overfitting problem may result.

The basic versions of sampling do not work well in practice, but using heuristic sampling methods may help. The idea is to eliminate special samples of the majority class and retain all rare class samples. The noisy, redundant, or borderline samples close to the boundary separating the positive and negative regions are discarded, and the concept of Tomek links is introduced into the algorithm to recognize them.

Two undersampling methods are EasyEnsemble and BalanceCascade. First, we create multiple subsets from the majority class. Next we use AdaBoost to train a classifier based on each subset and the rare class dataset. Finally, we combine the outputs of these classifiers. EasyEnsemble replaces samples from the majority class and BalanceCascade discards samples correctly classified by previous classifiers before subsequent sampling. Empirical results suggest BalanceCascade is more efficient on a highly skewed dataset; each new sample is generated in the direction of some or all of the nearest neighbors. We also know that synthetic minority oversampling technique (SMOTE) can improve the accuracy of classifiers in many rare class problems, and the combination of SMOTE and undersampling performs better than simply undersampling (Han et al., 2009).

3.3.2.2.1.1 Sampling Rate   While conducting a sampling, we may have a problem determining the proper sampling rate, and this directly affects the class distribution ratio. It is intuitive, given an imbalanced dataset problem, that a balanced distribution is likely to yield the best or approximately best performance, but studies show that the frequently used "even distribution" is not optimal for rare events. A ratio of 2:1 or even 3:1 in favor of the majority class may result in superior classification performance (Han et al., 2009).

3.3.2.3 Cost-Sensitive Learning

Cost-sensitive learning is widely used in data mining. This technique assigns different levels of a misclassification penalty to each class and has been incorporated into classification algorithms by considering the cost information and trying to optimize the overall cost during learning. It has recently been applied to the rare class problem; a higher cost is given to the misclassification of rare objects than to the majority class. A cost/benefit-sensitive algorithm called the statistical online cost-sensitive classification (STOCS) proposes to classify rare events in online data. Results show STOCS outperforms many well-known cost-insensitive online algorithms. In practice, however, it may be difficult to set the cost information. Although a false-negative prediction is known to be more risky than a false-positive prediction, how to make a quantitative analysis between these two risks may require prior knowledge and/or the involvement of domain experts. It may be wise to vary the cost ratio until a satisfactory objective function value is obtained (Han et al., 2009).

3.3.2.4 Algorithms for Rare Class Mining

This section reviews machine-learning algorithms that are either proposed or modified for the rare class mining problem.

3.3.2.4.1 Boosting Algorithms

Boosting, a powerful sequential ensemble-learning algorithm, can improve the performance of weak base learners. A series of basic classifiers are built based on the weighted distributions of the training set. At the end of each iteration, the weight of each training sample is adaptively changed based on the training error of the present classifier. Thus, later classifiers are forced to emphasize those learning samples misclassified by former classifiers.

We can consider Boosting to be a generalized sampling method, as it changes the distribution of the original dataset. Since Boosting focuses on difficult-to-classify samples, it is a good idea to use it to detect the rare class. That being said, because the standard Boosting algorithm treats the two kinds of errors (false positive and false negative) equally, the majority class may continue to dominate the training set even after successive Boosting iterations.

To solve this problem, RareBoost updates the weights of positive and negative samples differently: it permits the algorithm to put equal focus on Recall and Precision. AdaCost, a variant of AdaBoost, adopts the cost-sensitive technique, imposing different costs for the two types of errors to update the distribution of the training set and, thus, reduce the cumulative misclassification cost.

These various algorithms all seek to adaptively alter the distribution of the original dataset so classifiers can focus on the samples that are difficult to classify, but there are, of course, differences RareBoost and AdaCost apply a modified weight-updating mechanism to change the distribution; SMOTEBoost generates synthetic samples for the rare class. The standard Boosting focuses on all misclassified samples equally in its weight-updating mechanism; RareBoost treats false-positive samples and false-negative samples differently; AdaCost updates the weights differently for all four types of classification outputs.

By analyzing the key components of Boosting, namely, the accuracy metric, the ensemble-voting process, the weight-updating mechanism, and the base learner, researchers find that for the rare class mining problem, Boosting's performance critically depends on the abilities of its base learner (Han et al., 2009).

3.3.2.4.1.1 Rule-Based Algorithms Traditional rule-induction techniques often fail to perform well in rare class classification. Therefore, some modified algorithms have been proposed. Existing sequential covering techniques may not detect the rare class; they try to achieve high Recall and highPrecision rates simultaneously, and, as a result, they may encounter two problems: splintered false positives and small disjuncts caused by sparse target samples. A two-phase rule-induction approach PNrule may solve these problems. The first phase finds rules with high support and reasonable accuracy; these may contain both positive and negative samples. The second phase develops rules able to remove the false-positive samples to increase accuracy. PNrule is, therefore, especially suitable for rare class mining. Emerging patterns (EPs) are new types of patterns, referring to itemsets whose supports in one class are significantly higher than in others and can capture significant multiattribute contrasts between classes. Emerging pattern rare-class is an EP-based rare class classification approach which seeks to increase the discriminating power of EPs in three stages: generate new undiscovered rare class EPs, prune low-support EPs, and increase the supports for rare class EPs. Division for mining EPs is another novel approach to mining EPs; it divides the majority class into subsets so that the unseen rare class EPs can be discovered. It also defines a strength function to evaluate the rare class EPs and, thus, minimize the effect of noisy EPs (Han et al., 2009).

3.3.2.5 Obtaining Additional Training Data

Obtaining additional training data can directly address the problems encountered in rare case mining. If we obtain additional training data from the original distribution, most will be associated with the common cases, but because some will be associated with the rare cases, this approach may help with the problem of "absolute rarity." Unfortunately, this approach does not address the problem of relative rarity, as the same proportion of the training data will continue to cover common cases. Only by selectively obtaining additional training data for the rare cases can we address relative rarity (*note*: this sampling scheme would also be able to handle absolute rarity). We can only identify rare cases for artificial domains, so this approach generally cannot be implemented and has not been used in practice. If we assume small disjuncts are manifestations of rare cases in the learned classifier, this approach can possibly be approximated by preferentially sampling examples that fall into the small disjuncts of some initial classifier. This approach needs more research (Weiss, 2005).

3.3.2.6 *Employing Nongreedy Search Techniques*

Most data-mining algorithms are greedy in that they make locally optimal decisions regardless of what may be best globally. This ensures the data-mining algorithms are tractable, but because rare cases depend on the conjunction of numerous conditions and any single condition in isolation is likely to provide little guidance, such greedy methods are often ineffective for rare cases. One approach to rare cases is to use more powerful, global, search methods. As they operate on a population of candidate solutions, not a single solution, genetic algorithms fit this description and cope well with attribute interactions. Therefore, they are increasingly used for data mining. Some systems can be adapted to use genetic algorithms to handle rare cases, including more conventional learning methods. For example, Brute, a rule-learning algorithm, performs exhaustive depth-bounded searches for accurate conjunctive rules. It seeks to find accurate rules, even if they cover relatively few training examples. Although Brute performs well compared to other algorithms, the lengths of the rules must be limited to make the algorithm tractable. One of Brute's advantages is its ability to locate "nuggets" of information that other algorithms may not be able to find. Associated rule-mining systems usually employ exhaustive search algorithms, but while these algorithms are theoretically able to find rare associations, they become intractable if the minimum level of support, or minsup, is set small enough to find rare associations. In other words, such algorithms are heuristically inadequate if we are looking for rare associations and suffer from the rare item problem discussed above. Modifying the standard Apriori algorithm so that it can handle multiple minimum levels of support may be a solution. In this approach, a user specifies a different minimum support for each item, based on the item's frequency in the distribution. This means the minimum support for an association rule is the lowest minsup value among all items in the rule. Empirical results suggest such enhancements allow the modified algorithm to find meaningful associations involving rare items, without producing meaningless rules involving common items (Weiss, 2005).

3.4 Chance Discovery

Chance discovery is the discovery of chance, not by chance. Here, "chance" is defined as an event or a situation with a significant impact on human decision making (*note*: "chance" can also mean a suitable time or occasion to do something [*Oxford Advanced Learner's Dictionary*]), if the situation or event occurring at a certain time is more significant than the time itself, and a decision to do something precedes the action of doing it (Ohsawa, 2002). In other words, here, chance is a new event or situation that can be conceived either as an opportunity or a risk. The word "discovery" also has some ambiguity. In the process of discovering a law ruling nature, we sometimes "learn" frequent patterns through observation. This restricted use of "discovery" to mean "learning" is established in the machine-learning community of artificial intelligence; yet the general meaning of discovery is to explain events that have never been explained in explicit theories or hypotheses.

Chance discovery is defined as the awareness of a chance and the explanation of its significance, especially if the chance is rare and its significance is unnoticed. By understanding explicitly what actions can be taken to turn an opportunity into a benefit, we can promote the desirable effects of opportunities, while explicit preventive measures will become discovered risks.

An essential aspect of chance is that it can be a seed of significant future change. Our discovery of a new opportunity may lead to an unusual benefit, because it is not known yet by anyone including our business rivals, who are accustomed to frequent past opportunities. The discovery of a new hazard risk is indispensable to minimize damage, because existing solutions that have worked for frequent past hazards may not work. Chance discovery aims at solving problems by noticing (becoming aware of) and explaining (Ohsawa, 2000), such diverse things as the following:

- Promoting new products to increase sales and consumption
- Explaining side effect risks of a new drug
- Noting signs of future earthquakes
- Using the World Wide Web (WWW) to attract attention promote products, and so on

3.4.1 Prediction Methods for Rare Events

Statistical studies make it clear that we need many samples to predict an event that occurs at a low frequency (i.e., a rare event). Recently, attention in statistics has turned to extremals (rare substances of variables) as rare events can lead, for example, to extreme events, such as severe economic depression or remarkable prosperity. Studies of extremals can explain the static distributions of rare events (e.g., expected maximal values of known variables) but cannot explain the effects of hidden causes of rare events.

In data mining, the prediction of rare events is attracting considerable interest. Methods of learning a high-accuracy rule with a complex condition (i.e., long, even though optimized to be short) and a rare event B in the conclusion have been proposed. A rule can provide not only the prediction but also the explanation of a rare event. However, these methods cannot discover chances. Even exceptional rules cannot determine the risk of a big earthquake at an active fault from the data of past earthquakes. A rule is ignored if its support or confidence (the probability of the co-occurrence of A and B, and the same co-occurrence under the occurrence of A, respectively) is extremely small. In this example, the previous earthquakes at the active fault are too rare. Exceptional rules may suggest appropriate treatments and diagnoses for certain medical conditions, including considerably rare cases, but will not clarify the value of medical decisions in real situations based on these rules. And in business, the usefulness of a piece of knowledge varies with who uses it, depending on the applicability of the details to the user's situation, rather than confidence or support values.

Complex knowledge can also be acquired to predict a fixed rare event, such as a breakdown in the signal transmission on a certain line, by means of an extended genetic algorithm. However, this method does not select the significant rare events to predict nor does it explain the meaning of the rare event for human decisions (Ohsawa, 2000).

3.4.2 Chance Discoveries and Data Mining

Chance discovery started with data-mining methods for detecting significant changes, for example, from the WWW. At the same time, methods were developed to help users become aware of their own unnoticed interests by visualizing a cluster of words occurring on the WWW near to their current query, or words relevant to each other in their search history. By looking at the clusters and the relationships between clusters, users can discover topics that are significant to them. At this point, they may decide to enter new queries based on the awareness of their own hitherto unnoticed interests. The human awareness of unnoticed significant topics was, thus, introduced as a chance discovery tool.

In the area of marketing, the most basic application of data mining has been to segment buyers by feature values, for example, age, income, job, education, buying history, and so on. If a segment of buyers is found to buy items the vendor wants to sell, new buyers of this segment are seen as potential loyal customers. This method has successfully carried out customer evaluations for credit cards and found suitable advertising targets. If we regard a sightseeing spot as a market to sell sights to visitors, segmenting customers can be an effective approach to promoting somewhere to go. In social filtering, a place a touring user has not yet visited will be recommended, if the place has been visited by other people with a history of visiting similar places to those where the current user often visits. The system entertains tourists seeking new places to go, but users cannot tell why the recommended places are attractive to them.

In another example, the sales data of a bookstore can be used to make recommendations of new books for customers. The method searches a book at the intersection of multiple basic interests (i.e., already established in a buyer's mind). For example, when a buyer has a history of buying books on *Star Trek* as well as books on "cooking," a book titled *Star Trek Cook Book* can be recommended. This example particularly satisfies the buyer, because the new book links two separate interests—a comfortable coherence in the user's mind. This realizes a personal chance discovery in its ability to choose a rare but attractive book and to explain why the book is attractive to the buyer. In marketing, generally, a qualitative understanding of "in what context people like to buy what" should be achieved before determining the quantitative understanding of "how many and when." A chance discovery marketing tool should help investigate underlying desires and behaviors of buyers (Ohsawa, 2000).

3.5 Novelty Detection

The detection of novel events in any scheme of signal classification is an important ability. Since we cannot train a system of machine learning on all possible object classes whose data are in the system, it is important to be able to differentiate between known and unknown objects. The detection of novelty is an extremely challenging task, but several models of novelty detection perform well on different data. There is no single best model for the detection of novelty; success not only depends on the type of method used but also on the statistical properties of the data processed. To act as a detector instead of a classifier, several applications require a classifier; that is, the requirement is to detect whether an input is part of the data that the classifier was trained on or is, in fact, unknown. In applications such as fault detection, radar target detection and hand-written digit recognition, Internet and e-commerce, statistical process control, and several others, this technique is useful.

There is increased interest in novelty detection, and a number of research articles have appeared on autonomous systems based on adaptive machine learning. Yet only a few surveys are forthcoming, largely because the systems of high integrity cannot use traditional classification: either the abnormalities are very rare or there are no data describing the conditions of fault. By modeling normal data and using a distance measure and a threshold for determining abnormality, the technique for the detection of novelty can solve this problem.

Novelty detection has recently been used in a number of other applications, especially in image analysis and signal processing (e.g., biometrics). In applications like these, with multiple classes, the problem becomes more complicated, with noisy features, high dimensionality, and, quite often, not enough sample. The methods of novelty detection have tried to find solutions for these real-world problems.

In what follows, we review some of the current methods of novelty detection using statistical methods. Several important issues related to the detection of novelty can be summarized according to the following principles (Markou and Singh, 2003):

1. Robustness and trade-off: Novelty detection must be capable of robust performance while minimizing the exclusion of known samples. This trade-off should be, to a limited extent, predictable and under experimental control.

2. Uniform data scaling: To assist novelty detection, all the test and training data after normalization should be within the same range.

3. Parameter minimization: A novelty detection method should minimize the number of parameters set by the user.

4. Generalization: The system should be able to generalize without confusing generalized information with novel information.

5. Independence: A novelty detection method should be independent of the number of features and available classes and should perform reasonably well in the context of an imbalanced dataset, low number of samples, and noise.

6. Adaptability: A system recognizing novel samples during the test should be able to use the information for retraining.

7. Computational complexity: A number of novelty detection applications are available online; therefore, the computational complexity of a novelty detection mechanism should be as minimal as possible.

3.5.1 Outlier Detection

The problem of statistical outlier detection is closely related to novelty detection. Although there is no precise definition of an outlier, most authors agree outliers are observations inconsistent with, or lying a long way from, the rest of the data. Outlier detection aims to handle rogue observations in a set of data, as these observations can have an enormous effect on data analysis (such data points are called influential observations) (Marsland, 2002). Unfortunately, it is not possible to find an outlier in multivariate data by examining variables one at a time.

The importance of outlier detection to statistical methods can be seen in Figure 3.3. As shown, an outlying datapoint can completely change the least-squares regression line of the data. Statistical methods generally ignore unrepresentative data rather than explicitly recognizing these points. Techniques that manage to avoid the problems of outliers are called robust statistics. There are also sets of tests to determine whether predictions from particular distributions are affected by outliers. The appearance of some outliers in two dimensions is shown in Figure 3.4. The next subsections describe statistical techniques used to detect and deal with outlying datapoints.

3.5.1.1 Outlier Diagnostics

For outlier diagnosis, the residual of a point is $r_i = y_i - \hat{y}i$, that is, the difference between the actual point (yi) and the prediction of a point ($\hat{y}i$). The linear model of statistical regression for data X is

$$y = x\theta + e \tag{3.2}$$

where θ is the vector of (unknown) parameters, with e being the vector of errors. We can define the hat matrix, H (so-called because $Hy = \hat{y}$) as

$$H = X(X^TX)^{-1}X^T \tag{3.3}$$

Then

$$\mathrm{cov}(\hat{y}) = \sigma^2 H, \quad \text{and} \quad \mathrm{cov}(\hat{y}) = \sigma^2(1 - H) \tag{3.4}$$

where r is the vector of residuals and σ^2 is the variance. Each element h_{ij} of H can be interpreted as the effect exerted by the *jth* observation on $\hat{y}_i$, and $h_{ii} = \partial \hat{y}i/\partial yi$, the effect an observation has on its own prediction. The average of this is p/n, where $p = \sum_{i=1}^{n} hi$, and, in general, points are considered to be outliers if $h_{ii} > 2p/n$.

Interestingly, H is the pseudoinverse if $(X^T X)^{-1}$ exists. This means the hat matrix method is related to the Kohonen and Oja approach, which, in turn, can be considered an implementation of the hat matrix.

The residuals can be scaled by the values along the diagonal of the hat matrix, as shown in the three following methods:

$$\text{Standardized: } \frac{r_i}{s}, \quad \text{where } s^2 = \frac{1}{n-p}\sum_{j=1}^{n} r_j^2 \tag{3.5}$$

$$\text{Studentized: } \frac{r_i}{s\sqrt{1 - h_{ii}}} \tag{3.6}$$

$$\text{Jackknifed: } \frac{r_i}{s_i\sqrt{1 - h_{ii}}}, (s_i = \text{without the ith case}) \tag{3.7}$$

Another method is the Mahalanobis distance of each point

$$D^2 = (x - \mu)^T \sum{}^{-1} (x - \mu), \tag{3.8}$$

where Σ is the covariance matrix and μ is the mean. The Mahalanobis distance is a useful measure of the similarity between two sets of values. See Figures 3.3 and 3.4 (Marsland, 2002).

3.5.1.2 Recognizing Changes in the Generating Distribution

Given n-independent random variables from a common, but unknown, distribution μ, does a new input X belong to the support of μ?

To answer this question, we need to consider the following. The support, or the kernel, of a set is a binary-valued function; it is positive in areas of the input space where there are data and negative elsewhere. The standard approach to the problem of outlier detection is to take independent measurements of new distributions, assuming them to have a common probability measure v, and to test whether $\mu' = v$, where μ' is the probability measure of μ, in other words, to see if the support of $v \in S$, where S is the support of μ. How, then, do we estimate the support S from the independent samples $X_1,...X_n$?

The obvious approach is to estimate S_n as

$$S_n = \bigcup_{i=1}^{n} A(X_{i,\rho n}) \tag{3.9}$$

where $A(x, a)$ is a closed sphere centered on x with radius a, and ρ^n is a number depending only on n. The probability of making an error on datapoint X, given the data to this point, is

$$L_n = P(X \varepsilon S \mid X_1,........X_n) = v(s) \tag{3.10}$$

The detection procedure is considered consistent if $L_n \to 0$ in probability and strongly consistent if $L_n \to 0$ with probability 1 (Marsland, 2002).

3.5.1.3 Extreme Value Theory

The extreme value theory (EVT) is used to detect outliers in data by investigating the distributions of data with abnormally high or low values in the tails of the distribution generating the data. In this case, let $Z_m = \{z_1, z_2,...,z_M\}$ be a set of m independent and identically distributed random variables $z_i \varepsilon R$ drawn from some arbitrary distribution D. In addition, let $x_m = \max(Z_m)$. When observing other samples, the probability of observing an extremum

$x \geq x_m$ may then be given by the cumulative distribution function (Marsland, 2002)

$$p(x_m \mid \mu_m, \sigma_m, \gamma) = \exp\left\{ -\left[1 + \frac{\gamma(x_m - \mu_m)}{\sigma_m} \right]^{1/\gamma} \right\}$$
(3.11)

where $\gamma \in R$ is the shape parameter. With the limit as $\gamma \to 0$, this leads to a Gumbel distribution

$$p(x_m \leq x \mid \mu_m, \sigma_m) = \exp\left\{ -\exp(-y_m) \right\}$$
(3.12)

where μ_m and σ_m depend on the number of observations m, and y_m is the reduced variate. This gives

$$y_m = \frac{(x_m - \mu_m)}{\sigma_m}$$
(3.13)

3.5.1.4 Principal Component Analysis

Principal component analysis (PCA) is a standard statistical technique for extracting the structure from a dataset. Essentially PCA performs an orthogonal basis transformation of the coordinate system where data are described, thus reducing the number of features required for effective data representation. PCA can also be used to detect outliers that are, at least in some sense, orthogonal to the general data distribution. By looking at the first few principal components, we can find any datapoints that inflate the variances and covariances to a large extent. But by looking at the last few principal components, we can see features not readily apparent with respect to the original variables (i.e., outliers). A number of test statistics have been suggested to find these points, including a measure of the sum of squares of the values of the last few principal components and a version weighted by the variance in each principal component (Marsland, 2002).

3.5.2 Novelty Detection using Supervised Neural Networks

A main use of artificial neural networks is classification by clustering data into two or more classes. Neural networks can be trained in two ways: in supervised learning, each training input vector is paired with a target vector, or desired output; in unsupervised learning, the network self-organizes to extract patterns from the data with no target information. This section concentrates on supervised neural networks, such as the perceptron, related multilayer perceptron, and the radial basis function (RBF) network. Basically speaking, these networks adapt the connection weights between layers of neurons to approximate a mapping function and model the training data.

In the trained network, every input produces an output. In classification, this usually becomes an identifier for the best-matching class, but there is no guarantee that the best-matching class is a good match, only that it is a better match than the other classes for the set of training data used. Here, novelty detection is useful. It can recognize inputs not covered by the training data, inputs the classifier cannot reliably categorize (Marsland, 2002).

3.5.2.1 Kernel Density Estimation

Even if a network is well trained, its predictions could be poor if the dataset used to train the network is not representative of the whole set of potential inputs. There are two possible reasons:

- Only a few examples of an important class
- Incomplete classification set

There may, for example, be a strong relationship between the reliability of the output of the network and the degree of novelty in the input data, which evaluated the sum-of-squares error function of the network. This can be expressed as

$$E = \sum_{j=1}^{m} \int \left[y_j(x,w) - t_j \right]^2 p(x,t_j) dx dt_j$$

$$= \sum_{j=1}^{m} \int \left[y_j(x,w) - \langle t_j \mid x \rangle \right]^2 p(x,t_j) dx + \sum_{j=1}^{m} \int \left(\langle t_j^2 \mid x \rangle - \langle t_j \mid x \rangle^2 \right) p(x) dx, \quad (3.14)$$

where $p(x, t_j)$ is the joint probability density function for the data, $j = 1,...,m$ are the output units, w represents the weights, x is network input, t_j is the associated target for unit j, and y_j is the actual output of unit j.

The conditional averages of the target data in Equation 3.14 are given by

$$\langle t_j \mid x \rangle \equiv \int t_j p(t_j \mid x) dt_j$$

$$\langle t_j^2 \mid x \rangle \equiv \int t_j^2 p(t_j \mid x) dt_j \qquad (3.15)$$

Only the first of the two parts of Equation 3.14 is a function of the weights w; if the network is sufficiently flexible (i.e., it has enough hidden units), the minimum error E is gained when

$$y_j(x,w) = \langle t_j \mid x \rangle \qquad (3.16)$$

which is the regression of the target vector conditioned on the input. The first term of Equation 3.14 is weighted by density $p(x)$; hence, the approximation is most accurate where $p(x)$ is large (i.e., data are dense).

Although we know little about the density $p(x)$, we can generate an estimate $\hat{p}(x)$ from the training data and use it to get a quantitative measure of the degree of novelty for each new input vector. We could use this to put error bars on the outputs or to reject data where the estimate $\hat{p}(x) < \rho$ for some threshold ρ, thereby generating a new class of "novel" data. The distribution of novel data is generally unknown, but if we estimate it as constant over a large region of the input space and zero outside this region, we can normalize the density function, as shown in Figure 3.5 (Marsland, 2002).

3.5.2.2 Extending the Training Set

We can extend the training set so the neural network can be trained to recognize data from regions not included in the original set (Figure 3.7). Suppose the training set for some problem spans the region R is Rn. Then, (Marsland, 2002)

$$p(\text{class1} \mid x) = \frac{p(x \mid \text{class1})p(\text{class1})}{p(x \mid R)}$$

$$P(x \mid R) = p(x \mid \text{class1})p(\text{class1}) + p(x \mid \text{class2})p(\text{class2}). \qquad (3.17)$$

$$p(R \mid x) = \frac{p(x \mid R)p(R)}{p(x \mid R)p(R) + p(x \mid R')p(R')} = 1 - p(R' \mid x)$$

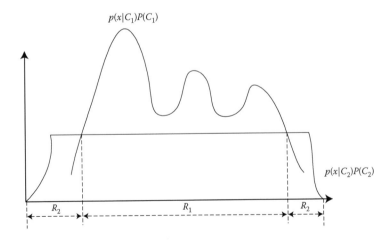

FIGURE 3.7
Novelty detection in the Bayesian formalism. The training data are used to estimate $p(x|C1)$ $P(C1)$ using $\hat{p}(x)$, with novel data (class C2) having a distribution that is assumed constant over some large region. Vectors that are in the regions labeled R2 are considered to be novel. (Redrawn from Marsland, S., 2002. *Neural Computing Surveys*, 3, 1–39.)

where R' is the missing class, separate from R. At this point, we can generate data in $R0$; we generate data and remove any data instances that are in R.

To consider the problem of density estimation, minimum mutual information can be used to factorize a joint probability distribution. If a Gaussian upper bound is put on the distribution, it can be used to estimate the density of the probability functions. Instead of the output density, the instability of a set of simple classifiers is measured. In this method, a number of classifiers are trained on bootstrap samples of the same size as the original training set. The output of all the classifiers is considered for new data. If the data are novel, the variation in responses from the different classifiers will be large. This approach can be applied to three different types of networks: a Parzen window estimator, a mixture of Gaussians, and a nearest neighbor method (Marsland, 2002).

3.5.3 Other Models of Novelty Detection

3.5.3.1 Hidden Markov Models

An HMM consists of a number of states, each with an associated probability distribution, along with the probability of moving between pairs of states (transition probabilities) each time instance. The actual state at any time is not visible to an observer (hence the name); instead, the observer can see an outcome or observation generated according to the probability distribution of the present state. A picture of an HMM is shown in Figure 3.8. HMMs are extremely useful in a number of different applications, especially speech processing (Marsland, 2002) (Figure 3.8).

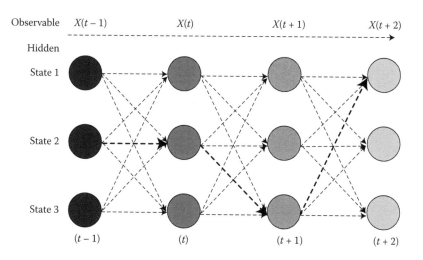

FIGURE 3.8
An example of an HMM. (Redrawn from Smyth, 1994. Hidden Markov models for fault detection in dynamic systems. *Pattern Recognition*, 27(1), 149–64.)

The standard HMM is not a useful technique for novelty detection because it has a predetermined number of states. Research into the problem of fault detection in dynamic systems using HMMs has addressed this issue. The faults that can occur need to be identified in advance and states generated for each model. The model assumes faults do not happen simultaneously, as this would cause problems with fault recognition. The technique is related to the density estimation method, but as the inputs are sequences, a modification allows extra states to be added while the HMM is being used. The model is expressed as follows. Let $w\{1,...,m\}$ be the event that the true system is in one of the states $w1,...,$ wm, and $p(w\{1,...,m\}|y)$ be the posterior probability that data are from a known state, given observation y. Then

$$p(wi|y) = pd(wi|y,w\{1,....m\})p(w\{1,....m\}|y), 1 \le i \le m \qquad (3.18)$$

where $pd(\cdot)$ is the posterior probability of being in state i, generated from some discriminative model. The second part can be calculated using Bayes' rule and the fact that

$$p(w_{m+1}|y) = 1 - p(w\{1,......,m\}|y) \qquad (3.19)$$

The probability of getting a novel state, for example, a machine fault, can be estimated from the mean time between failures.

3.5.3.2 Support Vector Machines

A support vector machine (SVM) is a statistical machine-learning technique; it performs linear learning by mapping data into a high-dimensional feature space. SVM selects the optimal hyperplane maximizing the minimum distance to the training points closest to the hyperplane. This is done in some high-dimensional feature space into which the input vectors are mapped using the kernel, a nonlinear mapping. The aim is to model the "support" of a data distribution, that is, a binary-valued function that is positive in those parts of the input space where the data lie, and negative otherwise. This means SVM can then detect inputs not in the training set, that is, novel inputs. This generates a decision function (Marsland, 2002) expressed as

$$\text{sign}(w \cdot \Phi(z) + b) = \text{sign}\left(\sum_i \alpha_j K(x_j, z) + b\right) \qquad (3.20)$$

where K is the kernel function (see Equation 3.20), $\Phi(\cdot)$ is the mapping into the feature space, b is the bias, z is the test point, x_i is an element of the training set, and w is the vector

$$w = \sum_{j} \alpha_{j}\Phi(x_{j}).$$ (3.21)

A hyperspherical boundary with minimal volume is put around the dataset by minimizing an error function containing the volume of the sphere using Lagrangian multipliers L (*note*: R is the radius of the hypersphere) shown as

$$L(R,a,\lambda) = R^2 - \sum_{i} \lambda\left[R^2 - (x_i - a)^2\right], \quad \lambda \geq 0$$

$$L = \sum_{i} \alpha_i(x_i \cdot x_i) - \sum_{i,j} \alpha_i\alpha_j(x_i \cdot x_i),$$ (3.22)

with $\alpha_i \geq 0, \sum_{i} \alpha_i = 1$

An object z is considered normal if it lies within the boundary of the sphere, that is,

$$(z - a)(z - a)^T = \left(z - \sum_{i} \alpha_i x_i\right)\left(z - \sum_{i} \alpha_i x_i\right)$$

$$= (z \cdot z) - 2\sum_{i} \alpha_i(z \cdot x_i) + \sum_{i,j} \alpha_i\alpha_j(x_i\ x_j) \leq R^2$$ (3.23)

The boundary can be made more flexible by replacing the inner products in the above equations by kernel functions $K(x, y)$. Using slack variables allows certain datapoints to be excluded from the hypersphere, so the task becomes minimizing the volume of the hypersphere and the number of datapoints outside it, that is,

$$\min\left[R^2 + \lambda\sum_{i} \xi_i\right] \text{ such that } (x_i - a) \cdot (x_i - a) \leq R^2 + \xi_i,\ \xi_i \geq 0 \quad (3.24)$$

3.5.3.3 Time to Convergence

One novelty detection method is based on an integrate-and-fire neuronal model. Generally, a training set of patterns known not to be novel are used to train the network and test patterns are evaluated for this training set. For this technique, it is the time the network takes to converge when an input is presented that suggests whether an input pattern is novel. The network architecture comprises a simple model of layer IV of the cortex. A two-dimensional sheet of excitatory and inhibitory neurons with recurrent connections is positioned according to a pseudorandom distribution. Neurons have local connections in a square neighborhood, and training takes the form of Hebbian learning. The state of each neuron is expressed as (Marsland, 2002)

$$S_i(t) = \begin{cases} 0 \\ H(U_i(t) - \theta) \end{cases} \quad \text{if } (t - t_{\text{spike}}) < \rho \tag{3.25}$$

where $H(\cdot)$ is the Heaviside function, $H(x) = 1$ for $x > 0$ and $H(x) = 0$ otherwise, and $U_i(t)$ is the control potential

$$U_i(t) = \sum_i C_{ij} S_j(t + 1) + U_i(t - 1) + s_i + f_i, \tag{3.26}$$

for connection strength C_{ij}, input s_i and variable-firing frequency function f_i. The network is applied to 7×5 images of numerical digits, together with noisy versions of the digits, as in Kohonen and Oja's novelty filter. For this task, it performs better than a back-propagation network.

3.6 Exception Mining

Patterns hidden in databases fall into three categories as follows:

- Strong patterns: Regularities for numerous objects
- Weak patterns: Reliable exceptions representing a relatively small number of objects
- Random patterns: Random and unreliable exceptions
- Weak "reliable-pattern" exceptions: Infrequent with high confidence

3.6.1 Confidence-Based Interestingness

When no other information is given, an event with lower-occurring probability gives more information than an event with higher probability. From information theory, we know that the number of bits required to describe the occurrence is defined as (Terano et al., 2000)

$$I = -\log_2 P$$

where P = the probability that the event will occur. Similarly, for a given rule $AB \rightarrow X$, with confidence $Pr(X|AB)$, we will require $-\log_2(Pr(X|AB))$ and $-\log_2(Pr(\neg X|AB))$ number of bits to describe the events X and $\neg X$, given AB. Thus, the total number of bits required to describe the rule $AB \rightarrow X$ is

$$I_C^{AB0} = (-Pr(X \mid AB)\log_2 Pr(X \mid AB)) + (-Pr(X \mid AB)\log_2 Pr(X \mid AB)) \quad (3.27)$$

where $I^{AB0}C$ = number of bits required to describe $AB \rightarrow X$ when no other knowledge has been applied. However, the difference in the number of bits to describe the rule $AB \rightarrow X$ in terms of $A \rightarrow X$ and $B \rightarrow X$ can cause a surprise. The bigger the difference in describing the rule $AB \rightarrow X$, the more interesting it is. Therefore, to estimate the relative interestingness in terms of $A \rightarrow X$ and $B \rightarrow X$, we need to know the number of bits required to describe event X when the probability of that event occurring given A and B, is $Pr(X|A)$ and $Pr(X|B)$, respectively.

Since the rule $AB \rightarrow X$ describes the event X in terms of A and B, to describe a similar event X, in terms of A and B using the rule $A \rightarrow X$ and $B \rightarrow X$, we need $-\log_2 Pr(X|A)$ and $-\log_2 Pr(X|B)$ number of bits. Now, in rule $AB \rightarrow X$, the probability of event X occurring is $Pr(X|AB)$.

Therefore, the expected number of bits required to describe all X events in rule $AB \rightarrow X$, in terms of A and B using the two rules is $-Pr(X|AB)$ $(\log_2 Pr(X|A) + \log_2 Pr(X|B))$. Similarly, for the event $\neg X$ in rule $AB \rightarrow X$, $Pr(\neg X| AB)(\log_2 Pr(\neg X|A) + \log_2 Pr(\neg X|B))$ number of bits will be required.

Thus, the total number of bits required to describe the events X and $\neg X$ in the rule $AB \rightarrow X$ by the rules $A \rightarrow X$ and $B \rightarrow X$ is

$$I_C^{AB0} = (-Pr(X \mid AB)\log_2 Pr(X \mid A)) + \log_2 Pr(X \mid B)$$
$$- Pr(X \mid AB)\log_2 Pr(X \mid A) + \log_2 Pr(X \mid B) \quad (3.28)$$

where $I^{AB1}C$ = number of bits required when $AB \rightarrow X$ is described by $A \rightarrow X$ and $B \rightarrow X$. Thus, the relative surprise, or relative interestingness, that comes from the difference between two descriptions for the given rule $AB \rightarrow X$ is

$$RI_C^{AB} = I_C^{AB1} - I_C^{AB0} = -Pr(X \mid AB)\left[\log_2 \frac{Pr(X \mid A)Pr(X \mid B)}{Pr(X \mid AB)}\right] \quad (3.29)$$

where $RI^{AB}C$ = the relative surprise or interestingness of the rule, considering the confidence in and knowledge about other rules. The interestingness of a rule that we have formulated in terms of confidence gives the exact impression of relative entropy. Here, the entropy of a rule is calculated relative to the other rules and is a measure of distance between two distributions. In statistics, this occurs as an expected logarithm of the likelihood ratio. The relative entropy $D(p(x)||q(x))$ is a measure of the inefficiency of assuming the distribution is $q(x)$, when the true distribution is $p(x)$. The relative entropy, or Kullback–Leibler distance, between two probability functions is defined as

$$D(p(x) \mid\mid q(x)) = \sum_{x \varepsilon X} p(x)\log \frac{p(x)}{q(x)} \quad (3.30)$$

In estimating the interestingness of the rule $AB - X$ with true confidence $Pr(X|AB)$, we approximate its confidence from the rules $A \rightarrow X$ and $B \rightarrow X$.

3.6.2 Support-Based Interestingness

By stating that the support of a rule $AB \rightarrow X$, we mean the frequency of the rule's consequent evaluation is A by AB, relative to the whole data-set. When we know the support of the two common sense rules $A \rightarrow X$ and $B \rightarrow X$, we know the relative frequency of the consequent X and $\neg X$ evaluated by A, and B, respectively. A similar relative entropy measure can be applied to estimate the surprise from the support. Now, for the newly discovered rule $AB \rightarrow X$, the true distributions of the consequent X and $\neg X$ evaluated by A and B are $Pr(ABX)$ and $Pr(AB \neg X)$, respectively. From the knowledge of one of our common sense rules, $A \rightarrow X$, for which the relative frequencies of X and $\neg X$ are $Pr(AX)$ and $Pr(A \neg X)$, respectively, we can find the distance between two distributions of consequence using relative entropy.

The relative entropy of $AB \rightarrow X$, relative to the rule $A \rightarrow X$ in terms of their support is, thus (Hussain et al., 2000),

$$D(AB \rightarrow X \| A \rightarrow X) = Pr(ABX)\log\frac{Pr(ABX)}{Pr(AX)}$$
$$+ Pr(AB \neg X)\log Pr(AB \neg X)\log\frac{Pr(AB \neg X)}{Pr(A \neg X)} \tag{3.31}$$

Hence, the total interestingness of rule $AB \rightarrow X$ relative to $A \rightarrow X$ and $B \rightarrow X$ is

$$RI = RI_c^{AB} + RT_s^{AB} \tag{3.32}$$

This includes support, confidence, and consideration of other rules in the estimation of the relative surprisingness.

3.6.3 Comparison with Exception-Mining Model

Patterns in a database can be divided into strong, weak, and random patterns. Strong patterns can be helpful for applications. As we have argued, weak patterns can also be very useful to applications, but most current data-mining techniques cannot effectively support weak pattern mining or "exception" mining as it is known. An algorithm for finding weak patterns uses reliable exceptions from databases and is written as the Exception Mining Model (EMM) model. Negative association rules are an important kind of weak pattern.

In the EMM model, the interestingness of an exceptional rule $AB \rightarrow X$ is measured by dependence on the composition of knowledge of the rules $A \rightarrow X$ and $B \rightarrow X$. The EMM model's search for interesting exceptional rules based on the χ-squared test, and requires eight probabilities, namely, supp(X), supp(Y), supp($\neg X$), supp($\neg Y$), supp($X \cup Y$), supp($X \cup \neg Y$), supp($\neg X \cup Y$), and supp($\neg X \cup \neg Y$), to construct the *contingency table* for the itemset $X \cup Y$ and to determine whether itemset $X \cup Y$ is the minimal dependent itemset using the χ-squared test.

For the PR model, we focus only on mining negative association rules of interest. The interestingness of a negative rule between itemsets X and Y is measured by four conditions as defined in Section 3.4. We need only five Probabilities, namely, supp(X), supp(Y), supp($X \cup Y$), supp($\neg X$), and supp($\neg X \cup Y$), to determine whether $X \rightarrow Y$ or $X \rightarrow \neg Y$ can be extracted as rules.

Note that the EMM model also generates negative association rules in databases, all of which are of interest. This often requires three steps: (1) testing confidence-based interestingness, (2) testing support-based interestingness, and (3) searching exceptional rules. However, if we wish to discover negative association rules by using the EMM model, it is not clear how we can identify which of $X \rightarrow \neg Y$, $\neg X \rightarrow Y$, $Y \rightarrow \neg X$, and $\neg Y \rightarrow X$ can be extracted, using the same facts as in the Chi-Squared test (CST) model. Therefore, it is clear that the PR model is better than the EMM model for finding negative association rules of interest (Zhang and Zhang, 2002).

3.6.4 Digging Out the Exceptions

Since exceptions are weak in terms of support, we must dig deeper into the data with lower support threshold to bring them out, but applying a lower support threshold for mining exceptions is not a cost-effective solution.

Moreover, a large number of rules will be generated and not all will be exceptions. In fact, we are going to mine those exceptions where the rules extracted as common sense will be used to alleviate the problem of dealing with a lower support threshold. In other words, we are searching for reliable exceptions starting from the common sense rules. To satisfy all the constraints defined in Table 3.1, we find exception $AB \rightarrow X$ from two common sense rules $A \rightarrow X$ and $B \rightarrow X$ (common sense infers $B \rightarrow X$ to be the reference for its obvious low support or/and low confidence). By doing this, we can estimate the amount of surprise the exception rule brings from the knowledge of the extracted rules.

TABLE 3.1

Rule Structure for Exceptions

$A \rightarrow X$	Common sense rule (strong pattern) (high support, high confidence)
$A, B \rightarrow \neg X$	Exception rule (weak pattern) (low support, high confidence)
$B \rightarrow \neg X$	Reference rule (low support and/or low confidence)

Source: Data from Terano, T., Liu, H., and Chen, A. L., 2000. Knowledge discovery and data mining. Current issues and new applications. In: *Lecture Notes in Artificial Intelligence; Subseries of Lecture Notes in Computer Science.* u.o.: Springer.

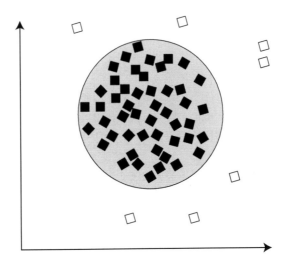

FIGURE 3.9
The principle of outlier detection. Empty squares are outliers to the dataset composed of the black squares. The circle surrounding the datapoints demonstrates a potential threshold, beyond which points are outliers. (Redrawn from Marsland, S., 2002. *Neural Computing Surveys*, 3, 1–39.)

From Figure 3.9, we can visualize how exceptions are mined by going deeper into the data. The threshold CS support is the minimum support to mine the common sense rules from the data and the EX support to assure the reliability of the exception rules. The following algorithm describes the way we mine interesting rules (Terano et al., 2000):

```
begin
LI = φ //list containing large item set
LC = φ //list containing common sense rules
LR = φ //list containing reference rules for a common sense
LE = φ //list containing candidate exception rules
LI ← GenerateLargeItemSet()          //running apriori [1]
LC ← GenerateAllCommonSenseSet(LI)
for each CS_i from LC do
    A ← GetAntecedent(CS_i)
    LR ← GetReferences(CS_i, LC)
    for each RR_j from LR do
            B ← GetAntecedent(RR_j)
            if (A ∪ B) is not in LI
                    insert(A ∪ B ∪ ¬Consequent(CSi), LE)
    end for
end for
LE ← GenerateExceptions(LE)          //Database scan once
EstimateInterestingness(LC, LE)      //Output interesting rules
end.                                 //according to the degree of
                                       surprise
```

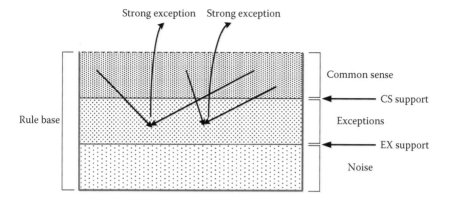

FIGURE 3.10
Rules in the data. (Redrawn from Hussain, F. et al. 2000. *Exception Rule Mining with a Relative Interestingness Measure*, Heidelberg, Berlin: Springer, pp. 86–97.)

The function *GetReferences(CS$_i$, LC)* returns all the candidate reference rules for *CS$_i$*, from *LC*. The reference rules are those common sense rules in *LC* with consequences similar to *CS$_i$*. Once we have inserted all the candidate exception rules into *LE*, we scan the database once to obtain the support and confidence of each candidate exceptions. We output those rules that satisfy the thresholds using *Generate Exceptions (LE)*. *Estimate Interestingness (LC, LE)* estimates the relative interestingness (Hussain et al., 2000) (Figure 3.10).

3.7 Noise Removal

The existing outlier detection techniques can be exploited to handle data with extremely high levels of noise. The objective is to improve data analysis by removing objects that may distort the analysis. By definition, the number of outliers in data is small, and outlier detection techniques traditionally remove only a small fraction of them. But if the amount of noise in the data is large from a data collection or data-analysis point of view, we will need data-cleaning techniques that remove large amounts of noise. In this case, we should turn to those outlier detection techniques that assign each object an outlier score characterizing the degree to which it is an outlier, as they can remove any specified percentage of noise. We sort the objects according to their "outlier score" and eliminate those with the highest outlier scores until we have eliminated the desired percentage of objects. In this section, we discuss three ways to do this, distance-based, density-based, and clustering-based methods, but any outlier detection technique that assigns a continuous outlier score to each object is viable (Xiong et al., 2006).

3.7.1 Distance-Based Outlier Detection Methods for Noise Removal

A simple way to detect outliers uses the distance measure, whereby an object in data set D is a distance-based outlier if at least a fraction α of the objects in D is at a distance greater than r. While this definition of an outlier is simple and easy to understand, we can have a problem if a data set has regions of varying density. More specifically, this approach is based on a criterion determined by the global parameters r and α; it cannot account for the fact that some objects are in regions of high density, while others are in low-density regions. Algorithm 3.1 expresses the pseudocode of our distance-based noise removal algorithm. For each object in the dataset, we record the number of objects lying within a distance r from it. According to the distance criteria, noise consists of those objects with the least number of neighbors within a specified radius. Therefore, all objects are sorted in ascending order according to the number of neighbors they have. The first $\varepsilon\%$ is declared noise and is removed from the data set. The complexity of this algorithm is $O(n^2)$, because nearest-neighbor sets must be constructed for each data object. *Note*: the cosine similarity measure is used instead of a distance measure, but this changes nothing essential (Xiong et al., 2006).

3.7.2 Density-Based Outlier Detection Method for Noise Removal

Another category of outlier detection methods identifies outliers in data sets with varying densities. One influential approach relies on the local outlier factor (LOF) of each object. This is based on the local density of an object's neighborhood, with an object's neighborhood defined by the MinPts nearest neighbors of the object. MinPts is a parameter specifying the minimum number of objects (points) in a neighborhood. Outliers, then, are objects with a high LOF. The use of the number of neighbors, rather than a specific distance or similarity, gives the approach its ability to handle data sets with varying densities.

Algorithm 3.2 shows the pseudocode of our implementation. As expressed by the algorithm, essentially, every object in a data set is considered an outlier to some extent, and this extent is measured using the LOF. The first part of Algorithm 3.2 computes this factor for each object. This algorithm has a computational complexity of $O(n^2)$, but can be reduced to $O(n\log(n))$ for low-dimensional data if we use efficient multidimensional access methods, such as the R* tree. As the LOF computation must be iterated over many values of MinPts, the associated constant in the complexity may be quite large (Xiong et al., 2006).

Since we use the cosine measure instead of a distance measure, the point with the lowest LOF value is determined to be the most unusual (noisy) point in the data set. To eliminate the required amount of noise from the data, all objects are sorted in ascending order based on their LOF values, and the first ε% is declared noise. *Note*: the sorting order differs from the case where distance measures are used to calculate LOF values.

While the LOF method does not suffer from problems of varying density, we may have a problem selecting parameters, such as MinPts. Indeed, since the LOF of each point may vary with the value of the MinPts parameter, it may be wise to calculate the LOF of each point for a range of values of MinPts and select one according to some criterion. To test this notion, we ran the LOF calculation algorithm for a range of values of MinPts, depending on the size of the data set.

For large data sets, this range was wide, for example, from 10 to 100, while smaller data sets had a smaller range, for example, from 5 to 25. In our work, the LOF of each point was determined to be the maximum of all LOF values calculated over this range. This approach, and the fact that the points with the least LOF are the most prominent outliers, suggests a point is labeled a local outlier only if it is a notable outlier for many values of MinPts (Xiong et al., 2006).

3.7.3 Cluster-Based Outlier Detection Methods for Noise Removal

As noted above, clustering algorithms can find outliers as a by-product of the clustering process. All objects in such clusters are treated as noise.

```
Data: Transaction set T, noise fraction ε. radius r
Result: Set of noise objects N, Set of non-noise objects p

for i = 1 to n_trans do
T[i].NumWithinDist ← 0;
    for i = 1 to n_trans do
        if ((j ≠ i)&&(CosineSimilarity(T[i],T[j] ≥ r) then
                T[i].NumWithinDist ++;
        end
    end
end
T_sorted ← Sort(T, NumWithinDist,ascending);
n_noise ← ε*n_trans;
N ← T_sorted [1...n_noise];
P ← T_sorted [n_noise +1...n_trans];
return N,P;
```

ALGORITHM 3.1
A distance-based noise removal algorithm.

Data: Transaction set *T*, noise fraction ε. Cluster label set *C* for *T*
Result: Set of noise points *N*, Set of non-noise points *P*

```
for i=1 to num_clusters do
        cluster_center[i][1...n_items]← avg(T[1...n_trans],i);
end

for i=1 to n_trans do
        T[i].ClusterCenterSimiliarity ←
        CosineSimiliarity(T[i],cluster_center [T[i]]);
end

T_sorted ← Sort(T,ClusterCenterSimiliarity,ascending);
n_noise ← ε*n_trans;
N← T_sorted [1...n_noise];
P← T_sorted [n_noise+1...n_trans];
return N,P;
```

ALGORITHM 3.2
A noise removal algorithm based on the LOF.

This method is sensitive to the choice of clustering algorithms, however, and has trouble determining which clusters should be classified as outliers. Another approach assumes that once data are clustered, noise objects are the farthest objects from their corresponding cluster centroids. Here, we consider a clustering-based data cleaner (CCleaner) based on this approach. Algorithm 3.3 shows the pseudocode of its implementation. Data are clustered using a *k*-means algorithm available in the CLUTO clustering package; then, the cosine similarity (distance) of each object from its corresponding cluster centroid is recorded. The top ε% objects obtained after sorting all objects in ascending (descending) order according to this similarity (distance) are the noise objects. The algorithm's overall complexity is the same as a *k*-means followed by a linear scan of the data, or $O(kn)$, where *k* represents the number of clusters and *n* is the number of points (Xiong et al., 2006).

CCleaner and other clustering-based approaches must consider how to select the number of clusters. If there is only one cluster, the cluster-based approach is very similar to the distance-based approach described earlier. If every object is a separate cluster, however, the cluster-based approach degenerates to randomly selecting objects as outliers. As our experimental results described in Section 3.6 show, CCleaner performs well only when the number of clusters is close to the "actual" number of clusters (classes) in the data set. Unfortunately, this limitation severely restricts the usefulness of this method (Xiong et al., 2006).

```
Data: Transaction set T, noise fraction ε. MinPtsLB, MinPtsUB, MinPtsStep
Result: Set of noise points N, Set of non-noise points P

for n = MinPtsLB; n ≤ MinPtsUB; n += MinPtsStep do
    MinPts ← n;
    for i = 1 to n_trans do
        InterSimilarity [1...n_trans] ← 0;
        for j = 1 to n_trans do
            InterSimilarity[j] ← CosineSimilarity(T[i],T[j]);
        end
        InterSimilarity[i] ← 0;
        UpdateKDistNeighbors(T[i], InterSimilarity)/*UpdateKD
        istNEighors finds the k nearest neighbors for
        transaction T[i] using the similarity vector
        InterSimilarity*/;
    end

    for i = 1 to n_trans do
        CalculateLRD (T[i])/*CalculateLRD calculates the
        local reachaility density (lrd) for transaction
        T[i] using its k nearest neighbor and their lrd
        values*/;
    end

    for i = 1 to n_trans do
        latestLOF ← CalculateLOF (T[i])/*LatestLOF computes
        the local outlier factor for T[i] using its lrd
        value and those of its k nearest neighbors, for
        the current value of MinPts*/;
    end
end

T_sorted ← Sort(T,lof,ascending); n_noise ← ε*n_trans;
N ← T_sorted [1...n_noise];
P ← T_sorted [n_noise+1...n_trans];
return N, P;
```

ALGORITHM 3.3
A cluster-based noise removal algorithm.

3.8 The Black Swan

A Black Swan event is an event in human history that is both unprecedented and unexpected. However, after evaluating the surrounding context, domain experts (and, in some cases, laypersons) can usually conclude: "it was bound to happen." Even though some parameters may differ (such as the event's time, location, or specific type), it is likely that similar incidents have had similar effects in the past.

The term "Black Swan" originates from the (western) belief that all swans are white because these were the only ones officially documented. However, in 1697, Dutch explorer Willem de Vlamingh discovered black swans in Australia. This was an unexpected event in (scientific) history and profoundly changed zoology. After the black swan was discovered, its existence seemed obvious, just as other animals with varying colors were known also to exist. In retrospect, then, the surrounding context (i.e., observations about other animals) seemed to imply the existence of the black swan, that is, the Black Swan assumption, and empirical evidence validated it. Detecting and analyzing Black Swan events helps us to gain a better understanding of why certain developments recur throughout history and what effects they have.

The Black Swan application aids users in finding Black Swan events throughout modern history. For this purpose, the application identifies outliers in statistical data and associates them with historic events. An outlier is a point in a statistic that does not "fit" into the overall trend, for example, an inflection point on a curve. The rules used to join events to outliers are automatically determined by using data-mining techniques, and the results can be explored using a web interface (Hasso Plattner Institut, u.d.).

3.8.1 Combining Statistics and Events

Throughout history, there have always been Black Swan events. These are defined as singular occurrences with high impact that were entirely unexpected, but in retrospect appear to be the logical consequence of certain preconditions. Examples include World War I and the burst of the dot-com bubble. The term originates in Vlamingh's discovery in 1697, discussed in the previous section. Until then, people had widely used the Black Swan to refer to impossibilities, as it was a common (western) belief that only white swans existed. With the discovery of black swans, the definition of the term changed radically.

In its combination of statistical data and information on events, the Black Swan is a tool helping domain experts (e.g., historians or econometricians) to identify important events throughout history. The goals of Black Swan analysis are first, to automatically detect outliers in global statistics and connect them with suitable events and second, to find patterns in the ways event types and statistical developments influence one another (Lorey et al., 2011).

To accomplish these goals, it is necessary to process statistical data, in which large deviations from expected statistical developments may suggest Black Swan events. Recently, a great deal of useful statistical data has been made available by governments and other organizations. Internet resources and databases also contain information on historical events which could be Black Swan events. See Figure 3.11.

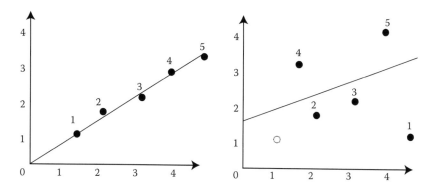

FIGURE 3.11

Why outlier detection is important in statistics. The five points represent the data; the line is the least-squares regression line. In the right-hand graph, point 1 has been misread. (Redrawn from Marsland, S., 2002. *Neural Computing Surveys*, 3, 1–39.)

3.8.2 Data

For our prototype, we must first create a database containing statistical and event data (Figure 3.12). To do so, we extract and merge information from a number of different internet sources (detailed in the next section). Events are characterized by a date, location, title, and event category. Instead of using

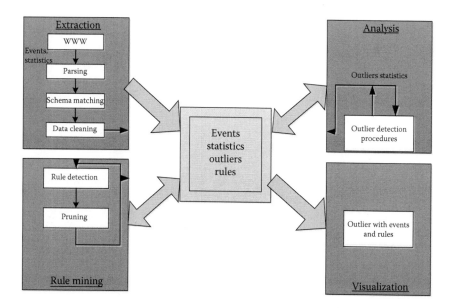

FIGURE 3.12

System architecture diagram. (From Lorey, J., Naumann, F., Forchhammer, B. et al., October 2011. Black swan: Augmenting statistics with event data. In: *Proceedings of the 20th ACM International Conference on Information and Knowledge Management*. ACM, pp. 2517–2520.)

a single date, we could chose to describe an event by selecting a time period with a start and end date. The event category provides a hierarchical taxonomy applicable to numerous and widely diverse types of events: natural disasters, political events, military conflicts, and so on.

A statistical data point is characterized by its numerical value and the type, location, and year of the statistic. For instance, one value might characterize the annual average income in the United Kingdom for 1969 using United States Dollar (USD). As discussed previously, statistical data outliers are extrema that deviate from a graph's underlying tendency. For statistical data, we turn to international organizations, such as the World Bank and the International Monetary Fund (IMF), but with the help of data provided by other projects, such as Gapminder or Correlates of War for event data, we also exploit DBpedia, EMDat, National Oceanic and Atmospheric Administration (NOAA), Correlates of War, Freebase, and the British Broadcasting Corporation (BBC) historical time line. While EMDat and NOAA are mostly limited to natural disasters, such as droughts or earthquakes, the BBC timeline pages provide a chronology of key events (mostly political) in a country's history. Not surprisingly, Correlates of War provides information on wars, while the DBpedia and Freebase data sets have structured data from user-generated content. With the help of the statistical outliers and event data, we can automatically detect patterns between these classes. The specific aim here is to decrease the number of events possibly causing an outlier in a statistic and to decrease the probable causes of an event (Lorey et al., 2011).

3.8.3 Extraction

The first step in our workflow to extract events and statistical data from our online sources. This process can be subdivided into parsing, schema matching, and data cleaning.

3.8.3.1 Parsing

First, we retrieve and parse event data from the sources noted in the previous section. We use a number of flexible parsers able to handle structured (e.g., CSV, HTML/XML, resource description framework [RDF]) and unstructured formats (i.e., plain text). Depending on the source, parsing is preceded by a preprocessing step which removes irrelevant parts of source data (e.g., HTML documents' header/footer). Our flexible design allows the existing parsers to be easily adapted to different formats and to be integrated with new parsers. We use Apache Tika to parse structured sources (e.g., HTML documents) and extract relevant text and the Jena Framework to load and query RDF data.

In this way, we can extract more than 43,000 distinct events from our sources. About half come from collections about natural disasters (most notably EMDat and NOAA). Our statistics include approximately 400 specie indicators (e.g., gross domestic product [GDP] or literacy rate) collected annually for about 200 countries and up to 200 years. While not all statistical data are available for all countries for such a long period, our final data include about 1,250,000 individual statistical values (Lorey et al., 2011).

3.8.3.2 Schema Matching

Next we map the parsed data from their source schema onto the unified event and statistics schemata. Because of the variety of data structures across sources, we configured the details for this mapping process manually for each source, using the following normalization steps for schema matching (Lorey et al., 2011):

- Attribute deduplication: Semistructured sources, as for example, DBpedia, may contain multiple attributes with essentially the same meaning. We merge these into a single attribute where applicable.
- Categorical classification of events and statistics into predefined hierarchical category structures: If a source already has taxonomy, we can perform classification using a static mapping between the existing classification and ours. Or we can take a machine-learning approach, a classification method particularly useful for unstructured sources (text classification).
- Geospatial classification of events and statistics using the GeoNames database and web service: In geospatial classification, multiple locations are permitted for events. In this step, we also generate titles for events if sources have not done so. Finally, generation is dynamic based on the values of other attributes (e.g., category, location).
- Date normalization: We map date and time values of the various data formats onto a unified schema.
- Value normalization: Here, we normalize statistical values to account for different value ranges and units. We also remove prognostic values, as our sole concern is analyzing the actual historical data.

Before we store an entity in the database, we perform a number of checks to ensure it is both valid and useful. For example, at the very least, events require a title, year, location, and category to be useful for rule mining. By the same token, we ignore missing or nonnumerical values in statistical time series data: they could cause the data-analysis algorithms to produce incorrect results or even no results at all.

3.8.3.3 Data Cleaning

The same event or similar events can most likely be found in several data sets. The following steps allow us to identify and then fuse about 2500 duplicate events.

First, we adapt the sorted neighborhood algorithm so that a window contains all events occurring during a period of five consecutive years: we only compare the events happening within this time frame. To find a duplicate, we compare all attribute values using a variety of similarity measures and weights. *Note*: the titles and locations have the highest impact in this case. The weight of the start date is increased for natural disasters, as a date is typically easier to pinpoint for such occurrences than, for example, political events.

Using a modified Monge Elkan distance metric that divides the similarity sum of the most similar words by the number of words from the shorter title, we can compare titles. We use the Porter Stemmer algorithm to stem all words. We eliminate stop words, and we define the similarity of two words as their Jaro Winkler distance.

These duplicates must now be clustered. We first considered clustering by deriving the transitive closure of individual duplicate candidates but rejected this when we found it to be both computationally expensive and likely to result in low-Precision values. Accordingly, we use the nearest neighbor technique for clustering. We use varying thresholds for different groups. Finally, for a cleaner event, we fuse each individual event cluster by fusing all its contained attributes. We select an appropriate title and start date using the source deemed to be most reliable, and we apply an end date if one is available. For the most part, the values of other attributes can be concatenated (Lorey et al., 2011).

3.8.4 Detection of Outliers

To identify interesting aspects in the statistical data, we look for outliers in the values, using a number of different methods to do so. For example, we can use linear regression to describe the relationship between a scalar variable (time, in our case) and a dependent variable, corresponding to a specific quantitative indicator (e.g., GDP in USD or the literacy rate in percent). We define an outlier as a point on the graph that differs noticeably from the estimated linear model; in other words, it has a large residual. To find the linear approximation of a curve, we draw on three algorithms: MM-estimation, least squares, and least median squares. For the latter method, the linear model is defined by points not regarded as outliers for this model.

Linear regression is only applicable to specific data sets, so we add two variants of nonparametric regression analyses: the Loess function and a generalized additive model (GAM).

To analyze the properties of the graph itself, we implement an algorithm that defines all extrema of a graph as its outliers. A more elaborate approach is to analyze the slope of the graph and define outliers as those points where the absolute change of the slope is above a certain threshold.

The final approach is an algorithm that defines a "global statistic" by calculating the mean of all country-specific statistics and analyzing the relationship of each statistic to the global statistic.

Each method has advantages and disadvantages and may be suitable only for specific data sets, for example, indicators expected to increase linearly over time. While the application's "default" setting should provide a good starting point, domain experts may opt for a more appropriate analysis of underlying data by selecting the corresponding algorithm in the web interface (Lorey et al., 2011).

3.8.5 Association Rule Mining

We want to detect interesting patterns of event–outlier combinations, such as: "in case of a major natural disaster, the annual GDP of a country declines." Here, we explain how to find these patterns using association rule mining.

3.8.5.1 Data Preparation

A rule consists of a premise X and a consequence Y, along with metrics describing the rule's quality, including support, confidence, and conviction. Support is a measure of the frequency of attribute combinations in the data set, while confidence tells us how often the consequence follows from a given premise, defined as

$$\text{Conf}(X \Rightarrow Y) = P(Y \mid X) = \frac{P(X, Y)}{P(X)} \tag{3.33}$$

Finally, conviction measures how strongly a rule holds when compared to purely random effects between the premise and consequence, defined as

$$\text{Conf}(X \Rightarrow Y) = \frac{P(X)P(\neg Y)}{P(X \wedge \neg Y)} \tag{3.34}$$

Association rule mining seeks to find dependencies between variables in a single-large database. Therefore, the first step in data preparation is to extract such a data set using the relational model. Since we are interested in the correlation between events and outliers, we join statistical outliers with events happening in a particular year and location (country) to form the basis for the data set. When we are finished, the data set contains information on the event category, the statistic category, the tendency of an outlier, and the statistical trend leading up to the outlier. For example, in one country in our data set, a change in government led to a local extremum in a previously ascending consumer price index in 383 cases with a conviction value of approx. 2.09. This means a change in government typically yielded a decline in the consumer price index.

The tendency of an outlier tells us whether it is a maximum or minimum compared to the expected statistical trend. The statistical trend in the years leading up to an outlier can help us identify Black Swans for which the statistical development is relevant to the event, instead of considering only the outlier. An example is the widespread increase of weapons production preceding WW I (Lorey et al., 2011).

3.8.5.2 Rule Generation

In our association rule mining, we use the open-source machine-learning software WEKA11 and employ the Apriori algorithm.

One challenge in rule mining generally is selecting the attributes we expect to find. In our problem, we decide every rule must contain at least the event category and the statistical category or indicator. Other attributes, such as outlier tendency or historical trend, may be permitted but are not required.

As our base data set for rule generation contains about 1.1 million combinations (by joining events, statistics, and outliers for each year), we want each determined rule to meet a minimum support of approximately 0.01% by considering only those implications with at least 100 occurrences.

Further, as the Apriori algorithm builds all combinations of item subsets, some generated rules do not comprise a useful rule that we can visualize, including rules without an event or statistic attribute and events with an unknown event class. We therefore prune the resulting set by removing all such rules. Each rule returns the conviction value for a given combination of an event and outlier; this value indicates the degree to which a rule applies to the given combination. The more rules that match the combination and the higher the conviction, the more likely the event for the given outlier (Lorey et al., 2011). In this way, we can determine the most fitting event(s) for an outlier.

References

Chandola, V., Banerjee, A., and Kumar, V., 2009. Anomaly detection: A survey. *ACM Computing Surveys (CSUR)*, 41(3), 15.

Chomboon, K., Kerdprasop, K., and Kerdprasop, N., 2013. Rare class discovery techniques for highly imbalanced data. In *Proc. International Multi Conference of Engineers and Computer Scientists (Vol. 1)*. Hong Kong.

Han, S., Yuan, B., and Liu, W., 2009. *Rare Class Mining: Progress and Prospect*. IEEE, pp. 1–5.

Hasso Plattner Institut, u.d. *What Are Black Swan Events?* [Online] Available at: http://blackswanevents.org/?page_id=26.

Holte, R. C., Acker, L. E., and Porter, B. W., 1989. *Concept Learning and the Problem of Small Disjuncts*. International Joint Conference on Artificial Intelligence (IJCAI), Vol. 89, pp. 813–818.

Hussain, F., Liu, H., Suzuki, E., and Lu, H., 2000. *Exception Rule Mining with a Relative Interestingness Measure*. Heidelberg, Berlin: Springer, pp. 86–97.

Joshi, M., 2002. *On Evaluating Performance of Classifiers for Rare*. Japan, Maebashi, pp. 641–644.

Lane, T. 1998. Machine learning techniques for the domain of anomaly detection for computer security (Doctoral dissertation, PhD thesis, Department of Electrical and Computer Engineering, Purdue University).

Liu, B., Hsu, W., and Ma, Y., 1999. Mining association rules with multiple minimum supports. In: *Proceedings of the Fifth ACM SIGKDD International Conference on Knowledge Discovery and Data Mining*, ACM, pp. 337–341.

Lorey, J., Naumann, F., Forchhammer, B. et al., October 2011. Black swan: Augmenting statistics with event data. In: *Proceedings of the 20th ACM International Conference on Information and Knowledge Management*. ACM, pp. 2517–2520.

Markou, M. and Singh, S., 2003. Novelty detection: A review—Part 1: Statistical approaches. *Signal Processing*, 83, 2481–2497.

Marsland, S., 2002. Novelty detection in learning systems. *Neural Computing Surveys*, 3, 1–39.

Medioni, G., 2001. Event detection and analysis from video streams. *IEEE Transactions on Pattern Analysis and Machine Intelligence*, 23(8), 873–889.

Ohsawa, Y., 2000. *Chance Discoveries for Making Decisions in Complex Real World*. [Online] Available at: http://ymatsuo.com/papers/NGCchance.pdf.

Ohsawa, Y., 2002. *A Scope of Chance Discovery*. AAAI Technical Report.

Riddle, P., Segal, R., and Etzioni, O., 1994. Representation design and brute-force induction in a Boeing manufacturing design. *Aritifical Intelligence*, 8, 125–147.

Seiffert, C., Khoshgoftaar, T. M., Hulse, J. V., and Napolitano, A., 2007. Mining data with rare events: a case study. In: *Tools with Artificial Intelligence, 2007. ICTAI 2007. 19th IEEE International Conference*, IEEE, Vol. 2, pp. 132–139.

Smyth, P., 1994. Hidden Markov models for fault detection in dynamic systems. Pattern Recognition, 27(1), 149–164.

Terano, T., Liu, H., and Chen, A. L. (Eds.), 2000. Knowledge discovery and data mining. Current issues and new applications. In: *Lecture Notes in Artificial Intelligence; Subseries of Lecture Notes in Computer Science*. Springer.

Weiss, G. M., 1995. Learning with rare cases and small disjuncts. In: *International Conference on Machine Learning (ICML)*, Morgan Kaufmann, pp. 558–565.

Weiss, G. M., 2005. Mining with rare cases. In: Maimon, O. and Rokach, L. (Eds.), *Data Mining and Knowledge Discovery Handbook*. USA: Springer, pp. 765–776.

Xiong, H., Pandey, G., Steinbach, M., and Kumar, V., 2006. Enhancing data analysis with noise removal. *IEEE Transactions on Knowledge and Data Engineering*, 18(3), 304–319.

Zhang, C. and Zhang, S. (Eds.), 2002. Association rule mining: Models and algorithms. In: *Lecture Notes in Computer Science; Vol. 2307: Lecture Notes in Artificial Intelligence*. Heidelberg, Berlin: Springer-Verlag, pp. 83–84.

4

Input and Output Data

The main function of machine learning is to design and develop algorithms that allow systems using empirical data, experience, and training to evolve and adapt to changes in their environment, that is, to changes in the state of the monitored assets. An important part in the investigation of machine learning is to automatically induce models, such as rules and patterns from the training data being analyzed. Figure 4.1 shows a combination of machine-learning techniques and approaches from different areas, including probability and statistics, psychology, information theory, and artificial intelligence.

An important part of any fault detection technique is the nature of the input data, where the input is usually a collection of data instances (also called an object, record, point, vector, pattern, case shows, watching, or entity) (Tan et al., 2005). Each data instance may be defined by a set of attributes (also called a variable, feature, function, field, or dimension). Attributes can be various types such as binary, categorical, or continuous. Each instance of data can be one of two types: it either has a single attribute (univariate) or multiple attributes (multivariate). For multivariate data entities, attributes may be all the same type of data or a mixture of different types (Chandola et al., 2007). The nature of the attributes determines the applicability of fault detection techniques. For example, statistical techniques include various statistical models for continuous and categorical data. Similarly, a distance measure is determined by the attributes of the data point's closest technical-based neighbor. In some cases, instead of the real data, the distance between pairwise cases may be provided as a distance or similarity matrix. In such cases, techniques requiring the consultation of original data are not applicable, for example, many statistical techniques based on classification. The input data can also be classified on the basis of the relationship between the various data instances (Tan et al., 2005).

Most of the existing techniques for detecting failures are responsible for recording data or data points, in which no relationship between data instances is assumed. In spatial data, each data instance is related to its neighboring instances, for example, data from vehicle traffic and ecological data. When data have a spatial (sequential) time component they are called spatio-temporal data, for example, climate data. When a graph format is used, data instances are represented as vertices and connected to other vertices by edges. Later in the chapter, we will discuss situations in which these kinds of relationships between data instances become relevant for fault detection.

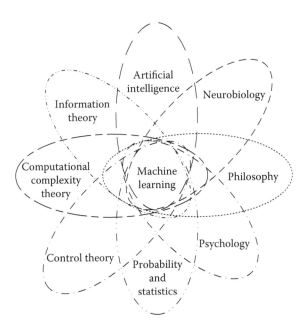

FIGURE 4.1
Machine learning.

The nature of data, as mentioned in previous chapters, is a major issue in condition monitoring. In addition, data collection must be contextually adaptive due to the changing environment; and algorithms for fault detection must be equally adaptive. To cite one example (Al-Karaki et al., 2004), wireless sensor network (WSN) applications operate in very challenging conditions in the field of condition monitoring and health management; they must constantly accommodate environmental changes, hardware degradation, and inaccurate sensor readings. To maintain operational correctness, a WSN application must frequently learn and adapt to changes in its running environment. Machine learning has been used in such cases.

Machine-learning algorithms can be classified into supervised and unsupervised learning, depending on whether instances of training data are labeled or not. In supervised learning, the student is provided with teaching-identified instances where both the input and the correct output are given. Unsupervised learning is the opposite of supervised learning since the correct output is not provided with the entry. Instead, the training program is based on other sources of feedback to determine whether the learning is done properly. There is a third class of machine-learning techniques called semisupervised learning, which uses a combination of both labeled and unlabeled data for training. Figure 4.2 shows the relationship between these three types of machine learning.

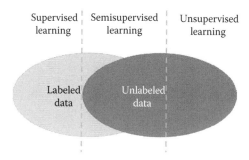

FIGURE 4.2
Classes of machine-learning algorithms.

In this chapter, we examine machine-learning algorithms in sensor networks from the perspective of the types of applications used. We provide more examples of the three types of machine learning and discuss their use in a number of sensor network applications. The machine-learning algorithms most commonly used include Clustering, Bayesian probabilistic models, Markov models, and decision trees. We also discuss the challenges, advantages, and disadvantages of the various machine-learning algorithms. Figure 4.3 shows the machine-learning algorithms introduced in this chapter.

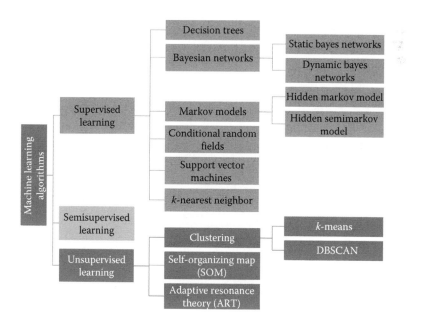

FIGURE 4.3
Classification of machine-learning algorithms used in WSN applications.

4.1 Supervised Failure Detection

In supervised learning, the student can be provided with labeled data entry. These data have a sequence of pairs of input/output $y_i > <x_i$, where x_i is an entry and y_i is the output associated with it. The main idea for the student in supervised learning is to learn the assignment of inputs to outputs. Therefore, the program is expected to identify function f that accounts for pairs of inputs/outputs seen so far: $f(x_i) = y_i$ for all i.

This function $f(x_i)$ can be called a classifier if the output is discrete and a regression function if the output is continuous. The distinction into function classifier/regression is to correctly predict the results of inputs that have not been seen before, for example, when the inputs are the redundancies of sensors and outputs are the activities causing them.

The execution of a supervised learning algorithm can be divided into five main steps (Figure 4.4). (Hu and Hao, 2013):

> *Step 1*: Determine which training data are needed and collect those data. In this case we have to ask two questions. First, what data are needed? Second, what amount of data are needed? Designers must make a decision and decide which training data best represent real life scenarios for a specific application. They also have to determine the amount of training data to be collected. The more data we have, the better we can train the learning algorithm, data collection training, and provide correct labels. However, this is usually both expensive and laborious. Therefore, an application designer always strives to find an algorithm large enough to provide sufficient training data, but small enough to avoid unnecessary costs associated with data collection and labeling.

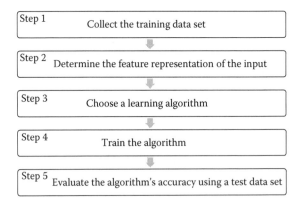

FIGURE 4.4
Stages of supervised machine learning.

Step 2: Identify the set of features, also called the feature vector, to represent the input. Each feature in the feature set represents a characteristic of the object, as well as events that are classified. There is a balance between the size of the feature vector and the classification accuracy of machine-learning algorithm. An array of many features can significantly increase the complexity of classification; however, the use of a vector of small features, which does not possess a sufficient description of the objects/events, may cause poor classification accuracy. Therefore, the feature vector must be large enough to represent the important features of object/event and small enough to avoid excessive complexity.

Step 3: Select a proper-learning algorithm. Certain factors must be considered when choosing a learning algorithm for a given task, including the content and size of the training dataset, the noise in the system, the accuracy of labeling, heterogeneity, and redundancy of the input data. Similarly, we must assess the needs and characteristics of the sensor network application itself. For example, for a recognition application activity, the duration of the sensor use plays a significant role in determining the activity being executed. Therefore, to achieve a high accuracy of activity recognition, machine-learning algorithms that can explicitly model state time are preferred.

Frequently used supervised machine-learning algorithms include support vector machines (SVMs), naive Bayes classifiers, decision trees, hidden Markov models (HMMs), conditional random field (CRF), and k-nearest neighbors (k-NN) algorithms. Likewise, a number of approaches have been used to improve the performance of the selected classifiers, such as bagging, boosting, or using a set of classifiers. Each algorithm has its advantages and disadvantages, making it suitable for some types of applications, but unsuitable for others.

Step 4: Train the learning algorithm using data collected. In this step, the algorithm treats the function that best fits the training instances of input/output.

Step 5: Evaluate the accuracy of the algorithm. The algorithm is tested with a data test set, that is, data that differ from the training data, to determine its accuracy.

Various supervised learning algorithms have been implemented and tested experimentally in a variety of sensor network applications. The most frequently used algorithms, WSN applications, are described in the remainder of this section.

4.1.1 Decision Trees

Decision trees are often characterized by a fast execution time, ease of interpretation of the rules, and scalability for large multidimensional data sets

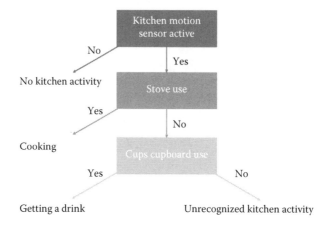

FIGURE 4.5
Sample decision tree for an activity detection application.

(Cabena et al., 1998; Han, 2005). The main function of decision tree learning is the elaboration of a model that predicts the value of the output variable based on the input variables in the feature vector. Each node corresponds to a variable feature vector. Each node has edges for each of the possible values (or range of values) of the input variable associated with the node. Each leaf in the decision tree represents a possible value for the output variable. The output variable can be calculated by following a path from the root upwards and is guided by the values of the input variables.

Figure 4.5 shows a decision tree for a sample application activity detection sensor network in a kitchen. In this scenario, it is assumed there are only two events of interest in the kitchen: cooking and getting a drink. The decision tree sensor node is used to differentiate between these two activities. For example, if there is movement in the kitchen and the stove is being used, the algorithm determines that the residents are cooking. However, if there is movement in the kitchen, the stove is not used, and someone opens the cupboard, the algorithm defines the activity as getting a drink. This is a simple example that illustrates how decision trees can be applied to sensor network applications. Actually, the decision trees learned from real applications are much more complex.

The C4.5 algorithm is currently known as top-down, a greedy search algorithm decision tree construction (Quinlan, 1993). The algorithm uses metric entropy and information gained to induce a decision tree. A decision tree is used for recognizing activity in the Place Lab project at MIT (Logan et al., 2007). Researchers monitored a home equipped with more than 900 sensors, including wire rod turns, current and water flow inputs, object detectors, people on the move, and radio-frequency identification (RFID) tags. Data from 43 typical household activities were collected. C4.5 was one of the classifiers used in its recognition approach.

C4.5 has also been used for target recognition in wireless underwater surveillance sensor systems (Cayirci et al., 2006). Each node in the network is equipped with multiple types of microsensors, including acoustic materials, magnetic radiation, and mechanical sensors. The readings from these sensors are used by the recognition algorithm decision trees to classify submarines, small delivery vehicles, mines, and divers.

In another instance, C4.5 was used as part of an algorithm to *automatically recognize physical activities and their intensities* (Tapia et al., 2007). More specifically, the algorithm monitored the readings of triaxial wireless accelerometers and wireless heart rate monitors and its efficacy was evaluated using datasets of 10 physical activities collected from 21 people.

4.1.2 Bayesian Network Classifiers

According to Hu and Hao (2013), Bayesian probability interprets the concept of probability as a degree of belief. A Bayesian classifier has the function of analyzing the feature vector describing a particular input instance and assigns the instance to the most likely class. This classifier is based on applying Bayes' theorem to evaluate the probability of particular events. Bayes' theorem gives the relationship between pre and postbeliefs of two events.

Assume $P(A)$ is the initial prior belief in A, and $P(A|B)$ is the posterior belief in A, after B has been found, that is, the conditional probability of A, given B. In a similar fashion, $P(B)$ is the initial belief before in A, and $P(B|a)$ is the posterior belief of B, given A. Supposing $P(B) \neq 0$, Bayes' theorem states

$$P(A|B) = \frac{P(B|A) \times P(A)}{P(B)}$$

The Bayesian network is a probabilistic model that represents a set of random variables and conditional dependencies through a direct acyclic graph (DAG). For example, a Bayesian network often represents the probabilistic relationships between the activities and the sensor readings. If set sensor readings are given, the Bayesian network can also be used to assess the likelihood of performing various activities.

Bayesian networks have a number of advantages. For example, a network of Bayes refers only to nodes that are probabilistically related by a causal dependency. This may lead to a huge savings in computing because there is no need to store all possible configurations of the states; we only need to store combinations of states relating to the sets of nodes between parents and children. Similarly, we can say that Bayes networks are also very adaptable, as they can start small, with limited knowledge about the domain, and grow as they acquire new knowledge.

Bayesian networks have been applied to a variety of sensor fusion problems where they must integrate data from various sources to build a complete picture of the current situation. In addition, they have been used to monitor and alert applications where the application must recognize whether certain events have occurred and decide whether to send a warning or notification. Finally, Bayesian networks have been applied to a number of activity recognition applications and numerous implementations have evaluated them using single source and multiple residents.

Bayesian networks are either static or dynamic, depending on whether groups are able to model the temporal aspects of events/activities of interest. In what follows, we present examples of these two classes: static naïve Bayes classifiers and dynamic naïve Bayes classifiers.

4.1.2.1 Static Bayesian Network Classifiers

A common representative of static Bayesian networks is the static naïve Bayes classifier. Learning Bayesian classifiers can be significantly reduced by making the naïve assumption that the characteristics describing a class are independent. The classifier usually assumes that the presence or absence of a class characteristic is not related to the presence or absence of any other features in the feature vector. The naive Bayesian classifier is one of the most practical learning methods and has been widely used in many sensor network applications, including activity recognition in nursing homes (Van Kasteren and Kröse, 2007), recognizing the PlaceLab activity in the project at MIT (Logan et al., 2007), outlier detection (Janakiram et al., 2006), and body sensor networks (Maurer et al., 2006).

Figure 4.6 shows a naïve Bayesian model used for the recognition of an activity. In this scenario, the activity at time t, *activity*, is independent of any previous activities. It is also assumed that the sensor data R_t depend only on the *activity*.

Naïve Bayes classifiers have the following advantages:

- Can be trained efficiently.
- Are well suited for categorical features.

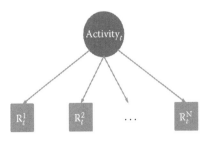

FIGURE 4.6
Static Bayesian network.

- Have performed well in complex real-world situations despite their naïve design and the independence assumptions and can work with more than 1000 features.
- Are good for combining multiple models and can be used in an iterative way.

Naïve Bayes classifiers have a disadvantage in that if the conditional independence is not true, that is, if there is a dependency between the features of the examined classes, then it is not a good model. In addition, naïve Bayes classifiers assume all corresponding attributes in a classification decision, which represent these observables. Despite these drawbacks, studies have shown that naïve Bayes classifiers are very accurate classifiers in a number of problem domains. Simple naïve Bayes networks have even been shown to be comparable to more complex algorithms such as decision trees (Tapia et al., 2004).

4.1.2.2 Dynamic Bayesian Network Classifiers

Another disadvantage of static Bayesian networks is that they cannot model the temporal aspect of sensor network events. However, dynamic Bayesian networks are capable of representing a sequence of variables when the sequence is a continuous reading from a sensor node. Therefore, dynamic Bayesian networks, although they are more complex to implement, can become more suitable for modeling events and activities in sensor network applications.

Figure 4.7 shows a dynamic naïve Bayesian model, where with the varying activity t, only one activity is directly influenced by the previous variable activity t. These models assume an event may cause another event in the future, but not vice versa. Therefore, directed arcs between events/activities must flow forward in time and cycles are not allowed.

4.1.3 Markov Models

A process is considered to be Markov if it exhibits the Markov property—lack of memory. That is, the conditional probability distribution of future

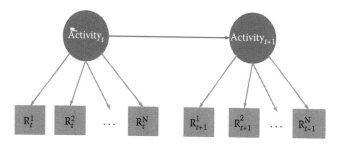

FIGURE 4.7
A naive dynamic Bayesian network.

states of the process depend only on the present state and not on events preceding it. Two types of Markov models are the HMM and the hidden semi-Markov model (HSMM).

4.1.3.1 Hidden Markov Model

The HMM can be considered a simple dynamic Bayesian network. When an HMM is used, the system is assumed to be a Markov process with unobserved states (hidden). Although the hidden state is sequenced, the output is dependent on the state that is visible. Therefore, in each time step, there is a hidden variable and an observable output variable. In sensor network applications, the hidden variable could be the event or activity being performed, and the observable output variable could be the vector of sensor readings.

Figure 4.8 shows an HMM where the states of the system are hidden, but the output variables X are visible. Two assumptions define the dependency model, as represented by the directional arrows in Figure 4.9:

1. Markov assumption: The hidden variable at time t, that is, Y_t, depends only on the previous variable Y_t hidden—1 (Rabiner, 1989).

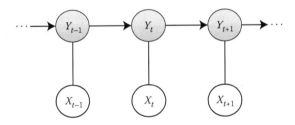

FIGURE 4.8
HMM example.

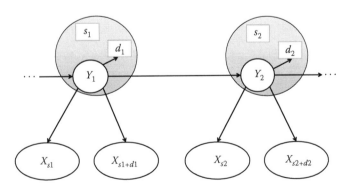

FIGURE 4.9
Hidden semi-Markov model.

2. The observable output variable at time t, that is, X_t depends only on the hidden variable Y_t.

With these assumptions, we can specify an HMM using three probability distributions:

1. The initial state distribution: The distribution in the initial states of $p(Y_1)$.
2. Distribution of transition distribution $p(Y_t|Y_t + 1)$, which represents the probability of moving from one state to another.
3. Distribution note: the distribution $p(X_t|Y_t)$, which indicates the probability of the hidden state $X_t Y_t$ generating observation.

Learning the parameters of these distributions corresponds to maximizing the joint probability distribution $p(X, Y)$ sequences paired with observations and labels in the training data. The ability to model the joint probability distribution $p(X, Y)$ makes a generative model.

HMMs have been widely used in many sensor network applications. Most of the previous work on activity recognition HMM was used to recognize activities from sensor data (Patterson et al., 2005; Van Kasteren et al., 2008; Wilson and Atkenson, 2005). An HMM was also used in the Smart Thermostat (Lu et al., 2010) project. This technology detected occupancy and sleep patterns in a home and used these patterns to automatically operate home heating, ventilation, and cooling (HVAC). The authors used an HMM to estimate the probability of being at home in each of three states: unemployed, employed with active residents, and residents occupied with sleeping. HMMs were also used in an application of biometric identification in the homes of several residents (Srinivasan et al., 2010). In this project, height sensors were mounted above the doors in a house and an HMM was used to identify the location of each resident.

A weakness of conventional HMMs is their lack of flexibility in the length of modeling the state. With HMMs, there is a constant probability of change of state, and this limits the ability of modeling. For example, the activity of preparing dinner normally extends at least for several minutes. To prepare dinner in less than a couple of minutes is not very common. The geometric distribution used by HMMs to represent the duration cannot be used to represent the distribution of events where shorter durations are less possible.

4.1.3.2 Hidden Semi-Markov Models

HSMM differs from an HMM in that HSMMs explicitly model the duration of hidden states. Therefore, the probability of a change in the hidden state depends on the amount of time elapsed since entry into the current state.

The recognition accuracy of an HSMM has been compared to that of an HMM (Van Kasteren et al., 2010). This example involved an evaluation of the recognition performance of the models using two fully annotated real-world databases consisting of several weeks of data. The first set of data was stored in a three-room apartment with one resident and the second data set was from a six-room house with one resident. When the results were analyzed, they showed that HSMM consistently outperformed HMM. This indicates that modeling the exact duration is important in applications for recognizing real-world activity, as it can result in significantly better performance. In many cases, the use of the duration of the classification process is especially helpful in scenarios where sensor data do not provide enough information to distinguish between activities.

4.1.4 Conditional Random Fields

CRFs are often considered an alternative to HMMs. The IRC is a method of statistical modeling that, given a sequence of particular observation, is a kind of probabilistic graphical model indirectly defining a single log-linear distribution of label sequences. This is used to encode the known relationships between observations and interpretations that are constantly being built.

CRFs are often considered an alternative to HMMs. The IRC is a statistical modeling method which is a type of probabilistic graphical model indirectly defining a single log-linear distribution of label sequences given a sequence of particular observation. CRFs are used to encode the known relationships between observations and interpretations that are consistently built.

The CRF model that most closely resembles an HMM is the linear chain CRF. As shown in Figure 4.10, the straight-chain model CRF is very similar to the HMM (Figure 4.8). The model still contains hidden and corresponding observable variables at each time step, but unlike the HMM, the CRF model is not directed. This means two connected nodes are no longer a conditional distribution. Moreover, we can talk about the potential between two connected nodes. Compared with HMM, the two conditional probabilities—the probability of observation $p(x_t|Y_t)$ and transition probabilities $p(y_t|Y_t 1)$—have been replaced by the corresponding potential. The essential difference lies in the way to learn the model parameters. In the case of HMM, the parameters are learned by maximizing the joint probability distribution $P(X, Y)$. CRFs

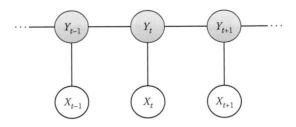

FIGURE 4.10
A linear-chain CRF model.

are discriminative models; their parameters are learned by maximizing the conditional probability distribution $p(Y|X)$, a member of the family of exponential distributions (Sutton and McCailum, 2006).

CRF models have been used for the recognition of activity in homes using video streams, where primitive actions, such as the "go-from -A -to- B" are recognized using a configuration of the room in question (Truyen et al., 2005). The results of these experiments show CRF performs much better than the generated HMMs even though a large portion of the labels are missing data. CRF has also been used for the modeling of concurrent activities and interleaving (Hu et al., 2008). These authors conducted experiments using one of the data sets at PlaceLab MIT (Logan et al., 2007), PLA1, comprising 4 h of sensor data.

Kasteren Van et al. used four different data sets, two sets of data from the bathroom and two from the kitchen, to compare the performance of HMM to CRF (Van Kasteren et al., 2010). Experiments showed that when applied to the activity recognition tasks, CRF models achieved higher accuracy than the HMM models. The authors applied the results to the flexibility of discriminant models such as CRF, to address violations of the model assumptions and found the higher accuracy achieved by models of CRF has a price:

1. Discriminative models take much longer to train than their generative counterparts.
2. Discriminative models are prone to overfitting. Overfitting occurs when a random noise pattern takes the place of the underlying relationship. The model is able to maximize performance on the training data; however, its effectiveness is not determined by how well it performs on training data, but by its generalizability and how well it performs on unseen data.

4.1.4.1 Semi-Markov CRF

Like HMMs, CRFs also have a semi-Markov variant: in this case, semi-Markov conditional random fields (SMCRF). A sample SMCRF model appears in Figure 4.11. The SMCRF inherits the following features from semi-Markov models and CRFs:

1. It explicitly models the duration of states (like HSMM).
2. Each hidden state is characterized by a start position and duration (like HSMM).
3. The model's graph is undirected (like CRF).

Hierarchical SMCRFs have been tested in an *activity recognition* application using a small laboratory dataset from the domain of video surveillance (Truyen et al., 2008). In this test, the task was to recognize indoor trajectories

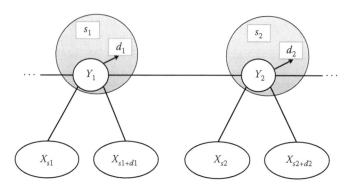

FIGURE 4.11
A semi-Markov CRF.

and activities of a person using the noise extracted from a video. The data had 90 sequences, each corresponding to one of three possible activities: preparing a short meal, preparing a normal meal, and having a snack. Results showed the hierarchical SMCRF outperformed both a conventional CRF and a dynamic CRF. Then, when SMCRFs were used for activity recognition by Van Kasteren et al. (Van Kasteren et al., 2010), the results indicated that contrary to the big improvements seen when using HSMMs over HMMs, SMCRFs only slightly outperformed CRFs. That being said, CRFs may be more robust in dealing with violations of modeling assumptions. Therefore, allowing them to explicitly model duration distributions might not yield the same significant benefits as using HSMM.

4.1.5 Support Vector Machines

SVM is a nonlinear probabilistic classifier binary organization, where the predicted output of an SVM is one of two classes. When there is a training set of instances, each is marked as belonging to one of two classes; one possible type of output is the SVM algorithm, which constructs a model N-dimensional hyperplane for future cases. As Figure 4.12 shows, an SVM model represents input instances as points in space, mapped so that separate instances of classes are clearly divided. New examples are assigned in the same space; and based on which side of the gap they fall on, their class is predicted. Put otherwise, the objective of SVM analysis is to find a line separating cases on the basis of class. As there is an infinite number of possible lines, finding the optimal line represents an essential challenge of using SVM models.

4.1.6 *k-NN* Algorithms

The *k-NN* algorithm is the easiest of all the machine-learning algorithms and is very accurate in many scenarios. The training examples are vectors

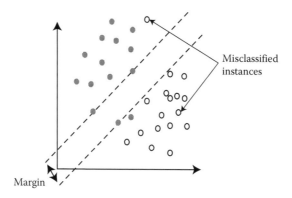

FIGURE 4.12
Two-dimensional SVM model.

in a multidimensional space of features, each having a class label. The training phase of the algorithm simply comprises storing the feature vectors and labels of class training samples. A new instance is divided by a majority vote of its neighbors and is ultimately assigned to the class found to be most common among its *k-NN*.

Figure 4.13 provides an example of a *k*-closest classification algorithm. The question mark represents the test sample and should be classified as either a star or a triangle. If $k = 3$, the sample is assigned to the class of triangles, as we have 2 and 1 star triangles in the inner circle. If $k = 7$, the test sample is assigned to the class of stars, as we have 4 stars and 3 triangles in the outer circle.

The best choice of k depends on the data, but k must be a positive integer and is generally small. If $k = 1$, the new instance is assigned to the class of its nearest neighbor. On the one hand, higher values of k reduce the effect of noise on the classification; on the other hand, they make boundaries between classes less distinct. Good k can be selected using a number of different heuristic techniques, for example, cross-validation. Although the *k-NN* is quite accurate, the time required to classify a sample may be higher because the algorithm must compute the distance (or similarity) of the instance to all

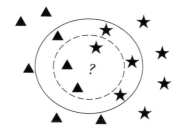

FIGURE 4.13
Example of *k*-nearest algorithm classification.

cases in the training set. In short, the classification time k-NN is proportional to the number of features and the number of training instances.

4.2 Semisupervised Failure Detection

Supervised learning algorithms use both semilabeled and unlabeled data for training. Labeled data are typically a small percentage of the training data. The function of semisupervised learning is to: (1) understand how combining labeled and unlabeled data may change the learning behavior, and (2) design algorithms to take advantage of this combination. Semisupervised learning is a very promising approach, as we can use readily available unlabeled data to improve supervised learning tasks when labeled data are scarce or expensive.

There are many different semisupervised learning algorithms. The following section explains some of the most common.

4.2.1 Expectation–Maximization with Generative Mixture Models

Expectation-maximization (EM) is an iterative method to find maximum likelihood estimates of parameters in statistical models that depend on unobserved latent variables (Dempster et al., 1977). Each iteration of the algorithm consists of Hope Step (E-step); this also has a maximization step (M-step). EM with a mixture of generative models is suitable for applications where the classes specified by the application data produced are tightly grouped.

4.2.2 Self-Training

Self-education can refer to a variety of schemes for using unlabeled data. For example, Ng and Cardie (2003) considered bagging and implementing self-training by majority voting. In their method, a group of classifiers is trained on instances of labeled data, and used to classify the unlabeled examples independently. Examples for which all classifiers are assigned the same label are added to the labeled training set and the classifier is retrained. The process continues until a stop condition is met.

A single classifier can also be self-taught. In this case, the classifier is trained on all the labeled data and then applied to the unmarked pattern. Only those cases that meet selection criteria are added to the labeled set and used for recycling.

4.2.3 Cotraining

Cotraining requires two or more views of the data, that is, disjoint sets of features that provide additional information about different instances

(Blum and Mitchell, 1998; Blum and Chawla, 2001). It is best if the two sets of features for each instance are conditionally independent. Moreover, each set of features should be able to accurately assign each instance in the respective class. The first step in cotraining is the use of all tagged data and training a separate classifier for each view. Then, the most accurate predictions of each classifier can be used in unlabeled data to build additional labeled training instances. Cotraining is an appropriate algorithm to use if the characteristics of the data set are naturally divided into two sets.

4.2.4 Transductive SVMs

Both transductive SVMs and extended SVMs can be used as partially labeled data for semisupervised learning (Gammerman et al., 1998) following the principles of supervised transduction. In inductive learning, the algorithm is trained on instances of specific training, but the goal is to learn the general rules, which are then applied to the test cases. By contrast, the transductive learning reasons from specific cases of training for specific test cases.

4.2.5 Graph-Based Methods

These are algorithms that use graph structure obtained by capturing pairwise similarities between the labeled and the unlabeled (Zhu, 2007) cases.

These algorithms define a graph structure where nodes are labeled, while the unmarked pattern and the edges, which can be weighted, represent the similarity of the connected nodes.

In sensor networks, semisupervised learning has been used to determine the location of moving objects. Pan et al. (2007) took a learning approach to semisupervised probabilistic models by reducing the calibration effort and increasing the tracking accuracy of the system. Their method is based on semisupervised CRF; the learned model, which contains a small number of training data, is improved with the addition of abundant unlabeled data. For better efficiency, it uses a generalized EM algorithm coupled with domain constraints. Yang et al. (2010) created a learning algorithm using a semimonitored collector to estimate the locations of mobile nodes in a WSN. Their learning algorithm calculates a subspace mapping function between the signal space and the physical space using a small amount of labeled data and a large amount of unlabeled data.

Finally, Wang et al. (2007) developed a learning algorithm based on semisupervised SVM. When this algorithm is applied to target classifications, experimental results show it can accurately classify targets in sensor networks.

Likewise semisupervised learning has also been applied to the detection and recognition of elements that are commonly displaced (Bulling and Roggen, 2011). For example, Xie et al. proposed a dual-sensor network camera that can be used as an auxiliary memory tool (Xie et al., 2008). The color characteristic of each new object is extracted and a semiclustering algorithm is used to classify the monitored object. The method provides the user with the option of reviewing the results of the classification algorithm and tagging the mislabeled images, thus providing real-time feedback for the system to filter the data model.

4.3 Unsupervised Failure Detection

It is difficult to collect labeled resources or to achieve accurate labeling. For example, obtaining training data sufficient for the recognition of the activity in a home may require several weeks of data collection and labeling. Moreover, labeling is difficult not only for remote and inaccessible areas but also for home and commercial building deployment. For any of these deployments, someone has to tag the data. In the instance of a house, the labeling can be done by the residents themselves. Although residents should have a record of what they are doing and when, previous experience shows these records are often incomplete and inaccurate. An alternative solution is to install cameras throughout the house and control the activities of residents. However, this approach is considered to be privacy invasive and, therefore, is not suitable.

In unsupervised learning, the learner is provided with input data that have not been tagged. The role of the student is to find the patterns inherent in the data and then apply these patterns to determine the correct value for new instances of data output. The assumption is that there is a structure to the input space, so that certain patterns occur more frequently than others. Overall, the learning wants to see what happens and what does not work. In statistics, this is known as density estimation.

In short, unsupervised learning algorithms are very useful for applications of sensor networks for a number of reasons:

- Data collection requires resources and time.
- Labeling is difficult to achieve accurately.
- As the applications of sensor networks are usually deployed in unpredictable and changing environments, applications must evolve and learn without guidance, using unlabeled patterns.

A variety of algorithms for unsupervised learning have been used in applications of sensor networks, including clustering algorithms such as k-means and mixture models, self-organizing maps (SOM), and the theory of adaptive resonance (ART). In the remainder of this section, some commonly used unsupervised learning algorithms are described.

4.3.1 Clustering

Clustering (or cluster analysis) is a form of unsupervised learning, often used in pattern recognition tasks and activity sensing applications. A clustering algorithm partitions the input case into a fixed number of subsets, called clusters, so that the instances in the same group are similar to each other with respect to a set of metrics (Hu and Hao, 2013).

Cluster analysis itself is not a given algorithm, but the general task to be solved. The grouping is done by algorithms, which differ significantly in their notion of what constitutes a cluster and how to find it efficiently. Choosing appropriate clustering algorithms and parameter settings, including the distance function, the threshold density, or the number of expected groups, depends on the dataset and the intended use of the results.

Figure 4.14 is a pictorial representation of a clustering algorithm. A clustering algorithm divides the set of input data instances into groups, called clusters. As the figure shows, the instances in the same group are more similar to each other than they are to instances in other clusters.

The idea of a *cluster* varies between algorithms, and the properties of the clusters found by different algorithms vary significantly. Typical cluster models include the following:

- *Connectivity models*: Hierarchical clustering, which builds models based on distance connectivity, is an example of a connectivity model.

- *Centroid models*: A representative is the *k*-means algorithm where each cluster is represented by a single mean vector.

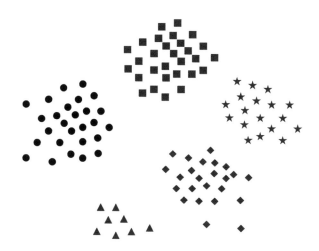

FIGURE 4.14
A clustering algorithm.

- *Distribution models*: In this case, clusters are modeled using statistics distributions.
- *Density models*: An example is density-based spatial clustering for applications with noise (DBSCAN) where clusters are identified as areas with a higher density than nonclusters.
- *Group models*: These clustering algorithms are not able to provide a refined model for results; they only generate the group information.

Of these, the two most commonly used in sensor network applications are *k*-means clustering and DBSCAN clustering.

4.3.1.1 k-Means Clustering

The function of *k*-means clustering is to separate the cases of entry into k groups, so that each instance belongs to the cluster with the nearest mean. When the problem is NP-hard, the common approach is to locate only approximate solutions. A number of efficient heuristic algorithms can rapidly converge to a local optimum, for example, the Lloyd algorithm (Lloyd, 1982). The algorithms find only local optima, which often run multiple times with different random initializations.

The *k*-means algorithm is simple and rapidly converges when the number of dimensions of the data is small, but *k*-means clustering also has certain drawbacks. First, *k* should be described in advance. Second, the algorithms prefer groups of similar sizes. This often leads to improperly reducing the boundaries between the groups; this is not surprising; as it is a model center of gravity, a *k*-means algorithm is optimized for the center of the cluster, not the cluster boundaries.

Figure 4.15 is an example of grouping where $k = 2$ and *k*-means is unable to precisely define the boundaries observed between the two groups. There are two sets of density in the figure: one is much larger and contains circles and the other is smaller and consists of triangles. Since *k*-means is optimized for the center cluster and tends to produce groups with similar sizes, it incorrectly splits the data instances in a green and a red cluster. These two groups, however, do not overlap with the original density of input data clusters.

4.3.1.2 DBSCAN Clustering

DBSCAN is a spatial clustering algorithm based on grouping the most popular density. In density-based clustering, the clusters are defined as areas of greater density than the rest of the data set. DBSCAN requires two parameters (Ester et al., 1996): threshold distance (Eps-neighborhood of a point) and the minimum number of points required to form a cluster (MinPts). DBSCAN is based on the points connected within a certain distance of each other, that is, in the same Eps-neighborhood. However, for a cluster, DBSCAN requires that each point in the cluster has at least MinPts number of points

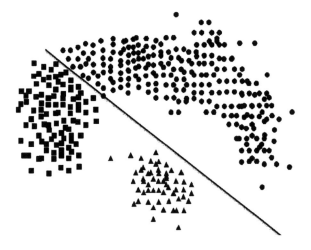

FIGURE 4.15
k-Means clustering incorrectly cutting borders between density-based clusters.

in the Eps-neighborhood. Figure 4.16 is an example of an observed DBSCAN grouping. The data set is the same as in Figure 4.15. The data are grouped correctly, but a clustering algorithm based on the density has been applied.

One advantage of DBSCAN is that, unlike many other clustering algorithms, it can form groups of any arbitrary shape. Moreover, its complexity is quite low and it essentially finds the same groups in each run. Therefore, in contrast to the *k*-means clustering, DBSCAN can be executed only once instead of multiple times. The main drawback is that a drop in DBSCAN has significantly

FIGURE 4.16
Density-based clustering with DBSCAN.

enough density to detect cluster boundaries. If cluster densities continuously decrease, DBSCAN can produce clusters whose borders look arbitrary.

In the applications of sensor networks, DBSCAN has been used as part of the FATS attack on security to identify the function of each room, including the bathroom, kitchen, or bedroom (Srinivasan et al., 2008). DBSCAN generates temporal activity groups, each of which forms a temporary block that is continuous with a relatively high shot density sensor. Experiments show DBSCAN works well because it stops and calculates the outliers' high-density clusters automatically. However, *k*-means clustering performs much better when identifying which sensors are in the same room. This is especially the case when all devices are highly correlated temporally, and there is no significant density drop in the boundary of clusters.

Apiletti et al. (2011) applied DBSCAN to *detecting sensor correlation* with data from a sensor network used in university labs. They found DBSCAN can identify different numbers of clusters based on the day of the week being analyzed. As a result, it can construct more accurate models for sensor use patterns in labs. Finally, DBSCAN can detect noisy sensors.

4.3.2 Self-Organizing Map

SOMs provide a way to represent multidimensional data spaces of much lower dimension, that is, typically one or two dimensions. A technique for data compression is vector quantization, the process of reducing the dimensionality of the feature vectors. SOM generates a map which is a representation of this function tablet space; a valuable feature of these maps is that information is stored so that the topological relationships are maintained within the set training.

SOM also contains components called nodes. Each node is associated with a position in the map space with a vector of weights; the size of this vector is the same as the input data instances. The nodes are regularly spaced on the map, typically having a rectangular or a hexagonal grid. A typical example of SOM is a color map (Figure 4.17). Each color is represented by a

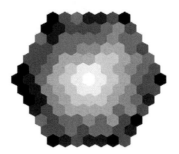

FIGURE 4.17
SOM representation for colors.

three-dimensional vector containing the values for red, green, and blue. The SOM colors represent two-dimensional space.

Placing an input data instance onto the map requires the following steps:

1. Initializing the node weights on the map.
2. Choosing an input training instance.
3. Finding the node with the closest vector to that of the input instance (termed the best matching unit [BMU]).
4. Calculating the radius of BMU's neighborhood. This value is frequently set to the radius of the whole map, but it decreases at each time step. Any node within this radius is considered inside BMU's neighborhood.
5. Assigning the values from the vector of the input instance and adjusting the weights of the nodes close to BMU toward the input vector; the closer a neighbor node is to BMU, the more its weight is altered.

4.3.3 Adaptive Resonance Theory

Existing learning algorithms tend to be *stable* (preserving previously learned information) or *plastic* (adapting to new input instances indefinitely). Typically, stable algorithms cannot easily learn new information, while plastic ones forget the old information they have learned. This is called the *stability–plasticity dilemma* (Carpenter and Grossberg, 1987).

The adaptive resonance theory (ART) architectures attempt to provide a solution to the stability–plasticity dilemma. This group of different neural architectures addresses the problem of how a learning system can maintain its previously learned knowledge while maintaining the ability to learn new patterns. A model of ART has the ability to distinguish familiar and unfamiliar events, and expected and unexpected events.

When we want to process familiar and unfamiliar events, the ART system contains two functionally complementary subsystems that allow us to perform this process: subsystems of attention and guidance. Family events are processed within subsystem attention; the objective of this subsystem is to constantly establish increasingly accurate internal representations and responses to family events. By itself, however, there is a drawback. The attention subsystem cannot simultaneously maintain stable representations of familiar categories and create new categories for family events. This is where help is needed from the guidance subsystem; it is used to restart the subsystem attention when a familiar event occurs. The orientation subsystem is essential to express that either a novel pattern is familiar and well represented by an existing code or an unknown code that requires recognition and a new recognition code.

Figure 4.18 shows the system architecture. The care system is classified into two successive stages, F1 and F2; these activation patterns encode the

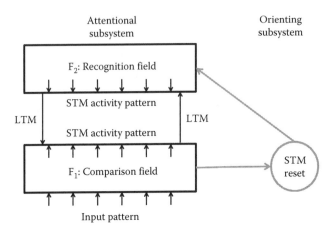

FIGURE 4.18
Architecture of an ART system.

short-term memory (STM). The input pattern is received in F1, and in F2, classification is performed. From top-to-bottom and from bottom-to-top, and the roads in between, the two stages of adaptation contain the long-term memory (LTM) traces. The guidance subsystem calculates the similarity between the vector input instance and the pattern produced by the fields in the attentional subsystem. As long as both are similar, that is, if the subsystem of attention has been able to recognize the input instance, the orientation subsystem does not interfere. However, if the two patterns are significantly different, the orientation subsystem resets the output of the recognition layer. The effect of the replacement is to force the system attention back to zero, thus making the system find a better match.

A disadvantage of some of the ART architectures is that the model results depend largely on the order in which the training instances are processed. The effect can be reduced to some extent when there is a slower rate of learning where the differential and degree of training in an entry depends on the time of the entry that is available. Even with the slow formation, however, the order of formation still affects the system, regardless of the size of the input data set.

4.3.4 Other Unsupervised Machine-Learning Algorithms

A wide variety of unsupervised learning algorithms, in addition to k-means clustering, for example, DBSCAN, SOM, and ART, have been used in WSN applications. The Smarthouse project uses a system of sensors to monitor the activities of a person in the household (Barger et al., 2005). The project objective is to recognize and detect different patterns of behavior. The authors use mixture models to develop a probabilistic model of behavior patterns.

The mixture model approach groups the observations into different types of events.

A number of project activity recognition algorithms have been developed from unsupervised learning models extracted from texts or the Web. One project used methods of an unsupervised learning guide to identify activities using RFID tags placed on objects (Philipose et al., 2003). This method is based on data mining techniques to extract activity patterns of the Web in an unsupervised way. For this project, the authors derived the term structure for about 15,000 activities at home.

An unsupervised approach based on the detection and analysis of the sequence of objects being used by residents has been described by Wu et al. (2007).

The method of recognizing activity is based on an RFID object correlated with video streams and the information obtained from how-to websites like about.com. Because the approach uses video streams, it provides high-grain recognition activity. For example, we can differentiate between tea and coffee. However, as mentioned above, collecting video data from household activities is difficult due to privacy concerns.

4.4 Individual Failures

A point of failure is when an instance can be considered as individually anomalous from the rest of the data. This is the simplest type of problem and is the focus of most research in anomaly detection. For example, in Figure 4.1, the points O1 and O2, and the points in the region of O3, are located outside the boundaries of normal regions, resulting in point anomalies, as they are different from the normal data points. As an example in real life, consider the detection of credit card fraud, wherein the data set corresponds to the credit card transactions of an individual. For the sake of simplicity, assume that the data are defined by a single feature: amount spent. A transaction for which the amount spent is very high compared with the normal range of costs for that person will be a point of failure. Either a point anomaly or a collective anomaly can be transformed to the contextual anomaly.

The most popular techniques for individual fault detection are classification, nearest-neighbor, clustering and anomaly-based techniques.

4.4.1 Classification-Based Techniques

These techniques work in two phases. During the first phase, learning, a prediction model (classificator) is built using the available labeled data. The classificator can distinguish between normal and anomalous data. During the second phase, testing, the tested data are classified into normal or anomalous classes. The learning phase can divide the normal data into several sets. When this occurs, it is called a multiclass technique. When only one normal class exists, it is a one-class technique. One group of classification techniques uses a classification algorithm based on neural networks. Such techniques can be used with either multiclass or one-class data. Other techniques use algorithms based on Bayesian networks, SVMs, or rule-based systems. The testing phase is generally very fast, because a predictive model has been built and instances are only compared to the model. These techniques also use algorithms that can distinguish between instances belonging to different normal classes. A disadvantage is that the techniques need labeled training data to build the predictive model.

4.4.2 Nearest Neighbor–Based Techniques

These techniques are based on the prediction that normal data instances form neighborhoods, while anomalous data instances do not. They compute distances to the nearest neighbors or use a relative density as the anomaly score. One group of techniques uses the distance to the k-nearest neighbors as the anomalous score. A second group computes the relative density in a hypersphere with the radius d. The advantage of these techniques is that they can work in the unsupervised mode, but if the semisupervised mode is used, the number of false anomaly detections is smaller (Pokrajac et al., 2007). Their computational complexity is relatively high, because the distance is computed between each pair of data instances. In addition, the rate of false anomaly detection is high if a normal neighborhood consists of only a few data instances.

4.4.3 Cluster-Based Techniques

These techniques are very similar to nearest neighbor techniques. The problem of detecting anomalies that form clusters can be transformed to the problem of nearest neighbor-based techniques; however, clustering-based techniques evaluate each instance with respect to the cluster it belongs to. There are three types of clustering-based techniques. The first assumes normal instances form clusters (Figure 4.19). These techniques apply known clustering-based algorithms and determine whether any data belong to the cluster. A disadvantage is that they are optimized to find clusters not anomalies. The second type assumes the normal data instances lie close enough to a closest cluster centroid (Figure 4.20). Such techniques are not applicable if the anomalies form clusters. Therefore, the third type of clustering technique assumes normal instances form large dense clusters, while anomalies form small sparse clusters (Figure 4.21). The first two types work in two phases: in the first phase, a clustering algorithm clusters data and in the second phase, it computes a distance as an anomalous score. Clustering-based techniques can work in the unsupervised mode, because clustering algorithms do not need labeled data.

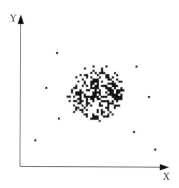

FIGURE 4.19
Normal instances as one big cluster.

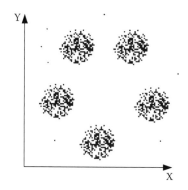

FIGURE 4.20
Normal instances forming more than one cluster.

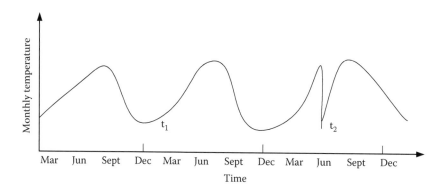

FIGURE 4.21
Contextual anomaly T_2 in a temperature time-series.

Once a model is built, the testing phase is quick, because it only compares tested instances to the model. A disadvantage is the high-computational complexity; furthermore, these algorithms are not optimized to find anomalies.

4.4.4 Anomaly Detection Techniques Based on Statistical Approach

The statistical methods of anomaly detection are based on the following assumption (Anscombe and Guttman, 1960): an anomaly is an observation suspected of being either partially or wholly irrelevant because it is not generated by a model assumed to be stochastic. Thus, normal data instances occur in high probability regions of the stochastic model, and anomalies occur in low probability regions. The statistical methods fit the statistical model to the normal data and then determine if a tested data instance belongs to y or x, as shown in Figure 4.21.

If a technique assumes the knowledge of the distribution, it is parametric (Eskin, 2000); otherwise, it is nonparametric (Desforges et al., 1998).

Nonparametric methods assume the model is determined from the given data. The most commonly used techniques are the kernel function-based and the histogram-based techniques. The former techniques use the Parzen windows estimation (Parzen, 1962). The latter are the simplest and are widely used in intrusion detection systems. The first step in using these techniques is to build a histogram based on different values taken from the training data. In the second step, a tested data instance is checked to determine whether it falls into one of the histogram bins.

Parametric methods assume data are generated by a parametric distribution and a probability density; they can be divided according to type of distribution (Kopka et al., 2010):

- Gaussian model based
- Regression model based
- Mixture of parametric distribution based

An advantage of statistical methods is their widespread use. If a good model is designed, the methods are very effective. They can also be used in the unsupervised mode with a lack of training data. The histogram-based techniques are not suitable for detecting the contextual anomalies, however, because they cannot record an interaction between the data instances. In addition, choosing the proper test method is nontrivial.

4.4.5 Other Detection Techniques

The aforementioned techniques are the most widely used. Some remaining methods use information theory techniques based on relative entropy or the Kolmogorov complexity. These techniques can operate in the unsupervised mode and do not need a statistical assumption about data. A final type, spectral anomaly detection techniques, tries to find an approximation of the data and determine the subspaces in which the anomalous instances can be easily identified. These techniques have high computational complexity.

4.5 Contextual Failures

A data instance is anomalous in a specific context; otherwise it is called a contextual anomaly (also called conditional anomaly) (Song et al., 2007). The notion of context is induced by the structure data set and must be specified as part of the problem formulation each data instance is defined by, using the following two sets of attributes:

Contextual features: Contextual attributes are applied to determine the context (or neighborhood) for an instance. For example, spatial data sets, the longitude and latitude of a location are contextual attributes. In time-series data, time is a contextual attribute that determines the position of the instance in the entire sequence.

Behavioral attributes: Behavioral attributes specify noncontextual features of an instance. For example, in a group of spatial data describing the average rainfall around the world, the amount of rain anywhere is an attribute of behavior. Anomalous behavior is determined using the values of the attributes of behavior within a given context. An example of data could be a contextual anomaly in a given context, but an instance of similar data (in terms of behavioral attributes) could be considered normal in a different context. Therefore, this property is key to the identification of contextual attributes and behavior in the detection technique of contextual anomalies. Contextual anomalies have been discussed most frequently in the data time series (Salvador et al., 2004; Weigend et al., 1995) and in spatial data (Kou and Lu, 2006; Shekhar et al., 2001). Figure 4.3 shows a time series of temperature with monthly temperatures of an area in recent years. It has to take into account the temperature at time $t1$ is the same as at time $t2$, but occurs in a different context and, therefore, is not considered a failure. A variety of anomaly detection techniques handle contextual anomalies (Figure 4.22). For these, the data must have a context attribute group in order to define a context, and a set of performance attributes for abnormalities in context. Song et al. (2007) used the terms of environmental attributes and indicators, and these are analogous to our terminology. Some attributes of the relevant data are discussed below.

1. *Spatial data*: The data may have attributes defining the spatial location of a particular data point and, therefore, a spatial neighborhood.

    A number of techniques for detecting anomalies is based on context (Kou et al., 2006; Lu et al., 2003; Shekhar et al., 2001; Sun and Chawla, 2004).

(a)   (b)   (c)

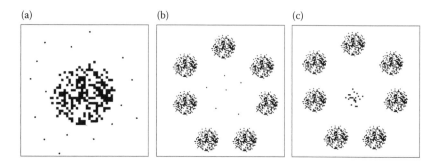

FIGURE 4.22
2-D Data sets. Normal instances are shown as circles and anomalies are shown as squares.
(a) Data set 1; (b) Data set 2; (c) Data set 3.

2. *Data based on graph*: The edges that connect the data nodes (instances) define the neighborhood for each node. Techniques for detecting contextual anomalies have been applied to graph-based data (Sun et al., 2005).

3. *Sequential data*: When data are sequential, the contextual attribute of a data instance is its position in the sequence. Time-series data have been explored extensively in the category of contextual anomaly detection (Abraham and Box, 1979; Abraham and Chuang, 1989; Bianco et al., 2001; Fox, 1972; Galeano et al., 2006; Rousseeuw and Leroy, 1987; Salvador et al., 2004; Tsay et al., 2000; Zeevi et al., 1997). Another form of sequence data developed for anomaly detection techniques is event data, in which each event has a timestamp (e.g., call data of the operating system or database of the Web) (Ilgun et al., 1995; Smyth, 1994; Vilalta and Ma, 2002; Weiss and Hirsh, 1998). The difference between the data time series and sequence of events is that for the latter, the time between arrivals of consecutive events is uneven.

4. *Profile*: Often data may not have an explicit spatial or sequential structure, but still can be segmented or grouped into parts using a set of contextual attributes. These attributes are typically used for profiling and to group users in activity monitoring systems, such as mobile phone fraud detection (Fawcett and Provost, 1999), CRM databases (He et al., 2004), and detection of credit card fraud (Bolton and Hand, 2001). Users are analyzed within their group of anomalies.

Compared to the literature on the point anomalies detection techniques, research on detecting contextual anomalies remains limited. In general, these techniques can be classified into two categories. The first category of techniques reduces the contextual detection problem to a problem of point-of-failure anomaly detection; the second category uses the model to detect anomalies.

Moreover, to find and identify faults in contextual techniques requires large and powerful computer data resources. In many cases, dividing the data into contexts is not a simple task. This is especially true for the data and time-series data of the sequence of events. In such cases, the time-series modeling and sequence-extending modeling techniques for detecting anomalies in the data context are used.

A generic technique in this category can be described as follows. When a model is learned from the training data, the expected behavior in a given context can be predicted. If the expected behavior is significantly different from the observed behavior, a fault is declared. A simple example of this technique is generic regression in which contextual attributes can be used to predict the behavior attribute by fitting a regression line to the data. The following models were developed for the detection of contextual anomalies: time-series data, multiple regression techniques for time-series modeling and robust regression (Rousseeuw and Leroy, 1987) based, autoregressive models (Fox, 1972), the auto-regressive moving averages (ARMA) models (Abraham and Box, 1979; Abraham and Chuang, 1989; Galeano et al., 2006; Zeevi et al., 1997) and auto-regressive integrated moving averages (ARIMA) models (Bianco et al., 2001; Tsay et al., 2000). Regression-based techniques have been extended to detect abnormalities in a contextual set of sequences of coevolution by modeling the regression and the correlation between the sequences (Yi et al., 2000).

One of the first works in anomaly detection time series was proposed by Fox (1972) who modeled time series as a stationary autoregressive process. In this process, a comparison is made between any observation test for the abnormal and the covariance matrix of the autoregressive process. If the observation is not in the modeling error for the process, we can say it is an anomaly. This technique was extended by Ma and Perkins (2003) in their Support Vector Regression to estimate regression parameters and use the learned model to detect novelties in the data. Their technique adopts a divide and conquer approach. The sequence is divided into two parts, and the Kolmogorov complexity is calculated for each event containing the anomaly. The sequence is divided recursively until we are left with one event that started the anomaly in the sequence.

4.5.1 Computational Complexity

The computational complexity of the training phase detection techniques based on contextual anomalies reduction depends on the reduction technique, just as the detection technique uses point of failure in each context. While segmentation techniques partition to have a rapid reduction stage, using clustering techniques or a mixture of the estimation models is slower. The anomaly detection techniques are usually used to accelerate the second step, as the reduction simplifies the problem of anomaly detection. The testing phase is relatively expensive, because with each test case, the context is determined and an anomaly or label score is assigned using a detection technique point of failure. The computational complexity is typically higher when the structure uses the data of the training phase anomaly detection techniques. An advantage of these techniques is the testing phase is relatively fast, as each instance is compared to a model and assigned a score of abnormality or an abnormality label.

The main advantage of the techniques of contextual anomaly detection is that they allow a natural definition of a problem in many real-world applications where data instances are often similar in context. Having a global view of the data, they can detect abnormalities that could not be detected by the techniques of detection of point anomalies.

However, contextual anomaly detection techniques are applicable only when a context can be defined and when detecting contextual anomalies with data instances that have contextual and behavioral attributes (Figure 4.22). The context between data can be defined using sequences, spaces, graphs, or profiles. Profiling is typically used to detect credit card frauds. A behavioral profile is built for each credit card holder (each holder denotes a separate context). Using the credit card abroad can be labeled the anomalous or the normal instance; this depends on the context, that is, the card owner. The problem of contextual anomaly detection can be transformed to the problem of point anomaly detection. In this case, it is necessary to identify the context and compute an anomaly score. Other methods utilize data structure and take a regression or divide-and-conquer approach. The advantage of these techniques is that they can identify an anomaly which would be undetectable using the techniques described in the previous section (Chandola et al., 2007).

4.6 Collective Failures

A collective failure occurs when a collection of related data instances is anomalous with respect to the entire data set. Individual instances of data in a group may not be abnormal themselves, but their appearance together as a collection is anomalous. Figure 4.4 shows an output of a human electrocardiogram (Goldberger et al., 2000). The highlighted region indicates a problem, because there is the same low value for an abnormally long time (corresponding to a premature atrial contraction). *Note*: Low value by itself is not an anomaly.

As another illustrative example, consider a sequence of actions in a computer . . . http–web, buffer–overflow, http–web, web–http, smtp–mail, ftp, http–web, ssh, smtp–mail, web–http, ssh, buffer–overflow, ftp, http–web, ftp, smtp–mail, http–web . . . The highlighted sequence of events (buffer–overflow, ssh, ftp) correspond to a typical web-based remote machine followed by a copy of data from the host computer to a remote destination via

FTP attack. *Note*: This collection event is an anomaly, but individual events are not anomalies when they occur elsewhere in the sequence.

Collective anomalies can occur only in cases (Forrest et al., 1999) when the data sets in data instances are related, while point anomalies can occur in any data set (Figure 4.23). In contrast, the occurrence of an abnormalities context depends on the availability of the data context attributes. Moreover, a point defect or anomaly collective is contextual with respect to failure if the context is analyzed. Therefore, a problem of an anomaly detection point or a collective anomaly detection problem can change and become a problem of contextual anomaly detection by incorporating context information. The techniques used to detect anomalies are very different from the point and detection techniques; contextual anomalies require detailed discussion and are not covered in this book. For a brief review of the research, the reader may consult Chandola et al. (2007).

Figure 4.24 shows different types of anomalies in a continuous sequence of real value. A series of aberrations is shown in black; these are located in the center of each series and correspond to a specific type of anomaly. The appropriate contexts for these anomalies are dark gray, while light gray elements are part of the context. The top panel contains a point defect—an abnormal point with respect to all other points of the series. The second panel contains a fault point—an abnormal point with respect to its context (in this case, a few leading and trailing points), but not necessarily a whole number. The third panel has a collective anomaly—a subsequence anomalous to the rest

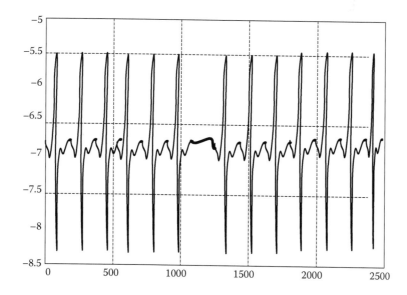

FIGURE 4.23

Collective anomaly corresponding to an atrial premature contraction in a human electrocardiogram output.

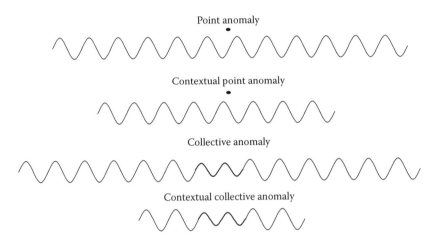

FIGURE 4.24
Different types of anomalies in a real-valued continuous sequence.

of the time series. The room contains a contextual collective anomaly—an anomalous subsequence with respect to context.

Collective anomaly detection in condition monitoring typically works with asset records. The data can have anomalies for several reasons, including abnormal condition, instrumentation errors, or recording errors. Several techniques have focused on detecting fault outbreaks in a specific area (Wong et al., 2003). In any event, anomaly detection is a critical problem in this domain and requires a high degree of accuracy. The data typically consist of records with several different types of features, such as age, pressure, vibration, temperature, and so on. The data may have both a temporal and a spatial aspect as well. Most anomaly detection techniques in this domain aim at detecting anomalous records (point anomalies). Typically, the labeled data are from healthy machines; hence, most techniques use a semisupervised approach. Time-series data, such as vibrations or acoustic emissions, represent another form of data handled by anomaly detection techniques.

Collective anomaly detection techniques have been used to find anomalies in such data (Lin et al., 2005). In this case, the most challenging aspect of the anomaly detection problem is the cost: classifying an anomaly as normal can be costly. Industrial units suffer damage due to continuous usage and normal wear and tear; this type of damage must be detected early to prevent damage escalation and losses. Data in this domain are usually termed sensor data, simply because they are recorded using sensors.

Anomaly detection techniques have been extensively used to detect damage. Industrial damage detection can be subclassified into a domain dealing with defects in mechanical components, including motors, engines, and so on (also called system health management), and a domain dealing with defects in physical structures; see Table 4.1.

TABLE 4.1

Anomaly Detection Techniques Used for Structural Damage Detection

Technique Used	Section	References
Statistical profit using histograms	Section 7.2.1	Manson et al. (2000, 2001), Manson (2002)
Parametric statistical modeling	Section 7.1	Ruotolo and Surace (1997)
Mixture of models	Section 7.1.3	Hickinbotham and Austin (2000), Hollier and Austin (2002)
Neural networks	Section 4.1	Brotherton and Johnson (2001), Brotherton et al. (1998), Nairac et al. (1999, 1997), Surace and Worden (1998), Surace et al. (1997), Sohn et al. (2001), Worden (1997)

4.6.1 Fault Detection in Mechanical Units

Anomaly detection techniques are used in mechanical units to monitor the performance of industrial products, such as engines, turbines, oil flow in pipes, or other mechanical components; they are also used for defects that may occur in circumstances due to wear or unforeseen incidents. The data in this field usually have a temporal aspect; time-series analysis is also used in some techniques (Basu and Meckesheimer, 2007, Keogh, et al., 2002, 2006). Abnormalities usually occur mainly because of an observation in a specific context (contextual anomalies) or an abnormal sequence of observations (collective anomalies). Normally, data on normal components without defects are readily available and semisupervised techniques are applicable. Abnormalities should be detected on an online form, as these require preventive measures before failure occurs. Some techniques for detecting anomalies in this domain are listed in Table 4.2.

TABLE 4.2

Anomaly Detection Techniques Used for Fault Detection

Technique Used	Section	References
Parametric statistical modeling	Section 7.1	Guttormsson et al. (1999), Keogh et al. (2002, 2004, 2006)
Nonparametric statistical modeling	Section 7.2.2	Desforges et al. (1998)
Neural networks	Section 4.1	Bishop (1994), Campbell and Bennett (2001), Diaz and Hollmen (2002), Harries (1993), Jakubek and Strasser (2002), King et al. (2002) Li et al. (2002), Petsche et al. (1996), Streifel et al. (1996), Whitehead and Hoyt (1995)
Spectral	Section 9	Parr et al. (1996), Fujimaki et al. (2005)
Rule-based systems	Section 4.4	Yairi et al. (2001)

TABLE 4.3

Anomaly Detection Techniques Used in Image Processing

Technique Used	Section	References
Mixture of models	Section 7.1.3	Byers and Raftery (1998), Spence et al. (2001), Tarassenko (1995)
Regression	Section 7.1.2	Chen et al. (2005), Torr and Murray (1993)
Bayesian networks	Section 4.2	Diehl and Hampshire (2002)
Support vector machines	Section 4.3	Davy and Godsill (2002), Song et al. (2002)
Neural networks	Section 4.1	Augusteijn and Folkert (2002), Cun et al. (1990), Hazel (2000), Moya et al. (1993), Singh and Markou (2004)
Clustering	Section 6	Scarth et al. (1995)
Nearest neighbor-based techniques	Section 5	Pokrajac et al. (2007), Byers and Raftery (1998)

TABLE 4.4

Anomaly Detection Techniques Used for Anomalous Topic Detection in Text Data

Technique Used	Section	References
Mixture of models	Section 7.1.3	Baker et al. (1999)
Statistical profiling using histograms	Section 7.2.1	Fawcett and Provost (1999)
Support vector machines	Section 4.3	Manevitz and Yousef (2002)
Neural networks	Section 4.1	Manevitz and Yousef (2000)
Clustering	Section 6	Allan et al. (1998), Srivastava and Zane-Ulman (2005), Srivastava (2006)

4.6.2 Structural Defect Detection

Structural defect detection techniques are used to find defects and structural damage abnormalities, for example, cracks in beams or tensions in airframes. The collected data have a temporal aspect, and anomaly detection techniques are similar to the techniques of detecting change point detection, as they try to detect the change in the data collected from a structure. The normal data and, therefore, the learned patterns are typically static in time. Some techniques used for detecting anomalies in this domain are listed in Tables 4.3 and 4.4.

References

Abraham, B. and Box, G., 1979. Bayesian analysis of some outlier problems in time series. *Biometrika*, 66(2), 229–236.

Abraham, B. and Chuang, A., 1989. Outlier detection and time series modeling. *Technometrics*, 31(2), 241–248.

Al-Karaki, J. N., Raza, U.-M. and Ahmed, E. K., 2004. *Data Aggregation in Wireless Sensor Networks—Exact and Approximate Algorithms.* pp. 241–245.

Allan, J., Carbonell, J. G., Doddington, G., Yamron, J. and Yang, Y., 1998. *Topic Detection and Tracking Pilot Study.* pp. 194–218.

Anscombe, F. and Guttman, I., 1960. Rejection of outliers. *Technometrics*, 2(2), 123–146

Apiletti, D., Elena, B. and Tania, C., 2011. Energy-saving models for wireless sensor networks. *Knowledge and Information Systems (Springer London)*, 28(3), 615–644.

Augusteijn, M. and Folkert, B., 2002. Neural network classification and novelty detection. *International Journal of Remote Sensing*, 23(14), 2891–2902.

Baker, D., Hofmann, T., McCallum, A. and Yang, Y., 1999. A hierarchical probabilistic model for novelty detection in text. In: *Proceedings of the International Conference on Machine Learning.*

Barger, T., Brown, D. and Alwan, M., 2005. Health status monitoring through analysis of behavioral patterns. *IEEE Transactions on Systems, Man, and Cybernetics, Part A: Systems and Humans*, 35(1), 22–27.

Basu, S. and Meckesheimer, M., 2007. Automatic outlier detection for time series: An application to sensor data. *Knowledge Information Systems*, 11(2), 137–154.

Bianco, A., Ben, M., Martinex, E. and Yohai, V., 2001. Outlier detection in regression models with ARIMA errors using robust estimates. *Journal of Forecast*, 20(8), 565–579.

Bishop, C., 1994. Novelty detection and neural network validation. In: *Vision, Image and Signal Processing*, IEE Proceedings, Vol. 141, No. 4. IET, pp. 217–222.

Blum, A. and Chwala, S., 2001. Learning from labeled and unlabeled data using graph mincuts. In: *Proceedings of the Eighteenth International Conference on Machine Learning.* C. E. Brodley and A. P. Danyluk, Eds., pp. 19–26.

Blum, A. and Mitchell, T., 1998. Combining labeled and unlabeld data with co-training, In: *Proceedings of the Eleventh Annual Conference on Computational Learning Theory.* ACM, pp. 92–100.

Bolton, R. J., and Hand, D. J. 2001. Unsupervised profiling methods for fraud detection. *Credit Scoring and Credit Control* VII, 235–255.

Brotherton, A. and Johnson, T., 2001. Anomaly detection for advanced military aircraft using neural networks. In *Aerospace Conference, 2001, IEEE Proceedings*, Vol. 6. IEEE. pp. 3113–3123.

Brotherton, A., Johnson, T. and Chadderdon, G., 1998. *Classification and Novelty Detection Using Linear Models and a Class Dependent-Elliptical Basis Function Neural Network.*

Bulling, A. and Roggen, D., 2011. *Recognition of Visual Memory Recall Precesses Using Eye Movemenr Analysis.* pp. 455–469.

Byers, S. and Raftery, A., 1998. Nearest neighbor clutter removal for estimating features in spatial point process. *Journal of American Statistics Association*, 93, 577–584.

Cabena, P. et al., 1998. *Discovering Data Mining: From Concept to Implementation.* New Jersey: Prentice-Hall, Inc.

Campbell, C. and Bennett, K., 2001. *A Linear Programming Approach to Novelty Detection.* Cambridge, UK: Cambridge Press.

Carpenter, G. and Grossberg, S., 1987. A massively parallel architecture for a self-organizing neural pattern recognition machine. *Journal on Computer Virion, Graphics, and Image Processing*, 37(1), 54–115.

Cayirci, E., Tezcan, H., Dogan, Y. and Coskun, V., 2006. Wireless sensor networks for underwater surveillance systems. *Ad Hoc Networks*, 4(4), 431–446.

Chandola, V., Banerjee, A., and Kumar, V. 2009. Anomaly detection: A survey. *ACM Computing Surveys (CSUR)*, 41(3), 15.

Chen, D., Shao, X., Hu, B. and Su, Q., 2005. Simultaneous wavelength selection and outlier detection in multivariate regression of near-infrared spectra. *Analytical Sciences*, 21(2), 161–167.

Cun, Y. et al., 1990. Handwritten digit recognition with a back-propagation network. In: *Advances in Neural Information Processing Systems*. Morgan Koufamann, pp. 396–404.

Davy, M. and Godsill, S., 2002. Detection of abrupt spectral changes using support vector machines, an application to audio signal segmentation. In: *Proceedings of the IEEE International Conference on Acoustics, Speech and Signal Processing*.

Dempster, A. P., Laird, N. M. and Rubin, D. B., 1977. Maximum likelihood from incomplete data via the EM algorithm. *Journal of the Royal Statistical Society. Series B (Methodological)*, 39(1), 1–38.

Desforges, M., Jacob, P., and Cooper, J., 1998. *Applications of Probability Density Estimation to the Detection of Abnormal Conditions in Engineering*. pp. 687–703.

Diaz, I. and Hollmen, J., 2002. *Residual Generation and Visualization for Understanding Novel Process Conditions*. IEEE, pp. 2070–2075.

Diehl, C. and Hampshire, J., 2002. Real-time object classification and novelty detection for collaborative video surveillance. In: *Proceedings of the IEEE International Joint Conference on Neural Networks*. IEEE.

Eskin, E., 2000. *Anomaly Detection Over Noisy Data Using Learned Probability Distributions*. pp. 255–262.

Ester, M., Kriegel, H.-P., Sander, J. and Xu, X., 1996. *A Density-Based Algorithm for Discovering Clusters in Large Spatial Databases with Noise*. pp. 226–231.

Fawcett, T. and Provost, F., 1999. Activity monitoring: Noticing interesting changes in behavior. In: *Proceedings of the 5th ACM SIGKDD International Conference on Knowledge Discovery and Data Mining*. ACM Press, pp. 53–62.

Forrest, S., Warrender, C. and Pearlmutter, B., 1999. Detecting intrusions using system calls: Alternate data models. In: *Proceedings of the IEEE ISRSP*. IEEE Computer Society, pp. 133–145.

Fox, A., 1972. Outliers in time series. *Journal of Royal Statistical Sovial Series B*, 34(3), 350–363.

Fujimaki, R., Yairi, T. and Machida, K., 2005. An approach to spacecraft anomaly detection problem using kernel feature space. In: *Proceedings of the 11th ACM SIGKDD International Conference on Knowledge Discovery in Data Mining* ACM Press, pp. 401–410.

Galeano, P., Peña, D., and Tsay, R. S. 2006. Outlier detection in multivariate time series by projection pursuit. *Journal of the American Statistical Association*, 101(474), 654–669.

Gammerman, A., Vovk, V. and Vapnik, V., 1998. Learning by transduction. In: *Proceedings of the Fourteenth Conference on Uncertainty in Artificial Intelligence*. Chicago: Morgan Kaufmann Publishers Inc. pp. 148–155.

Goldberger, A. et al., 2000. PhysioBank, physioToolkit, and physioNet: Components of a new research resource for complex physiologic signals. *Circulation*, 101(23), e215–e220.

Guttormsson, S., Marks, R., El-Sharkawi, M. and Kerszenbaum, I., 1999. Elliptical novelty grouping for online short-turn detection of excited running rotors. *IEEE Transactions of Energy Conversions*, 14(1), 14.

Han, J., 2005. *Data Mining: Concepts and Techniques.* San Francisco: Morgan Kaufmann Publisher Inc.

Harries, T., 1993. *Neural Network in Machine Health Monitoring.* Professional Engin.

Hazel, G., 2000. Multivariate gaussian MRF for multi-spectral scene segmentation and anomaly. *GeoRS*, 38(3), 1199–1211.

He, Z., Xu, X., Huang, J. and Den, S., 2004. *Mining Class Outliers: Concepts, Algorithms and Applications.* Heidelberg, Berlin: Springer.

Hickinbotham, S. and Austin, J., 2000. *Novelty Detection in Airframe Strain Data.* pp. 536–539.

Hollier, G. and Austin, J., 2002. *Novelty Detection for Strain-Gauge Degradation Using Maximally Correlated Components*, pp. 257–262.

Hu, D. H. et al., 2008. *Real World Activity Recognition with Multiple Goals*, pp. 30–39.

Hu, F. and Hao, Q., 2013. *Intelligent Sensor Networks; The Integration of Sensor Networks, Signal Processing and Machine Learning.* Boca Raton, FL: CRC Press, Taylor & Francis Group.

Ilgun, K., Kummerer, A. and Porras, P., 1995. State transition analysis: A rule-based intrusion detection approach. *IEEE Transactions of Software Engineering*, 21(3), 181–199.

Jakubek, S. and Strasser, T., 2002. *Fault-Diagnosis Using Neural Networks with Ellipsoidal Basis Functions.* pp. 3846–3851.

Janakiram, D., Adi Mallikarjuna Reddy, V. and Phani Kumar, A. V. U., 2006. *Outlier Detection in Wireless Sensor Networks Using Bayesian Belief Networks.* New Delhi, pp. 1–6.

Keogh, E., Lin, J., Lee, S.-H. and Herle, H., 2006. Finding the most unusual time series subsequence: Algorithms and applications. *Knowledge Information Systems*, 11(1), 1–27.

Keogh, E., Lonardi, S. and Chi'Chiu, B., 2002. Finding surprising patterns in a time series database in linear time and space. In: *Proceedings of the 8th ACM SIGKDD International Conference on Knowledge Discovery and Data Mining.* ACM Press, pp. 550–556.

Keogh, E., Lonardi, S. and Ratanamahatana, C., 2004. Towards parameter-free data mining. In: *Proceedings of the 10th ACMSIGKDD International Conference on Knowledge Discovery and Data Mining.* ACM Press, pp. 206–215.

King, S. et al., 2002. *The Use of Novelty Detection Techniques for Monitoring High-Integrity Plant.* pp. 221–226.

Kopka, J., Reves, M. and Giertl, J., 2010. *Anomaly Detection Techniques for Adaptive Anomaly Driven Traffic Engineering.*

Kou, Y., Lu, C. and Chen, D., 2006. *Spatial Weighted Outlier Detection.* Bethesda, Maryland, pp. 613–617.

Kou, Y. and Lu, C.-t., 2006. Spatial weighted outlier detection. In: *Proceedings of SIAM Conference on Data Mining.* pp. 614–618.

Li, Y., Pont, M. and Jones, N., 2002. Improving the performance of radial basis function classifiers in condition monitoring and fault diagnosis applications where unknown faults may occur. *Pattern Recognition Letters*, 23(5), 569–577.

Lin, J., Keogh, E., Fu, A. and Herle, H., 2005. Approximations to magic: Finding unusual medical time series. In: *Proceedings of the 18th IEEE Symposium on Computer-Based Medical Systems.* IEEE Computer Society, pp. 329–334.

Lloyd, S., 1982. Least squares quantization in PCM. *IEEE Transactions on Information Theory*, 28(2), 129–137.

Logan, B., Healey, J., Philipose, M., Tapia, E. M. and Intille, S., 2007. A long-term evaluation of sensing modalities for activity recognition. In: *Lecture Notes in Computer Science, UbiComp*, vol. 4717, pp. 483–500.

Lu, C., Chen, D. and Kou, Y., 2003. *Algorithms for Spatial Outlier Detection*. pp. 597–600.

Lu, J. et al., 2010. *The Smart Thermostat: Using Occupancy Sensors to Save Energy in Homes*. pp. 211–224.

Ma, J. and Perkins, S., 2003. Online novelty detection on temporal sequences. In: *Proceedings of the 9th ACM SIGKDD International Conference on Knowledge Discovery and Data Mining*. ACM Press, pp. 613–618.

Manevitz, L. and Yousef, M., 2000. Learning from positive data for document classification using neural networks. In: *Proc 2nd Workshop on Knowledge Discovery and Learning*. Jerusalem.

Manevitz, L. and Yousef, M., 2002. One-class SVMS for document classification. *Journal of Machine Learning Research*, 2, 139–154.

Manson, G., 2002. Identifying damage sensitive, environment insensitive features for damage detection. In: *Proceedings of the Third International Conference on Identification in Engineering Systems*. pp. 187–197.

Manson, G., Poerce, S. and Worden, K., 2001. *On the Long-Term Stability of Normal Conditions for Damage Detection in a Composite Panel*. Cardiff, UK.

Manson, G. et al., 2000. *Long-Term Stability of Normal Condition Data for Novelty Detection*. pp. 323–334.

Maurer, U., Smailagic, A., Siewiorek, D. P. and Deisher, M., 2006. *Activity Recognition and Monitoring Using Multiple Sensors on Different Body Positions*. Washington, DC: IEEE Computer Society, pp. 113–116.

Moya, M., Koch, M. and Hostetler, L., 1993. One-class classifier networks for target recognition applications. In: *Proceedings of the World Congress on Neural Networks, International Neural Network Society*. pp. 797–801.

Nairac, A. et al., 1997. *Choosing an Appropriate omdel for Novelty Detection*. pp. 227–232.

Nairac, A. et al., 1999. A system for the analysis of jet engine vibration data. *Integrated Computer Aided Engineering*, 6(1), 53–56.

Ng, V. and Cardie, C., 2003. *Bootstrapping Coreference Classifiers with Multiple Machine Learning Algorithms*. pp. 113–120.

Pan, R., Zhao, J., Zheng, V. W. and Pan, J. J., 2007. *Domain-Constrained Semi-Supervised Mining of Tracking Models in Sensor Networks*. pp. 1023–1027.

Parr, L., Deco, G. and Meisbach, S., 1996. Statistical independence and novelty detection with information preserving nonlinear maps. *Neural Computing*, 8(2), 260–269.

Patterson, D. J., Fox, D., Kautz, H. and Philipose, M., 2005. *Fine-Grained Activity Recognition by Aggregating Abstract Object Usage*. pp. 44–51.

Petsche, T. et al., 1996. *A Neural Network Autoassociator for Induction Motor Failure Prediction*. pp. 924–930.

Philipose, M., Fishkin, K. P., Fox, D., Kautz, H., Patterson, D., and Perkowitz, M. 2003. Guide: Towards understanding daily life via auto-identification and statistical analysis. In: *Proc. of the Int. Workshop on Ubiquitous Computing for Pervasive Healthcare Applications (Ubihealth)*.

Pokrajac, D., Lazarevic, A. and Latecki, L., 2007. Incremental local outlier detection for data streams. In: *Proceedings of the IEEE Symposium on Computational Intelligence and Data Mining*.

Quinlan, R., 1993. *C4.5: Programs for Machine Learning.* San Mateo, CA: Morgan Kaufmann Publishers Inc.

Rabiner, L. R., 1989. *A Tutorial on Hidden Markov Models and Selected Applications in Speech Recognition.* pp. 257–286.

Rousseeuw, P. and Leroy, A., 1987. *Robust Regression and Outlier Detection.* New York: John Wiley and Sons, Inc.

Ruotolo, R., and Surace, C. 1997. A statistical approach to damage detection through vibration monitoring. *Applied mechanics in the Americas*, 314–317.

Salvador, S., Chan, P., and Brodie, J. 2004. Learning states and rules for time series anomaly detection. In: *FLAIRS Conference.* pp. 306–311.

Scarth, G., McIntyre, M., Wowk, B. and Somorjai, R., 1995. Detection of novelty in functional images using fuzzy clustering. In: *Proceedings of the 3rd Meeting of the International Society for Magnetic Resonance in Medicine.*

Shekhar, S., Lu, C. and Zhajng, P., 2001. Detecting graph-based spatial outliers: Algorithms and applications (a summary of results). In: *Proceedings of the 7th ACM SIGKDD International Conference on Knowledge Discovery and Data Mining.* ACM Press, pp. 371–376.

Singh, S. and Markou, M., 2004. An approach to novelty detection applied to the classification of image regions. *IEEE Transactions of Knowledge Data Engineering*, 16(4), 396–407.

Smyth, P., 1994. Markov monitoring with unknown states. *IEEE Journal of Selected Areas of Communication*, 12(9), 1600–1612.

Sohn, H., Worden, K. and Farrar, C., 2001. *Novelty Detection Under Changing Environmental Conditions.* In: *SPIE's 8th Annual International Symposium on Smart Structures and Materials.* International Society for Optics and Photonics. pp. 108–118.

Song, Q., Hu, W. and Xie, W., 2002. Robust support vector machine with bullet hole image classification. *IEEE Transactions of System Man Cybernetics. Part C: Applications and Reviews*, 32(4), 440–448.

Song, X., Wu, M., Jermaine, C. and Ranka, S., 2007. Conditional anomaly detection. *IEEE Transactions of Knowledge data Engineeing*, 19(5), 631–645.

Spence, C., Parra, L. and Sajda, P., 2001. Detection, synthesis and compression in mammographic image analysis with a hierarchical image probability model. In: *Proceedings of the IEEE Workshop on Mathematical Methods in Biomedical Image Analysis.* IEEE Computer Society.

Srinivasan, V., Stankovic, J. and Whitehouse, K., 2008. *Protecting your Daily In-Home Activity Information from a Wireless Snooping Attack.* pp. 202–211.

Srinivasan, V., Stankovic, J. and Whitehouse, K., 2010. *Using Height Sensors for Biometric Identification in Multi-resident Homes.* pp. 337–354.

Srivastava, A., 2006. *Enabling the Discovery of Recurring Anomalies in Aerospace Problem Reports Using High-Dimensional Clustering Techniques.* pp. 17–34.

Srivastava, A. and Zane-Ulman, B., 2005. *Discovering Recurring Anomalies in Text Reports Regarding Complex Space Systems.* pp. 3853–3862.

Streifel, R., Maks, R. and El-Sharkawi, M., 1996. Detection of shorted-turns in the field of turbine generator rotors using novelty detectors–development and field tests. *IEEE Transactions of Energy Conversion*, 11(2), 312–317.

Sun, J., Qu, H., Chakrabarti, D. and Faloutsos, C., 2005. *Neighborhood Formation and Anomaly Detection in Bipartite Graphs.* pp. 418–425.

Sun, P. and Chawla, S., 2004. On local spatial outliers. In: *Data Mining, 2004. ICDM'04. Fourth IEEE International Conference.* IEEE. pp. 209–216.

Surace, C. and Worden, K., 1998. *A Novelty Detection Method to Diagnose Damage in Structures: An Application to an Offshore Platform.* Colorado, pp. 64–70.

Surace, C., Worden, K. and Tomlinson, G., 1997. *A Novelty Detection Approach to Diagnose Damage in a Cracked Beam.* pp. 947–953.

Sutton, C. and McCailum, A., 2006. An introduction to conditional random fields for relational learning. In: *Introduction to Statistical Relational Learning.* Cambridge, MA: MIT Press, pp. 93–129.

Tan, P.-N., Steinbach, M. and Kumar, V., 2005. *Introduction to Data Mining.* Addison-Wesley.

Tapia, E. M., Intille, S. S., and Larson, K. 2004. Activity recognition in the home using simple and ubiquitous sensors. In: Lecture Notes in Computer Science Volume 3001, Heidelberg, Berlin: Springer, pp. 158–175.

Tapia, E., Intille, S., Haskell, W. and Larson, K., 2007. *Real-Time Recognition of Physical Activities and Their Intensities Using Wireless Accelerometers and a Heart Rate Monitor.* Boston, IEEE, pp. 37–40.

Tarassenko, L., 1995. *Novelty Detection for the Identification of Masses in Mammograms.* pp. 442–447.

Torr, P. and Murray, D., 1993. Outlier detection and motion segmentation. In: *Proceedings of the SPIE. Sensor Fusion VI.* pp. 432–443.

Truyen, T. T., Bui, H. H. and Venkatesh, S., 2005. *Human Activity Learning and Segmentation using Partially Hidden Discriminative Models.* pp. 87–95.

Truyen, T. T., Phung, D. Q., Bui, H. H. and Venkatesh, S., 2008. *Hierarchical Semi-Markov Conditional Random Fields for Recursive Sequential Data.* pp. 1–56.

Tsay, R., Pea, D. and Pankratz, A., 2000. Outliers in multi-variate time series. *Biometrika,* 87(4), 789–804.

Van Kasteren, T., Englebienne, G. and Kröse, B., 2010. Activity recognition using semiMarkov models on real world smart home datasets. *Journal of Ambient Intelligence and Smart Environments,* 2(3), 311–325.

Van Kasteren, T. and Kröse, B., 2007. *Bayesian Activity Recognition in Residence for Elders.* pp. 209–212.

Van Kasteren, T., Noulas, A., Englebienne, G. and Kröse, B., 2008. *Accurate Activity Recognition in a Home Setting.* pp. 1–8.

Vilalta, R. and Ma, S., 2002. Predicting rare events in temporal domain. In: *Proceedings of the IEEE.* IEEE Computer Society.

Wang, X., Wang, S., Bi, D. and Ding, L., 2007. Hierarchical wireless multimedia sensor networks for collaborative hybrid semi-supervised classifier learning. *Sensors,* 7(11), 2693–2722.

Weigend, A., Mangeas, M. and Srivastava, A., 1995. Nonlinear gated experts for time-series: Discovering regimes and avoiding overfitting. *International Journal of Neural Systems,* 6(4), 373–399.

Weiss, G. and Hirsh, H., 1998. Learning to predict rare events in event sequences. In: *Proceedings of the 4th International Conference on Knowledge Discovery and Data Mining.* AAAI Press, pp. 359–363.

Whitehead, B. A. and Hoyt, W. A. 1995. Function approximation approach to anomaly detection in propulsion system test data. *Journal of Propulsion and Power,* 11(5), 1074–1076.

Wilson, D. H. and Atkenson, C., 2005. Simultaneous tracking and activity recognition (STAR) using many anonymous, binary sensors. In: *Pervasive Computing, Third International Conference.* Munich, Germany: Pervasive 2005, pp. 62–79.

Wong, W.-K., Moore, A., Cooper, G. and Wagner, M., 2003. Bayesian network anomaly pattern detection for disease outbreaks. In: *Proceedings of the 20th International Conference on Machine Learning*. AAAI Press, pp. 808–815.

Worden, K., 1997. Structural fault detection using a novelty measure. *Journal of Sound Vibrations*, 201(1), 85–101.

Wu, J., Osuntogun, A., Choudhury, T., Philipose, M., and Rehg, J. M. 2007. A scalable approach to activity recognition based on object use. In: *Computer Vision, 2007. ICCV 2007. IEEE 11th International Conference*. IEEE, pp. 1–8.

Xie, D., Yan, T., Ganesan, D. and Hanson, A., 2008. *Design and Implementation of a Dual-Camera Wireless Sensor Network for Object Retrieval*. pp. 469–480.

Yairi, T., Kato, Y. and Hori, K., 2001. Fault detection by mining association rules from housekeeping data. In: *Proc. of International Symposium on Artificial Intelligence, Robotics and Automation in Space*. Vol. 3, No. 9.

Yang, B., Xu, J., Yang, J. and Li, M., 2010. Localization algorithm in wireless sensor networks based on semi-supervised manifold learning and its application. *Cluster Computing*, 13(4), 435–446.

Yi, B.-K. et al., 2000. Online data mining for co-evolving time sequences. In: *Proceedings of the 16th International Conference on Data Engineering*. IEEE Computer Society, San Diego, CA.

Zeevi, A., Meir, R. and Adler, R., 1997. Time series prediction using mixtures of experts. *Advances in Neural Information Processing*, 9, 309–315.

Zhu, X., 2007. *Semi-Supervised Learning Literature Survey*, Madion: University of Wisconsin–Madison [Online] Available at: http://pages.cs.wisc.edu/~jerryzhu/pub/ssl_survey.pdf.

5

Two-Stage Response Surface Approaches to Modeling Drug Interaction

5.1 Classification-Based Techniques

Classification is defined as the act of forming into a class or classes; distribution into groups as classes, orders, families, and so forth, in accordance with some common relations or affinities Thus, a classifier is, "a theme which creates classifications" (Michie et al., 1994, p. 1).

Many popular research areas include automatic categorization of patterns, for example, Machine learning methods for classification and learning from data that includes classified instances known as training sets. They try to develop models that, given a set of attributed values, will predict a class for classification. In supervised learning problem, we are given a sample of input–output data also known as training samples. Here, the task is to find a deterministic function that maps any input to an output. Current classification techniques include; neural networks (NNs), classification trees, variants of naive Bayes, k-nearest neighbors, classification through association rules, logistic regression, function decomposition, and support vector machines. The performance of a classification method is task dependent. The use of a classifier depends on the application and its efforts to solve an exclusive categorization problem (Michie et al., 1994).

Classification is a statistical method used to construct predicative models that distinguish and classify new data points. For example, to distinguish between junk emails and necessary emails, existing emails can be categorized as spam, and real emails thus creating a total N that can be used for training to make email spam free. We refer to the training data as: $X = \{x_1, x_2, ..., x_N\}$.

Now, from our training data, detecting spam emails depends on features selected. Features can be continuous or discrete depending on the context, that is, if m features are measured, it will consist of each email that contains an $m \times 1$ row vector x_i, for $i \in \{1, ..., N\}$ of data, this is denoted by R^m as the feature space. The training data is an $N \times m$ matrix, where entry x_{ij} characterizes the j-th feature of the i-th email. Hence, using training data X, classification will create a decision function, $D(x)$; a function that precedes a new data point and predicts its population (Kim et al., 2010).

The aim of classification is to create a rule whereby a new observation can be classified into an existing class. There are two distinct types of classification, it can refer to a set of observations for the purpose of establishing presence of classes or it can mean clusters in data and division of data into classes. The former is known as unsupervised learning (or clustering), and the latter, supervised learning. In statistics, supervised learning may be referred as discrimination, or creating classification rule from appropriately classified data (Michie et al., 1994).

In statistics, properties of observations are termed explanatory variables (or independent variables, regressors, etc.), and predicted classes are outcome, or likely values of a dependent variable.

The existence of appropriately classified data presupposes that someone (a supervisor in a work situation) is able to classify without error (Michie et al., 1994). If this is so, why is approximation necessary? In fact, there are many reasons for establishing a classification procedure, such as:

1. Mechanical classification procedures may be far faster, for example, postal code reading machines may be able to categorize majority of letters, leaving hard cases to human readers.

2. These procedures are unbiased. For example, a mail order firm may have to make a decision on permitting credit, based on the information provided in the application form whereas human operators may have a biased opinion.

3. They are useful in the medical field, for example, they can review external symptoms to make an exact diagnosis and possibly avoid surgery.

4. They can be used in forecasting, as in meteorology or stock-exchange transactions or investments, and loan decisions.

The probable classifier should be concerned with the following issues:

Accuracy: Some errors are more crucial than others; therefore, it is necessary to control error rate for some classes.

Speed: The speed of a classifier can be a major issue depending on the task at hand.

Comprehensibility: It is important that human operators trust the system. An often-quoted example is the Three-Mile Island nuclear power plant case, where the programmed devices recommended a shutdown, but this recommendation was not processed by the human operators; they did not trust the system. The Chernobyl disaster was similar.

Time to Learn: In a dynamic environment, in particular, it may be essential to learn a classification rule rapidly, or make changes to the existingrule in time. "Rapidly," implies establishing a rule with fewer samples.

At one extreme, the naive 1-nearest neighbor rule, training set is searched for the "nearest" datum whose class is then assumed for new cases. This rule is quick to learn but is very slow in practice. At another extreme, some cases need a faster method or require verification for using another method (Michie et al., 1994).

The nature of classes and their definitions are important concerns. The cases described below are common, but only the first is considered a classification; in practice, datasets combine these types:

1. Classes resemble labels for different populations: membership of various populations is not important. For example, animals and humans constitute separate classes or populations, and it is known with certainty whether a being is an animal or a human. Membership in a class or population may be assigned by an independent authority, such as a supervisor, as the distribution into a class is determined independently of any specific attributes or variables.

2. The class attribute is essentially a result of knowledge prediction, and statistically speaking, class is a random variable. A typical example is prediction of interest rates. Will interest rates rise (class = 1) or not (class = 0)?

3. Classes are predefined by a divider of a sample space, that is, attributes. Since attributes are a class function, an industrial item may be classed as faulty if some attributes are outside programmed limits. A rule classifies data based on the attributes. The trick is to create a rule that impersonates the actual rule. Many credit datasets are of this sort.

There are two main stages of work in classification. The "classical" stage focuses on derivatives of work on linear discrimination while the "modern" stage exploits flexible models, by joint distribution of features of each class to create a classification rule (Michie et al., 1994).

Statistical approaches are categorized as an explicit probability model that affords a probability of presence in each class rather than a classification. It is assumed that the techniques will be used by statisticians, and, hence, some human factor intervention is expected in terms of variable selection and transformation. There are three main strands of research: statistical, machine learning, and NN common objectives. They have all originated procedures that are able to (Michie et al., 1994):

- Be equal or not greater than a human decision-maker's performance, but have benefit of consistency up to adjustable extent
- Manage a wide range of problems, and given sufficient data
- Be broad thinking with established success

In machine learning and statistics, classification is a problem of recognizing a set of categories in a new observation, based on source of a training set of data, comprising observations of membership categories. The discrete observations are examined and placed into a set of quantifiable properties as various explanatory variables, features, and so on. These various properties may be categorical, ordinal, integer-valued, or real-valued. Some algorithms concentrate on discrete data and need real-valued or integer-valued data discretized into groups.

An algorithm that carries out classification in a concrete implementation is known as a classifier. "Classifier" occasionally refers to mathematical function formulated by a classification algorithm that maps the input data into a category.

In machine learning, observations are regularly recognized as instances, explanatory variables are called features, and probable categories to be predicted are classes. Some classification methods do not include a statistical model.

To improve classification performance, first selection and then forward feature-selection methods can be employed. Principal component analysis is used to lessen *dimensionality* of features while taking advantage of classification accuracy (Tato et al., 2002).

The accuracy of a training set is different than that of an accuracy of hidden data (test set). In machine-learning applications, the training set may be fitted with out flaws, but may not perform well on a test set. The important challenge is to accurately classifying hidden data. It is generally assumed that class memberships can overcome this issue. The procedure is, first, a considerable amount (training set) of a given data is used to train a procedure. This rule is then tested on the residual data (test set) and the results are associated with the known classifications. The correct proportion in a test set is an unbiased evaluation of accuracy of the rule, if random sampled data are used for a training set (Michie et al., 1994).

5.1.1 Neural Network–Based Approaches

NNs have varied sources, ranging from imitating a human brain replicating human skills such as speech and language, to practical commercial, scientific, and engineering disciplines of pattern recognition, modeling, and prediction. The hunt for new technology is a strong motivating force in many areas of research.

NNs are made up of layers of interconnected nodes, each node comprising a nonlinear function of its input. The input to a node may originate from other nodes or straight from an input data. Certain nodes are also recognized in output of a network. The complete network signifies a very intricate set of interdependencies, which may integrate any degree of nonlinearity modeled across all functions.

In the simplest networks, output from one node is provided from another node so that "messages" are broadcast by means of layers of interconnecting nodes. More complex actions may be modeled by networks in which last output nodes are connected with earlier nodes; this gives a system features of an extremely nonlinear system.

NNs, to a degree reflects, conduct of neuron networks in a brain. NN methods use the complex statistical techniques of machine-learning to emulate human intelligence at "unconscious" level, and supplementing these with learned concepts apparent to the operator (Michie et al., 1994).

Linear discriminants were introduced by Fisher in 1936, as a statistical procedure for classification. Here, the space of attributes can be divided with a set of *hyperplanes* by a linear arrangement of the variables of an attribute. An analogous model for logical processing was proposed by McCulloch and Pitts in 1943, as bearing resemblance to neurons in a human brain. These researchers demonstrated that the model could be used to build any finite logical expression. The McCulloch–Pitts neuron (see Figure 5.1) contains a weighted sum of its inputs, monitored by a nonlinear function called the em activation function. Officially:

$$y_k = \begin{cases} 1 & \text{if } \sum_j w_{kj}x_j - \mu_k \geq 0 \\ 0 & \text{otherwise} \end{cases}, \tag{5.1}$$

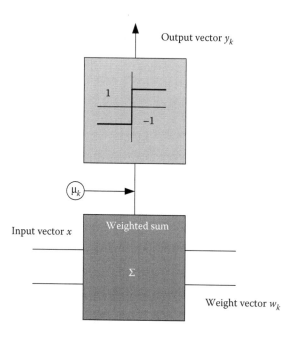

FIGURE 5.1
McCulloch and Pitts neuron. (From McCulloch, W. and Pitts, W., 1943. A logical calculation of the ideas in nervous activity forms. *Bulletin of Methematical Biophysics, 7*, pp. 127–147.)

otherwise, Other neural models are used very broadly; for example, Section 5.1.1.3 discusses the radial basis function in detail. Networks neurons for arbitrary logical expressions, with the capacity to learn by strengthening behavior, were derived later. This is established in Hebb's *The Organisation of Behaviour* (Hebb, 1949). It was recognized that functionality of NNs depended on strength of networks between neurons. Hebb's learning rule advises that if network links an essential method to a specified input, the weights should be adjusted to increase probability of a similar reaction to analogous inputs in future. Equally, if a network reacts uncomfortably to input, weights should be adjusted to lessen probability of a similar reaction (Michie et al., 1994).

There is a difference in pattern recognition between supervised learning and unsupervised learning. In supervised learning, training data measurements of surroundings are conveyed by labels representing class of event characterized by measurements as an anticipated reaction to measurements. Supervised learning networks, including Perceptron and Multilayer Perceptron (MLP), Cascade Correlation learning architecture, and radial basis function networks, are discussed later in the chapter.

Unsupervised learning refers to circumstances when measurements are not conveyed by class labels. Some networks can model structure samples in measurement, and also attribute space either in terms of a probability density function or by demonstrating data in relationships of cluster centers and widths. These include Gaussian mixture models and Kohonen networks.

Once a model is developed, a classifier can be used in two ways. First, is to decide which class of pattern in training data, that is, each node or neuron in model, reacts most powerfully, most often. Unseen data can then be classified, class label of neuron determines patterns. The Kohonen network or mixture model can be applied as first layer of a radial basis function network, with a succeeding layer of weights used to compute a set of class probabilities. Weights in this layer are designed using a linear *one-shot* learning algorithm providing radial basis functions. The first layer of a radial basis function network can be adjusted by selecting a subset of training data points as centers (Michie et al., 1994).

5.1.1.1 Introduction to NNs

An artificial neural network (ANN) is an information processing standard based on biological nervous system. The term *artificial* is used to distinguish these networks from biological neural systems, but it is unstated within computational environment, thus, they can also be considered NNs (Tato et al., 2002).

The most important aspect of an ANN is original construction of information processing. An ANN comprises many highly interrelated processing elements (neurons) employed in harmony to resolve precise problems. An input is accessible to some of its input units; this input vector is spread across whole network. Fundamentally, these inputs are *functions*. Since, input and output can involve various units or mechanisms, they are constructed as vectors.

ANN's real power is its capability to learn. Although, the function is variable, but it can be changed dynamically. ANNs learn by examples and an ANN is arranged for a particular application, for example, in pattern recognition or data classification, through a learning process. Learning in biological systems comprises alterations to synaptic networks that occur among the neurons.

ANNs are arrangement of a multiprocessor computer system, with the following elements:

- Simple processing elements (neurons or nodes)
- A high degree of interconnection (links between nodes)
- Simple scalar communications
- Adaptive interaction among elements

The basic processing element, artificial neuron, or node (Figure 5.2), is based on biological neuron model with numerous inputs and one output. Each input comes through a connection that has an asset (or weight); these weights resemble synaptic effectiveness in a biological neuron. Each neuron has a sole threshold value. Weighted sum of inputs is formed, and threshold subtracted, to comprise activation of neuron. Then, activation signal traverses through an activation function to generate the output of a neuron, as shown in Figure 5.3. The activation function, but it can be altered and even self-programmed for improved enactment in a definite assignment.

An artificial neuron has a training mode and, a use (testing) mode. In a training mode, a neuron can be trained to fire (or not) for specific input configurations. In a use mode, when a trained input pattern is perceived upon input, the

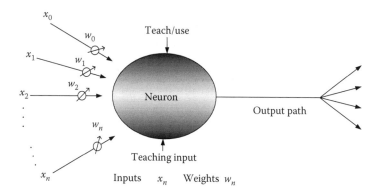

FIGURE 5.2
Artificial neuron model. (From Tato, R. et al., 2002. Classifiers. In: *Emotion Recognition in Speech Signal*. Sony.)

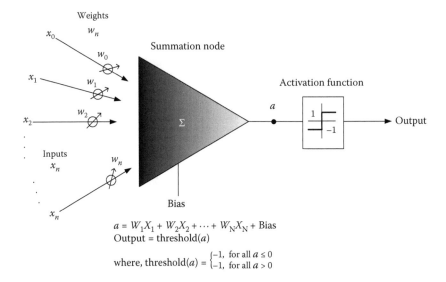

$$a = W_1X_1 + W_2X_2 + \cdots + W_NX_N + \text{Bias}$$
$$\text{Output} = \text{threshold}(a)$$
$$\text{where, threshold}(a) = \begin{cases} -1, & \text{for all } a \leq 0 \\ -1, & \text{for all } a > 0 \end{cases}$$

FIGURE 5.3
Artificial neural neuron activation process. (From Tato, R. et al., 2002. Classifiers. In: *Emotion Recognition in Speech Signal*. Sony.)

corresponding output converts into current output. If the input configuration does not appear on a trained list of input configurations, the firing rule is used to decide whether to fire or not. Depending on their function in the net, we can differentiate three types of units, portrayed in Figure 5.4. The units with activations are problematic input for the net and are called input units; the output units characterize the output of the net output units. The remaining

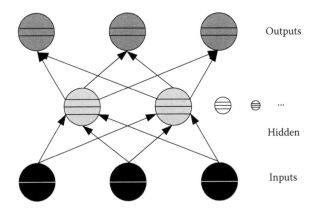

FIGURE 5.4
Types of units within an ANN. (From Tato, R. et al., 2002. Classifiers. In: *Emotion Recognition in Speech Signal*. Sony.)

units are called hidden units, because they are not noticeable from outside. An ANN necessity is to have both input and output units, with no hidden units (single-layer), found in one or numerous layers (multilayer). By relating these basic units and linking them, several network configurations can be created.

An ANN is categorized by its:

- Architecture, or its pattern of networks between the neurons
- Learning algorithm, or its method of defining weights on connections
- Activation function governing its output; common activation functions are step, ramp, sigmoid, and Gaussian.

There are two categories of ANNs, regardless of the number of layers (single-layer or multilayer),

1. *Feed-forward networks* permit signals to traverse one way, only, from input to output. There is no feedback (loops); that is, the output of any layer does not disturb that layer. Feed-forward ANNs lean toward direct networks that relate inputs *with* outputs. They are widely applied in pattern recognition. This type is also known as bottom–up or top–down (Figure 5.5).
2. *Feedback networks* can have signals traversing in both directions by presenting loops in the network. Feedback networks can become very intricate. They are dynamic; their "state" varies, uninterrupted, until they touch an equilibrium point. They continue to be at this point until input fluctuates and a new equilibrium requirement is initiated. Feedback architectures are also mentioned as interactive or recurrent to denote single-layer feedback connections.

5.1.2 Supervised Networks for Classification

Nominally supervised learning has data, i, encompassing an attribute vector X_i and a target vector Y_i. It processes X_i, with a network to generate an output y_i, with the same form as the target vector Y_i.

The parameters of the network w are changed to improve connection between outputs and targets by diminishing the sum-squared error

$$E = \frac{1}{2}\sum_i (y_i - Y_i)^2, \tag{5.2}$$

It might appear as a percentage misclassification error measure in classification problems, but the total squared error has accommodating smoothness and differentiability properties. The total squared error has been used in the *StatLog* trials for training, whereas percentage misclassification in the trained networks has been used for assessment (Michie et al., 1994).

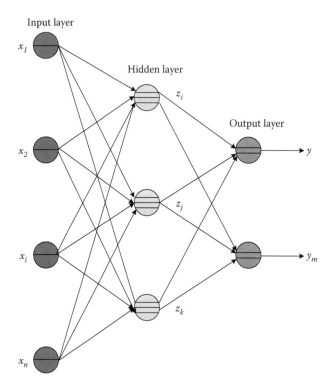

FIGURE 5.5
MLP employing feed forward fully connected topology. (From Tato, R. et al., 2002. Classifiers.
In: *Emotion Recognition in Speech Signal.* Sony.)

5.1.2.1 Perceptrons and Multilayer Perceptrons

The activation of the McCulloch–Pitts neuron takes the form:

$$y_j = f_j \left(\sum_i w_{ji} * X_i \right), \tag{5.3}$$

where the activation function, f_j can be any nonlinear function. The nodes
have been separated into an input layer I and an output layer O. The threshold level or bias of Equation 5.1 is involved in the sum, with an assumption of
an additional constituent in the vector X whose value is fixed at 1.

   Rosenblatt studied competences of clusters of neurons in a single layer, a
structure called perceptron. Rosenblatt proposed the Perceptron Learning
Rule for learning appropriate weights for classification complications. When
f is a difficult threshold function, Equation 5.3 defines a nonlinear function, *transversely* a hyperplane in attribute space; with a threshold activation
function, on one side of the hyperplane the neuron output is 1 and on the

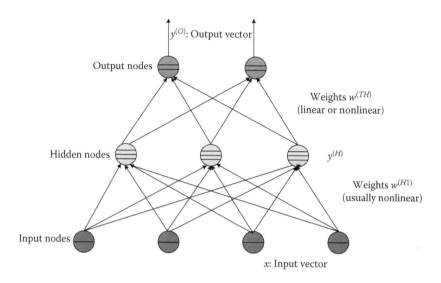

FIGURE 5.6
MLP structure. (From Michie, D., Spiegelhalter, D. and Taylor, C., 1994. *Machine Learning, Neural and Statistical Classification*. [Online] Available at: http://www1.maths.leeds.ac.uk/~charles/statlog/whole.pdf.)

other, it is 0. When united in a perceptron structure, neurons can divide, the attribute space into areas. This practice is the foundation of aptitude of perceptron networks to accomplish classification.

Citing the problem as the simplest specimen Minsky and Papert stated that numerous physical domain problems do not take this simple framework. Here, it is essential to separate two convex regions, composing them into a single class. They also showed this was not probable with a perceptron network but could be done with a two-layer perceptron structure. Even though the Perceptron Learning Rule (also called the Delta Rule) cannot be used universally to calculate weights for this structure, yet, resulting MLP is extensively used today.

A rule which permits MLP for learning was proposed in 1985. The generalized Delta rule (Section 5.1.4.2) defines a notion of back propagation of error derivatives, from side to side of a network, and qualifies a large class of models with dissimilar joining structures for training. This research triggered academic curiosity about NNs; later, it became of considerable interest to industry as well (Figure 5.6).

5.1.2.2 MLP Structure and Functionality

Figure 5.7 shows the arrangement of a standard two-layer perceptron: The inputs form the input nodes of the network, and the outputs are the output nodes. The middle layer of nodes is labeled as the hidden layer; it cannot be

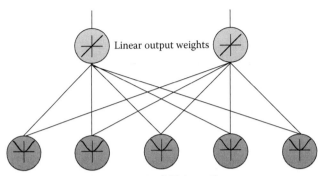

Nonlinear receptive fields in attribute space

FIGURE 5.7
Radial basis function network. (From Michie, D., Spiegelhalter, D. and Taylor, C., 1994. *Machine learning, neural and statistical classification*. [Online] Available at: http://www1.maths.leeds.ac. uk/~charles/statlog/whole.pdf.)

seen by either the inputs or outputs and its size is variable. A hidden layer is usually used to make a blockage, requiring the network to make a simple model of a system producing the data with facility to oversimplify earlier unseen configurations.

The process is quantified by

$$y_i^{(H)} = f^{(H)} \left(\sum_j w_{ij}^{(HI)} * x_j \right), \tag{5.4}$$

$$y_i = f^{(T)} \left(\sum_j w_{ij}^{(TH)} * y_j^{(H)} \right), \tag{5.5}$$

This stipulates how the input pattern vector x is charted into the output pattern vector $y^{(0)}$, via the hidden pattern vector $y^{(H)}$, in a fashion parameterized by the two layers of weights and $w^{(HI)}$ and $w^{(TH)}$. The univariate functions $f^{()}$ is set to,

$$f_{(x)} = \frac{1}{1 + R^{-x}}, \tag{5.6}$$

which differs smoothly from 0 at $-\infty$ to 1 at ∞, as a threshold function would do. If the number of hidden layer nodes is less than the amount of degrees of freedom characteristic in a training data, the activations of hidden nodes are inclined to arrange an orthogonal set of variables, either linear or nonlinear arrangements of the attribute variables, of which the duration is, as big a subspace of the problem, as conceivable. With a slight additional restraint on a network, these internal variables form a linear or nonlinear principal constituent depiction of an attribute space. If statistics have noise, it is not an intrinsic part of the producing structure. The main constituent network changes as a strainer of a lower-variance noise signal on a condition that signals to noise ratio, of the data, is adequately high. This property gives MLPs capacity to oversimplify formerly unnoticed patterns, by modeling only significant fundamental structure of the creating system. The hidden nodes can be viewed as sensors of intangible features of the attribute space (Michie et al., 1994).

5.1.2.2.1 Universal Approximators and Universal Computers

In general, inMLPs of the two-layer types indicated in Equations 5.4 and 5.5, the output-layer node values y are functions of the input-layer node values X (and the weights w). The two-layer MLP can estimate a random variable as if there is no boundary to hidden nodes. In this context, the MLP is an approximate common function. This proposition does not suggest that added intricate MLP architectures are worthless; it can be more efficient to use different numbers of layers for dissimilar problems. There is a lack of rigorous ideologies based on a select architecture, but various experimental principles have been developed. These include symmetry principles and constructive algorithms (Michie et al., 1994).

MLP is a feedforward network of the output vector y and a function of the input vector X and particular parameters w, therefore:

$$y = F(x; w), \tag{5.7}$$

for some vector function F given in detail by Equations 5.11 and 5.12, in the two-layer case. Similarly it is possible to describe a recurrent network by feeding outputs back to inputs. The common arrangement of a recurrent perceptron is:

$$y_i(t + 1) = f\left(\sum_j wi_j * y_j(t)\right), \tag{5.8}$$

This can be transcribed as,

$$y(t + 1) = F(y(t); w), \tag{5.9}$$

This is a discrete-time model. Continuous-time models administered by a differential equation of analogous structure can also be considered.

Recurrent networks use logic that assumes an unbound quantity of nodes, they can imitate any design that can be finished on a Universal Turing machine. This outcome is easily demonstrated for hard-threshold recurrent perceptrons in one-node network, that performs not–AND. Or one that functions as a FLIP–FLOP. These elements are all that is needed to construct a computer (Michie et al., 1994).

5.1.2.2.2 Training MLPs by Nonlinear Regression

In NN terminology, training is an application of appropriate network bounds (its weights) to given data. The training data comprises a set of examples of equivalent inputs and anticipated outputs, or "targets." In this case, let the i-th specimen be given by input X_{ji} for input element j, and target Y_{ji} for target element j. A least-squares fit is attained by result of the parameters which lessen degree of error (Michie et al., 1994).

Thus,

$$E = \frac{1}{2} \sum_i \sum_j (y_{ji} - Y_{ji})^2, \qquad (5.10)$$

where y_{ji} is the output value found by replacing the inputs X_{ji} for x_j in Equations 5.4 and 5.5; if the fitting is impeccable, $E = 0$, otherwise $E > 0$.

5.1.2.2.3 Probabilistic Interpretation of MLP Outputs

If there is one-to-many relationship among inputs and targets in a training data, no plotting of the formula 5.7 can be achieved effortlessly. If a probability density $P(Y|X)$ designates a data; Equation 5.10 is accomplished by mapping X as an average target, such that,

$$\int dY P(Y|X) Y, \qquad (5.11)$$

Classification problems are denoted using *1-out-of-n* output coding. One output node is assigned for each class, and the target vector Y_i. For example, i is all 0s except for a 1 on the node demonstrating the accurate class. In this case, the value calculated by the j-th target node can be understood as a probability that the input pattern is appropriate to class j and, cooperatively, the outputs prompt $P(Y|X)$. This affords helpful information and also provides a principle with which NN models can be pooled with supplementary probabilistic models.

The probabilistic explanation of output nodes indicate a common error measure for classification problems. If the value y_{ji} output by the j-th target node specified the ith training input X_i, is $P(Y_{ji} = 1)$, and $1 - y_{ji}$ is $P(Y_{ji} = 0)$, then the probability of a complete assortment of training outputs Y is,

$$P(Y) = \prod_{ji} y_{ji}^{Y_{ji}} (1 - y_{ji})^{1 - Y_{ji}}, \qquad (5.12)$$

In general, this is the exponential of a cross-entropy,

$$E = \sum_i \sum_j (Y_{ji}\log y_{ji} + (1 - Y_{ji})\log(1 - y_{ji})) \qquad (5.13)$$

So, the cross-entropy can be applied as an error measure for an alternative of a sum of squares 5.10.

The probabilistic explanation of MLP outputs in classification problems should be prepared with extreme care. It only spreads over a network if trained to its minimum error, and then only if the training data precisely characterizes original probability density $P(Y|X)$. The second condition is challenging, if X belongs to a continuous space or a large discrete set, because, theoretically, a huge or unbounded volume of data is essential. This problem is correlated to overtraining and generalizing concerns.

Aimed at theoretical details, cross-entropy is another fitting error measure used in classification problems, most recommend using it with slight modifications. The sum of squares was mostly used in the StatLog NN trials (Michie et al., 1994).

5.1.2.2.4 Minimization Methods

NN models are trained by altering their weight matrix factors, w, to decrease an error measure, such as, in Equation 5.10. In ordinary circumstances, a network outputs are linear in weights, creating quadratic Equation 5.10. Then, the minimal error can be calculated by solving a linear system of equations. This special case is further discussed in Section 5.1.2.3, within the framework of radial basis function networks. In a standard, nonlinear case, minimization is done using a particular type of gradient descent, thus generating a local minimum, w from which any miniscule variation creates E, but not inevitably the global minimum of $E(w)$ (Michie et al., 1994).

5.1.2.2.5 First-Order Gradient-Based Methods

The gradient $\nabla E(w)$ of $E(w)$ is the vector field of derivatives of E such that:

$$\nabla E(w) = \left(\frac{dE(w)}{dw_1}, \frac{dE(w)}{dw_2}, \ldots \right)$$

A linear approximation of $E(w)$ g in the miniscule neighborhood of a random weight matrix w^0 is specified by,

$$E(w) = E(w^0) + \nabla E(w^0) * (w - w^0), \qquad (5.14)$$

At any point w of a constraint space (weight space) of a network, vector ∇E points in course of the strongest surge of E; that is of all the minuscule variations δw (of a given magnitude) which we could make to, w, a change in the course of ∇E surges, E the greatest. Therefore, a modification of w in the course of $-\nabla E$ is responsible for the maximum potential reduction in E. A straightforward approach in gradient descent is to calculate the gradient and alter the weights in the opposite direction.

A concern is that the theorem on maximal descent relates only to minuscule modifications. The gradient deviates as well as the error; therefore, the best direction for (minuscule) descent deviations is when w is attuned. The pure gradient descent algorithm needs a small step size parameter η selected for a $\eta\nabla E$ to be minuscule as attaining descent is difficult, otherwise as large as possible in terms of speed. The weights are continually modified by,

$$w \leftarrow w - \eta\nabla E(w), \tag{5.15a}$$

until the error E fails to descend.

The trial and error is an aspect used for major step, size η, that works. With large step sizes, a gradient has a tendency to vary with each step. A common experimental approach is using an average of the gradient vector to determine a systematic inclination. This is possible by adding a momentum term to Equation 5.15b, connecting a parameter $\alpha \le 1$, so that:

$$w \leftarrow w - \eta\nabla E(w) + \alpha\delta\, w_{\text{old}}, \tag{5.15b}$$

where $\alpha\delta\, w_{\text{old}}$ denotes the furthermost current weight modification.

These methods seem straightforward, but their presentation is contingent on the parameters η and α. Various principles give an impression of being suitable for dissimilar complications, however, not for the same stages of training in a particular problem. This situation is relevant to an overabundance of heuristics for adaptive, adjustable, and step size algorithms (Michie et al., 1994).

5.1.2.2.6 Second-Order Methods

Basically, in first-order gradient-centered methods, the linear approximation Equation 5.14 disregards the curvature of $E(w)$. This can be equalized by applying Equation 5.14 to the quadratic approximation,

$$E(w) = E(w^0) + \nabla E(w^0) * \delta w + \delta w \nabla\nabla E(w^0)\delta w$$

where $\nabla\nabla E$ is the matrix with constituents $(d^2E/dw_i dw_j)$, called the inverse Hessian (or the Hessian, contingent on pacts), and $\delta w = w - w^0$. The change $\delta w = (1/2)H\nabla E$, where $H^{-1} = \nabla\nabla E$, carries w to a fixed point of this quadratic form. This may be a minimum, maximum, or saddle point. If it is minimum, then a step in that direction appears to be a good idea, if not, then a positive or negative step in a conjugate gradient's direction, $H\nabla E$, is preferable.

Hence, a class of algorithms has been established connecting a conjugate gradient.

A record number of these algorithms need explicit calculation or assessment of the Hessian H. Number of components of H are roughly half the square of number of constituents of w in huge grids with numerous weights; such algorithms are directed toward unreasonable computer memory requests. But, one algorithm usually called, the conjugate gradient algorithm or the *memoryless* conjugate gradient algorithm. This algorithm gives an approximation of conjugate direction without signifying H directly.

In general, a conjugate gradient algorithm uses an array of *linesearches*, one-dimensional searches for a minimum of $E(w)$, beginning with the furthermost current estimate of minimum, and ending at minimum in direction of the current estimate of the conjugate gradient. Linesearch algorithms are relatively easy since the subject of direction selection condenses to a binary choice. The linesearch in an inner loop of a conjugate gradient algorithm is very efficient. Therefore, a linesearch is characteristically used to find complex components of conjugate gradient applications. Numerical round-off problems are additional uses of linesearch applications; as the conjugate gradient is approximately orthogonal to a gradient, creating a variation of $E(w)$ laterally, with an exclusive minor conjugate gradient.

The updated rule for a conjugate gradient direction is

$$s \leftarrow -\nabla E(w) + \alpha s_{\text{old}}, \tag{5.16}$$

where

$$\alpha = \frac{(\nabla E - \nabla E_{\text{old}}) * \nabla E}{\nabla E_{\text{old}} * \nabla E_{\text{old}}}, \tag{5.17}$$

Note: This is a Polak–Ribiere variant; there are others.

Rather complicated evidence suggests, if E is quadratic in w, s is adjusted to the gradient, and the linesearches are achieved precisely, then s meets on the conjugate gradient and E meets on its minimum, after as many repetitions of Equation 5.16 as there are constituents of w. Much broader functions can also be attained using linesearches. It is essential to expand Equation 5.17 with a rule to reset s to $-\nabla E$, whenever s develops too orthogonally to a gradient for growth to continue.

The conjugate gradient algorithm has numerous parameters that regulate the particulars of a linesearch, and others which describe precisely when to reset s to $-\nabla E$. But, with dissimilar step sizes and momentum parameters of weaker methods, the enactments of the conjugate gradient method are more or less oblivious to its parameters; if they are set inside rational arrays. All algorithms are susceptible to methods of choosing early weights (Michie et al., 1994).

5.1.2.2.7 *Gradient Calculations in MLPs*

It remains debatable whether, in the case of an MLP, NN model calculation of the gradient $\nabla E(w)$ will generate an error quantity such as shown in Equation 5.10. The calculation is prearranged as a back transmission of error. For a network with a single layer of hidden nodes, this calculation transmits node output values y, that advance from an input to output layers for each training example, and then transmitting magnitudes δ connected to the output errors at the rear, with a linearized form of a grid. Products of δs and ys, then, contribute to the gradient. In the case of a network with an input layer (I), a single hidden layer (H), and an output or target layer (T), the calculation is:

$$y_i^{(H)} = f^{(H)}\left(\sum_j w_{ij}^{(HI)} * x_j\right)$$

$$y_i = f^{(T)}\left(\sum_j w_{ij}^{(TH)} * y_j^{(H)}\right)$$

$$\delta_{ji}^{(T)} = \left(y_{ji} - Y_{ji}\right)$$

$$\delta_{ji}^{(H)} = \sum_k \delta_{ki}^{(T)} f_{ki}'^{(T)} w_{kj}^{(TH)}$$

$$dE/dw_{jk}^{(TH)} = \sum_i \delta_{ji}^{(T)} f_{ji}'^{(T)} y_{ki}^{(H)},$$

$$(5.18)$$

$$dE/dw_{jk}^{(HI)} = \sum_i \delta_{ji}^{(H)} f_{ji}'^{(H)} X_{ki} \qquad (5.19)$$

The index i is summed over training examples, while js and ks refer to nodes:

$$f_{ki}'^{(\cdot)} = \frac{d}{dx} f(x)\Big|_{x=f^{-1}\left(y_{ki}^{(\cdot)}\right)}$$

5.1.2.2.8 *Online versus Batch*

The error E in Equation 5.10 and the gradient ∇E in Equations 5.18 and 5.19 comprise a summation of examples, projected by arbitrarily choosing a subsection of examples to be added together. In extreme, use of a sole example influence for every gradient estimate, is called a stochastic gradient method. If an analogous policy is used devoid of random choice, but with data engaged in the directive it originates, the technique is an online one. If a sum above all training data is executed for a gradient calculation, the technique is of a batch variety.

Several online and stochastic gradient methods have benefit of speed, if an estimate is functional. These are preferable for problems with huge volumes of training data. However, such approximations cannot be used in the conjugate gradient method, because it is constructed on procedures and theorems; whereby E is a given function of w, which can be assessed accurately, allowing for judgments at neighboring points. Thus, the stochastic gradient and online methods are more likely to be used with simple step-size and momentum methods. Some research has looked at finding a negotiation method (Møller, 1993).

5.1.2.3 Radial Basis Function Networks

The radial basis function network consists of a layer of components without linear or nonlinear functions of the attributes, trailed by a layer of weighted networks to nodes whose outputs have the same arrangement as target vectors. It has an arrangement much like an MLP with one hidden layer, except each node of a hidden layer calculates a random function of inputs, and transfer function of each output node is the insignificant identity function. In its place of "synaptic strengths" hidden layer has parameters suitable for any functions used, for example, Gaussian widths and positions. In certain circumstances, this network has a number of advantages over an MLP, even if the two models are same, computationally.

These advantages include a linear training rule which evades problems related to local minima; for example, it is able to determine better, the accuracy of the probabilistic elucidation of the outputs discussed in Section 5.1.2.2.

Figure 5.8 indicates the construction of a radial basis function. The nonlinearities include a location in attribute space with the function situated in the center, and a nonlinear function of the distance of an input point from that center by any function at all. Common selections comprise a Gaussian response function, $\exp(-x^2)$ and inverse multiquadrics ($[z^2 + c^2])^{-(1/2)}$, and nonlocal functions, such as, thin plate splines ($z^2 \log z$) and multiquadrics ($[z^2 + c^2])^{(1/2)}$. Though, it appears counterintuitive to attempt and produce an interpolating function using nonlocalized functions, but, these are known to have well-interpolating properties in region occupied by a training data.

The radial basis function network method, includes increase or preprocessing of input vectors into a high-dimensional space. This works in achieving a theorem (Cover, 1965) that infers that, a classification problem in a high-dimensional space is more likely to be linearly independent than one in a low-dimensional space (Michie et al., 1994).

5.1.2.3.1 *Training: Choosing Centers and Nonlinearities*

Many approaches can be used to select centers for a radial basis function network. It is imperative that dissemination of centers in attribute space be

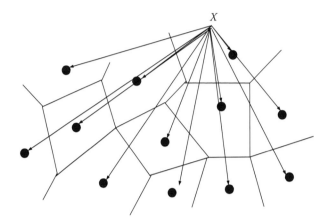

FIGURE 5.8
k-Means clustering: The center is moved to the mean position of the patterns within each patch. (From Michie, D., Spiegelhalter, D. and Taylor, C., 1994. *Machine Learning, Neural and Statistical Classification*. [Online] Available at: http://www1.maths.leeds.ac.uk/~charles/statlog/whole. pdf.)

alike and in same region as training data. The training data is assumed to be illustrative of the problem.

A primary technique for selecting centers is to consider points on a square grid that cover region of an attribute space, enclosed by training data. There might be predictable in improved performance, if centers were sampled at random from training data, because extra densely populated regions of an attribute space probably have a greater resolution model than sparser regions. In this situation, it is important that at least one sample from each class is used as a pattern center. When center locations are elected for radial basis function networks with localized nonlinear functions for Gaussian receptive fields, it is necessary to compute appropriate variances for those functions. This ensures large regions of space do not occur among centers; where no centers react to patterns, and conversely, no two centers react almost identically to patterns. This problem mainly occurs in large dimensional attribute spaces.

Several other methods use a "principled" clustering technique to spot centers, such as a Gaussian mixture model or a Kohonen network. These models are considered as unsupervised learning (Michie et al., 1994).

5.1.2.3.2 Training: Optimizing Weights

As explained in Section 5.1.2.2, radial basis function networks are trained by resolving a linear system. There is a similar problem in normal linear regression; the only difference being that input to a linear system is an output of a hidden layer of a network, not the attribute variables. To illustrate: let $y_{ki}^{(H)}$ be the output of the k-th radial basis function in the i-th example. The output of each target node j is calculated using the weights w_{jk} as:

$$y_{ji} = \sum_k w_{jk} * y_{ki}^{(H)},$$

(5.20)

Let the anticipated output, for example, i on target node j be Y_{ji}. The error measure (Equation 5.10) occupied is written out as:

$$E(w) = \frac{1}{2} \sum_{ji} \left(\sum_k w_{jk} * y_{ki}^{(H)} - Y_{ji} \right)^2,$$

(5.21)

which has its minimum where the derivative,

$$\frac{dE}{dw_{rs}} = \sum_k \sum_i w_{rk} * y_{ki}^{(H)} * y_{ji}^{(H)} - \sum_i Y_{ri} * y_{si}^{(H)},$$

(5.22)

vanishes. Let R be the correlation matrix of the radial basis function outputs,

$$R_{jk} = \sum_i y_{ki}^{(H)} * y_{ji}^{(H)},$$

(5.23)

The weight matrix w^*, which minimizes E, lies where the gradient vanishes so that:

$$w^*_{jk} = \sum_r \sum_i Y_{ji} * y_{ri}^{(H)} * (R^{-1})_{rk},$$

(5.24)

Thus, the problem is solved by inverting the square $H \times H$ matrix R, where H is the number of radial basis functions.

The matrix inversion can be calculated by standard methods, namely, LU decomposition, if R is neither singular nor nearly so. If two radial basis function centers are adjacent to each other, it will result in a singular matrix, and this is a certainty, if number of training samples is, not as great as H. There is no applied method to confirm a nonsingular correlation matrix. Hence, the best choice is, an additional, computationally expensive, singular value decomposition method. Such methods deliver an approximate inverse by diagonalizing a matrix, and inverting only eigenvalues that exceed zero by a parameter-specified margin, and converting back to the initial coordinates. This creates an optimal minimum-norm approximation to an inverse in least-mean-squares.

Let n be the number of training examples, and let the $H \times H$ linear system be given by the derivatives of E Equation 5.22; these are rooted in the error formula Equation 5.21 as shown by

$$\sum_k w_{jk} * y_{ki}^{(H)} = Y_{ji},\qquad(5.25)$$

Unless $n = H$, this is a rectangular system. Broadly speaking, a precise solution is not required, but an optimal solution in a least-squares sense is provided by pseudo-inverse $y^{(H)+}$ of $y^{(H)}$, the matrix with elements $y_{ji}^{(H)}$:

$$w^* = Yy^{(H)+},\qquad(5.26)$$

The identity $Y^+ = \tilde{Y}(Y\tilde{Y})^+$, where $\sim$ denotes a transposed matrix, can be applied to Equation 5.26 to show that pseudo-inverse method gives the same result as Equation 5.24:

$$w^* = Y\tilde{y}^{(H)}(y^{(H)} * \tilde{y}^{(H)})^+,\qquad(5.27)$$

The ability to invert or pseudo-invert a matrix, reliant on a whole data-set, denotes this as a batch method. An online variant is, known as Kalman Filtering (Scalero and Tepedelenlioglu, 1992). A countenance occurs for the inverse correlation R^{-1} is, if another example is supplementary to the sum (Equation 5.23) and does not require inverse to be computed.

5.1.2.4 Improving the Generalization of Feed-Forward Networks

5.1.2.4.1 Constructive Algorithms and Pruning

Recently, efforts have been made to improve the perceptron and MLP training algorithms by altering architecture of networks while training continues. These techniques include pruning useless nodes or weights, and constructing algorithms with extra nodes as necessary. The benefits include smaller networks, faster training times on serial computers, increased simplification capability, andresulting resistance to noise. In addition, it is much easier to understand what a trained network is undertaking. As mentioned previously, a minimalist network uses its hidden layer to model as much of a problem as possible in partial number of degrees of freedom existing in its hidden layer. We can compare this with other pattern classifying techniques, such as, decision trees and expert systems.

For a network to have good simplification capability, it must be able to calculate a suitable number of hidden nodes. If nodes are too rare, the network may not learn at all. While excessively hidden nodes lead to over learning of individual samples in the beginning, that results in a close optimal model of data distributions comprising training data.

Later methods eliminate some problems, and are appropriate for statistical classification problems. They often build a single hidden layer and include stopping conditions that permit them to find solutions for statistical problems. Cascade correlation is an example of such a network algorithm.

Pruning has been carried out on networks in three ways. The first is a heuristic approach grounded on recognition of nodes or weights that contribute to mapping. After these have been determined, additional training results in an improved network. An alternative technique is to contain terms in an error function, so that, weights tend to zero under certain circumstances. Zero weights can then be detached without degrading a network performance. Finally, if sensitivity of a global network error to removal of a weight or node is defined, it is possible to detach those weights or nodes to which global error is least sensitive. The sensitivity measure is not restricted to training, and includes only a minor amount of spare computational effort (Michie et al., 1994).

5.1.2.4.2 *Cascade Correlation: A Constructive Feed-Forward Network*

Cascade correlation is a way to construct a feed-forward network as training continues in a supervised mode (Fahlman and Lebiere, 1990). Instead of fine-tuning weights in a fixed architecture, it uses a small network, and supplements new hidden nodes one by one, thus creating a multilayer structure. Once a hidden node has been added to a network; its input-side weights are stationary and it develops a perpetual feature-detector in a network, obtainable for output or for making other, more complex feature detectors in later layers. Cascade correlation can compromise condensed training time, and it automatically concludes size and topology of networks.

Cascade correlation combines two ideas: first, cascade architecture, where hidden nodes are added one at a time by using outputs of all other input nodes, and the second, maximizing correlation between a new unit's output and residual classification error of parent network.

Nodes added into net may be of any kind. Examples include, linear nodes that might be trained using linear algorithms, threshold nodes such as single perceptrons; possibly using simple learning rules such as the Delta rule or the Pocket Algorithm, or nonlinear nodes such as sigmoid or Gaussian functions, requiring Delta rules or new progressive algorithms such as Fahlman's Quickprop (Fahlman, 1988). Standard MLP sigmoids were used in the StatLog trials.

At every stage of training, each node from a group of candidate nodes is trained on residual error of a main network. Those nodes that have output of the highest correlation with error of parent nodes are chosen. The error function minimized in this scheme is S—summation of output units of the degree of correlation (or the covariance) between V, a candidate unit's value, and $E_{i,o}$ residual error detected at output unit o, for example, i. S is defined by:

$$S = \sum_o \left| \sum_i (V_i - \bar{V})(E_{i,o} - \bar{E}_o) \right| \qquad (5.28)$$

where the quantities $\bar{V}$ and $\bar{E}_o$ are the values of V and E_o averaged over all patterns. In order to maximize S, partial derivative of the error is considered with respect to each of the weights coming into the node, w_j. Thus:

$$\frac{dS}{dw_j} = \sum_{i,o} \sigma_o (E_{i,o} - \bar{E}_o) * f_i' * I_{j,i}, \qquad (5.29)$$

where, σ_o is the sign of a correlation between a candidate's value and the 'output O, f_i' is the derivative for pattern i of the candidate unit's activation function, with respect to the sum of its inputs, and $I_{j,i}$ is the input the candidate unit obtains for pattern i.

The partial derivatives are used to achieve gradient ascent that will maximize S. When S no longer recovers in training for any of candidate nodes, the best contender is added to a network, and others are abandoned (Michie et al., 1994).

In a sample problem, connecting classification of data points and creating two interlocked spirals, cascade correlation is described as 10–100 times faster than conservative back-propagation of error derivatives in a fixed architecture network. Empirical tests on a range of actual problems have found that it speeds up one to two orders of scale with minimal degradation of classification accuracy. These outcomes were only attained after numerous experiments in order to define suitable values for various parameters which essentially are set in a cascade correlation application. Cascade correlation can also be applied in computers with partial precision and in recurrent networks (Hoehfeld and Fahlman, 1992).

5.1.3 Unsupervised Learning

Recently, increased attention has been paid to unsupervised learning. It has the advantage of being able to determine the structure of a data without requiring class information and can disclose unpredictable features or those unknown beforehand. This can include segregation of data previously assumed to be a single uniform cluster into a number of lesser clusters; each with distinct recognizable properties. The clusters create a model of a data in terms of cluster; centers, sizes, and shapes. These can often be defined using less information, and with fewer parameters than previously required for training datasets. This has clear benefits for storing, coding, and transmitting stochastically generated data. If their distribution in attribute space is identified, corresponding data can be extracted from a model as required.

Unsupervised learning methods, such as Boltzmann machines, are computationally very expensive. However, Iterative clustering algorithms such as Kohonen networks, k-means clustering, and Gaussian mixture models compromise similar modeling with abridged training time. As class labels are not used in models, freedom from this constraint, along with cautious initialization of models using any previous information accessible about a data, can yield very rapid and effective models. These models, known as vector quantizers, can be used as nonlinear part of supervised learning models. In this case, a linear part is added and trained far enough to link the mapping derived from activating various pieces of a model to likely classes of events engendering a data (Michie et al., 1994).

5.1.3.1 k-Means Clustering Algorithm

The principle of clustering involves a depiction of a set of data to be initiated in a model of distribution of samples in an attribute space. The k-means algorithm (Krishnaiah and Kanal, 1982), as a model with a fixed number of cluster centers attains this swiftly and efficiently flagged by the user in advance. The cluster centers are selected from a data; each center forms code vector for that section (patch) of input space, where whole points are nearer to center than any other place. This division of space into patches is known as a Voronoi tessellation. Since the preliminary distribution of centers may not yield a decent model of the probability distribution function (PDF) of an input space, a series of repetitions, where each cluster center is stimulated to mean location of all training patterns in its tessellation region is required (Michie et al., 1994).

The generalized variant of the k-means algorithm is the Gaussian mixture model, or Adaptive k-means clustering. In this arrangement, Voronoi tessellations substitute changes from one center's field to another by allocating a variance to each center, thereby outlining a Gaussian kernel at each center. These kernels are brought together by a set of mixing weights to approximate PDF of an input data; an algorithm iteratively calculates a set of mixing weights and variances for centers. While the number of centers for these algorithms is fixed in advance, in more general implementations, certain methods seem to permit new centers to be auxiliary as training proceeds (Wynne-Jones, 1993).

5.1.3.2 Kohonen Networks and Learning Vector Quantizers

Kohonen's network algorithm (Kohonen, 1984) also results in a Voronoi tessellation of input space into patches with conforming code vectors. It has an extra feature whereby centers are organized in a low dimensional structure with nearby points, in a topological structure (i.e., map of points are close in an attribute space). It is assumed that structures of this type occur in nature, for example, in retinotopic maps and in mapping an ear to an auditory cortex, from retina to visual cortex or optic tectum.

In training, winning node of a network is the nearest node in input space to an agreed-upon training pattern. This indicates a good distribution of a network topology in a nonlinear sub-space of a training data.

Vector quantizers that preserve topographic relations between centers are beneficial in communications, especially when noise added to coded vectors cause some corruption. The topographic mapping confirms that a minor alteration in code vector is decoded as a minor alteration in attribute space and, thus, a minor alteration at output. Though, it is basically an unsupervised learning algorithm, learning vector quantizer can be used as a supervised vector quantizer, where network nodes have class labels. The Kohonen Learning Rule is used when winning node signifies the same class as a fresh training pattern. Change in class between winning node and a training pattern causes a node to move away from a training pattern by the same distance. Learning vector quantizers perform extremely well in readings of statistical and speech data (Kohonen et al., 1988).

5.1.3.3 RAMnets

The use of neurally inspired classification algorithms is still common. The n-tuple recognition method devised by Bledsoe and Browning (1959) and Bledsoe (1961) laid the foundation for the subsequent development of Wisard. The patterns to be classified are bit strings of a given length. Several (say N) sets of n bit locations are nominated arbitrarily. These are the n-tuples. The constraint of a pattern to an n-tuple can be observed as an n-bit number which establishes a "feature" of a pattern. A pattern is classified as fitting the class if it has features in common with at least one pattern in a training data.

To be precise, class assigned to unclassified pattern u is

$$\arg\max_{c} \left(\sum_{i=1}^{N} \Theta \left(\sum_{v \in C_c} \delta_{\alpha i(u)\alpha i(v)} \right) \right), \tag{5.30}$$

where C_c is the set of training patterns in class c, $\Theta(x) = 0$ for $\Theta \le 0$, $\Theta(x) = 1$ for $\Theta > 0$, $\delta_{i,j}$ is Kronecker delta ($\delta_{i,j} = 1$ if $i = j$ and 0 otherwise) and $\alpha_i(u)$ is the i-th feature of pattern u:

$$\alpha_i(u) = \sum_{j=0}^{n-1} u_{\eta i(j)} 2^i, \tag{5.31}$$

where, u_i is the i-th bit of u and $\eta_{i(j)}$ is the j-th bit of the i-th n-tuple.

With C classes to discriminate, the system can be realized as a set of N C RAMS, in which the memory content m_{ci} α at address α of the i-th RAM allocated to class c is

$$m_{ci\alpha} = \Theta\left(\sum_{v \in Cc} \delta_{\alpha 1 \alpha 2(v)}\right), \tag{5.32}$$

Thus, $m_{ci\alpha}$ is set if any pattern of C_c has feature α and is unset.

Recognition is accomplished by totaling set bits in RAMS of each class for the specified features of an unclassified pattern.

RAMnets can be trained faster than MLPs or radial basis function networks by order of magnitude, but delivers analogous results (Michie et al., 1994).

5.2 SVM-Based Approaches

SVM was first used in 1992 by Boser, Guyon, and Vapnik in COLT-92. SVM is a classification and regression prediction tool that uses machine learning theory to maximize predictive accuracy while avoiding overfit to the data, automatically (Jakkula, 2011). SVMs are a set of connected supervised learning methods used for classification and regression. In other words, SVMs might be defined as systems using hypothesis space of linear functions in a high-dimensional feature space. It is trained with a learning algorithm from optimization theory that implements a learning bias resulting from a statistical learning theory. SVM was originally used by neural information processing systems (NIPS) but is now an essential aspect of machine learning research everywhere in the world. SVMs commonly use pixel maps as input; their accuracy is similar to sophisticated NNs with explained features in a hand-writing recognition task (Moore, 2003). They are also used for numerous applications, particularly pattern classification and regression-centered applications, such as, hand writing analysis, face analysis, and so forth.

After its original formulation, SVM quickly gained acceptance because of many promising features, such as, improving empirical performance and so on. It uses the structural risk minimization (SRM) principle, shown to be better than the empirical risk minimization (ERM) principle used by conventional NNs. SRM minimizes an upper bound on predictable risk, where ERM minimizes an error in a training data. It is this difference that gives SVM better ability to generalize, the objective of statistical learning. SVMs were established to resolve classification problem, but lately they have been extended to solve regression problems (Vapnik et al., 1997).

5.2.1 Statistical Learning Theory

Statistical learning theory provides a framework for learning problem of gaining knowledge and making predictions and decisions from a set of data. In simple terms, it allows selection of a hyperplane space while carefully signifying the essential function in a target space (Evgeniou et al., 2000).

In statistical learning theory, problem of supervised learning is expressed as follows: we assume a set of training data $\{(x_1,y_1)...(x_l,y_l)\}$ in an $R^n \times R$ sampled rendering to unidentified probability distribution $P(x,y)$, and loss function $V(y,f(x))$ that measures the error, for a given x, with $f(x)$ "predicted" in place of the actual value y. The problem requires finding a function f that minimizes anticipation of an error on new data; that is, outcome is a function f that minimizes expected error: $\int V(y,f(x)) P(x,y) \, dx \, dy$.

In statistical modeling, we would select a model from hypothesis space, which borders on essential function in a target space. More on statistical learning theory can be found in (Bousquet et al., 2004).

5.2.1.1 Learning and Generalization

Machine-learning algorithms were initially intended to learn representations of simple functions. Hence, the objective of learning was to output a hypothesis that executed precise classification of a training data, and early learning algorithms were intended to find an accurate fit to data. The ability of a hypothesis to acceptably classify data not in training set is known as its generalization. SVM does not over-generalize, whereas NNs may effortlessly over-generalize. For information on trade-offs with complexity, see Figure 5.9. *Note*: the illustration is prepared from class notes (Jakkula, 2011).

5.2.1.2 Introduction to SVM: Why SVM?

Employing NNs for both supervised and unsupervised learning shows good results. MLPs use feed forward and recurrent networks. MLP properties comprise collective approximation of continuous nonlinear functions and embrace learning with input–output patterns. MLPs also contain radical network architectures with multiple inputs and outputs (Skapura, 1996). Figure 5.10 is a simple visualization of NNs.

Some NNs have numerous local minima; having numerous neurons might constitute an additional task, affecting optimality of that NN. Note that even if NN solutions are inclined to converge, outcome may not be a unique solution. When we plot the data, as shown in Figure 5.11, and attempt to classify it, we see several hyperplanes that can classify it. But which is better?

As the above diagram shows, there are many linear classifiers (hyperplanes) that isolate data. But, only one of these classifiers achieves maximum

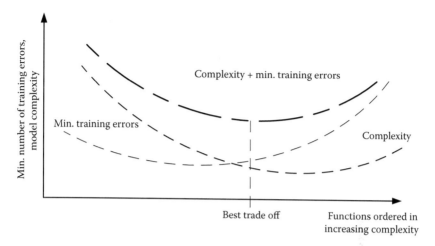

FIGURE 5.9
Number of epochs versus complexity. (From Cristianini, N. and Shawe-Taylor, J., 2000. *An Introduction to Support Vector Machines and Other Kernel-Based Learning Methods.* Cambridge University Press; Burges, C., 1998. A tutorial on support vector machines for pattern recognition. In: *Data Mining and Knowledge Discovery.* Boston: Burges C.)

separation. If we use a hyperplane to classify, it might end up nearer to one set of datasets than others and we do not need this to occur. Thus, a maximum margin classifier or hyperplane is a special solution; Figure 5.12 gives the maximum margin classifier example that solves the problem (Jakkula, 2011).

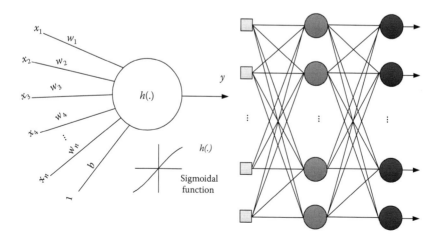

FIGURE 5.10
(a) Simple NN, (b) MLP. (From Skapura, D. M., 1996. *Building Neural Networks.* Addison-Wesley Professional. New York: ACM Press; Mitchell, T., 1997. *Machine Learning.* Boston, MA: McGraw-Hill Computer Science Series.)

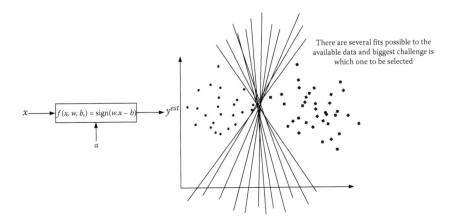

FIGURE 5.11
Many hyperplanes can classify data, but which one is best? SVM is required. (From Moore, A., 2003. Andrew W. *Moore's Home Page*. [Online] Available at: http://www.cs.cmu.edu/~awm.)

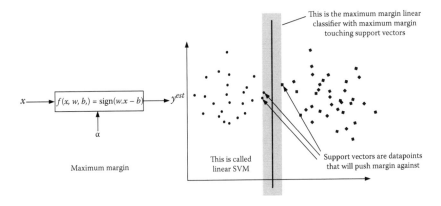

FIGURE 5.12
Illustration of Linear SVM. (From Moore, A., 2003. Andrew W. *Moore's Home Page*. [Online] Available at: http://www.cs.cmu.edu/~awm.)

The maximum margin is expressed as:

$$\text{margin} \equiv \arg\min_{x \in D} d(x) = \arg\min_{x \in D} \frac{|x \cdot w + b|}{\sqrt{\sum_{i=1}^{d} w_i^2}}$$

Figure 5.12 shows the maximum linear classifier with the maximum rank. This situation is an example of a simple linear SVM classifier. An interesting question is, why maximum margin? Empirical performance is the answer.

In addition, even if we have made a small error in position of a boundary, this error has little chance of causing a misclassification. Another benefit is, evasion of local minima and enhanced classification. The goals of SVM are to separate data with a hyperplane and spread them to nonlinear boundaries using the kernel trick. The objective to calculate SVM is to appropriately classify all data. The mathematical calculations are:

a. If $Y_i = +1$; $wx_i + b \geq 1$
b. If $Y_i = -1$; $wx_i + b \leq 1$
c. For all i; $y_i (w_i + b) \geq 1$

In this equation x is a vector point and w is weight and also a vector. To separate the data [a] should always be greater than zero. Among all possible hyperplanes, SVM picks the one where distance of the hyperplane is as great as possible. If training data is good, every test vector is positioned in radius r from training vector. Now, the special hyperplane is placed at the farthest possible point from the data. The hyperplane that maximizes the margin also intersects the lines in the middle of the closest points on the convex body of the two datasets. Thus, we have [a], [b], and [c] (Jakkula, 2011) (Figure 5.13).

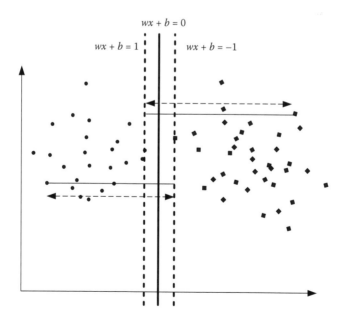

FIGURE 5.13
Representation of hyperplanes. (From Cristianini, N. and Shawe-Taylor, J., 2000. *An Introduction to Support Vector Machines and Other Kernel-Based Learning Methods.* Cambridge University Press.)

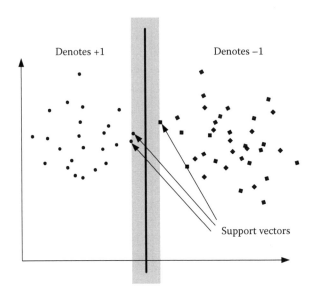

FIGURE 5.14
Representation of support vectors. (From Moore, A., 2003. Andrew W. *Moore's Home Page.*
[Online] Available at: http://www.cs.cmu.edu/~awm.)

The distance of the closest point on the hyperplane can be found by maximizing x, as x is on the hyperplane. We have a similar development for further side points. Therefore, resolving and subtracting the two distances, we get the summed distance from the separating hyperplane to the closest points. Maximum margin = $M = 2/||w||$.

Now, maximizing the margin is the same as minimizing. We have a quadratic optimization problem and we have to solve for w and b. To resolve this, it is essential to optimize the quadratic function with linear constraints. The solution involves building a dual problem, using a Langlier's multiplier α_i. We need to find w and b such that $\Phi(w) = 1/2|w'||w|$ is minimized and for all $\{(x_i,y_i)\}$: $y_i(w^*x_i + b) \geq 1$.

This gives us $w = \Sigma\alpha_i^*x_i$; $b = y_k - w^*x_k$ for any x_k such that $\alpha k \neq 0$.

The classifying function will have the resulting form:

$f(x) = \Sigma\alpha_i y_i x_i^*x + b$ (Figure 5.14).

5.2.2 SVM Representation

A superficial depiction of the QP formulation for SVM classification is shown in this section. SV classification:

$$\min_{f,\xi_i} \|f\|_K^2 + C\sum_{i=1}^{l} \xi_i \; y_i f(x_i) \geq 1 - \xi_i, \quad \text{for all } i \; \xi_i \geq 0$$

SVM classification, dual formulation:

$$\min_{\alpha_i} \sum_{i=1}^{1} \alpha_i - \frac{1}{2} \sum_{i=1}^{1} \sum_{j=1}^{1} \alpha_i \alpha_j y_i y_j K(x_i, x_j) 0 \leq \alpha_i \leq C, \quad \text{for all } i; \sum_{i=1}^{l} \alpha_i y_i = 0$$

Variables ξ_i are called slack variables, and they measure the error made at point (x_i, y_i). Training SVM is interesting when number of training points is large.

5.2.2.1 Soft Margin Classifier

In an actual domain problem, it is not possible to get a precise and distinct line isolating data surrounded by space. We may have a rounded decision boundary. We could also have a hyperplane which might separates data precisely, but this may not be necessary if data have noise in them. It is better for smooth boundary to discount a few data points than be curved or go in loops near outliers. This is not figured in the same way; in this case, we have slack variables. Now we have $y_i(w'x + b) \geq 1 - s_k$. This permits a point to be a small distance s_k on wrong side of a hyperplane. We might end up having huge slack variables which permit any line to separate data; in such situations, we use the Lagrangian variable:

$$\min L = \tfrac{1}{2} w'w - \sum \lambda_k (y_k(w'x_k + b) + s_k - 1) + \alpha \sum s_k$$

where plummeting α lets more data lie on the wrong side of the hyperplane where they would be preserved as outliers which give a smoother decision boundary (Lewis, 2004).

5.2.3 Kernel Trick

The section begins by defining kernel and feature space,

5.2.3.1 Kernel

If data are linear, a separating hyperplane may be used to divide them; It is frequently the case. that data are not linear and datasets are inseparable. Kernels are used to map input nonlinearly to a high-dimensional space. The new charting is linearly separable [1]. A simple illustration of this situation is presented in Figure 5.15.

This mapping is defined by the kernel: $K(x,y) = \Phi(x) \cdot \Phi(y)$

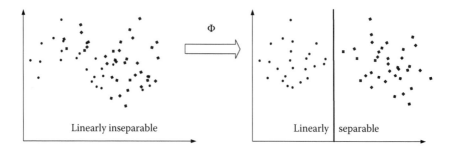

FIGURE 5.15
Why use kernels? (From Cristianini, N. and Shawe-Taylor, J., 2000. *An Introduction to Support Vector Machines and Other Kernel-Based Learning Methods.* Cambridge University Press; Mitchell, T., 1997. *Machine Learning.* Boston, MA: McGraw-Hill Computer science series; McCulloch, D., 2005. *An Investigation into Novelty Detection.*)

Feature space: It possible to define a similar measure on basis of a dot product by transforming data into feature space. The pattern recognition may be easy if feature space is chosen properly [1] (Figure 5.16):

$$\langle x_1 \cdot x_2 \rangle \leftarrow K(x_1, x_2) = \langle \Phi(x_1) \cdot \Phi(x_2) \rangle$$

Note, the legend is not described as these are sample plots to understand the concepts.

Now getting back to the kernel trick, we see that when w, b is found, the problem is resolved for a linear situation in which data are separated by a hyperplane. The kernel trick allows SVMs to form nonlinear boundaries. Steps of the kernel trick are given below (Lewis, 2004; Schölkopf et al., 1999):

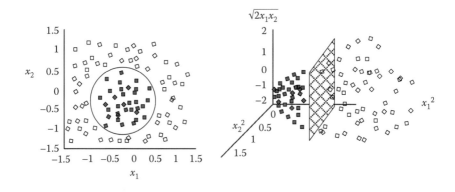

FIGURE 5.16
Feature space representation. (From Mitchell, T., 1997. *Machine Learning.* Boston, MA: McGraw-Hill Computer Science Series.)

a. The algorithm is expressed using solitary inner products of data sets; This is also called a dual problem.

b. Original data are approved through nonlinear maps to arrange different data with respect to new dimensions by tallying a specific pair-wise product of original data dimension for each data vector.

c. Somewhat more than an inner product for these new, larger vectors is to store data in tables and later do a table lookup; we can characterize a dot product of data after doing nonlinear mapping on them. This function is the kernel function. More on kernel functions follows in the next section.

5.2.3.2 Kernal Trick: Dual Problem

First, we change a problem with optimization to dual form in which we try to eradicate w, and a, Lagrangian now is only a function of λ_i. To resolve the issue, we should maximize the LD with respect to λ_i. The dual form abridges optimization; the major accomplishment is dot product attained (Jakkula, 2011).

5.2.3.3 Kernal Trick: Inner Product Summarization

The dot product of nonlinear mapped data can be costly, but the kernel trick performs a suitable function resembling the dot product of certain nonlinear mapping. A particular kernel is chosen by trial and error on the test set; selecting the right kernel centered on the problem or application improves SVM's performance (Jakkula, 2011).

5.2.3.4 Kernel Functions

The idea of using the kernel function is to allow processes to be achieved in an input space rather than in possibly high-dimensional feature space. Hence, inner product should not be estimated in feature space. We need the function to achieve mapping of attributes of input space to feature space. The kernel function plays a pivotal role in SVM and its performance. It is constructed by replicating Kernel Hilbert Spaces (Jakkula, 2011), as follows:

$$K(x,x') = \langle \phi(x),\phi(x')\rangle,$$

If K is a symmetric positive definite function, that contains Mercer's conditions,

$$K(x,x') = \sum_{m}^{\infty} a_m\phi_m(x)\phi_m(x'), \quad a_m \geq 0,$$

$$\iint K(x,x')g(x)g(x')dxdx' > 0, \quad g \in L_2$$

and the kernel represents a genuine inner product in feature space. The training set is not linearly separable in an input spacebut is linearly separable in feature space named "the kernel trick."

Various kernel functions are listed below. More descriptions on kernel functions can be found in literature. The following are mined from several studies:

1. *Polynomial:* A polynomial mapping is a popular method for nonlinear modeling. The second kernel is desirable as it avoids problems of hessian becoming zero.

$$K(x,x') = \langle x,x' \rangle^d,$$

$$K(x,x') = (\langle x,x' \rangle + 1)^d,$$

2. *Gaussian radial basis function:* Radial basis functions are done most frequently with a Gaussian form:

$$K(x,x') = \exp\left(-\frac{\|x-x'\|^2}{2\sigma^2}\right)$$

3. *Exponential radial basis function:* A radial basis function creates a piecewise linear solution that can be attractive when discontinuities are acceptable:

$$K(x,x') = \exp\left(-\frac{\|x-x'\|^2}{2\sigma^2}\right)$$

4. *MLP:* The long-established MLP, with a single hidden layer, also has a valid kernel representation:

$$K(x,x') = \tanh\left(\rho\langle x,x' \rangle + e\right)$$

There are numerous additional kernel methods including Fourier, splines, B-splines, additive kernels, and tensor products (see Cristianini and Shawe-Taylor, 2000).

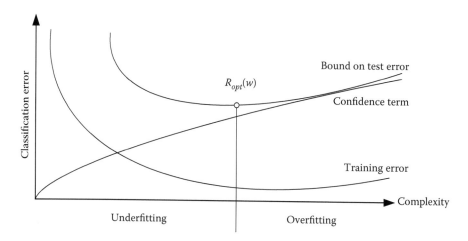

FIGURE 5.17
How to control complexity. (From Moore, A., 2003. Andrew W. *Moore's Home Page*. [Online] Available at: http://www.cs.cmu.edu/~awm.)

5.2.3.5 Controlling Complexity in SVM: Trade-Offs

SVM is used to estimate any training data and generalize enhancements of specified datasets. Complexity of the kernel affects performance of new datasets. SVM supports parameters for monitoring complexity; above all, SVM does not tell us how to set these parameters. We should be able to fix these parameters by cross-validation on given datasets. The diagram given in Figure 5.17 is a better illustration (Jakkula, 2011).

5.2.4 SVM for Classification

SVM is a useful method for data classification. While NNs are easier to use than SVMs, they occasionally have unacceptable consequences. A classification task contains training and testing data with specific data instances. Every one of the instances in training set covers one target value and many attributes. The goal of SVM is to produce a model that predicts target value of data instances in a testing set, but only attributes.

Classification in SVM is an example of supervised learning. Recognized labels designate whether a system is performing in a factual way or not. This information points to an anticipated response authenticating accuracy of the system, or is used to assist system to perform correctly. A step in SVM classification includes identification. which is closely associated with known classes and termed as a feature selection or feature extraction. Feature selection and SVM classification are used when prediction of unfamiliar samples is not essential. They can be used to recognize key sets in procedures that differentiate classes.

SVM has been established for pattern classification problems. The application of support vector approach to a specific practical problem encompasses: determining a number of questions centered on problem definition, and strategy intertwined with it. One of the major challenges is selecting a fitting kernel for a specified application. Generally, Gaussian or polynomial kernel is a default choice , but if this does not work or if inputs are discrete structures, more intricate kernels are needed. Through a feature space, kernel delivers description language used by a machine for inspecting data. The main constituents of the system are in place, once choice of kernel and optimization condition has been decide.

The job of text categorization is classification of normal text documents into a fixed quantity of predefined categories that are centered on their content. Since a document can be allocated to more than one category, this is not a multilass classification problem, but a series of binary classification problems, one for every category. One of the regular demonstrations of text for resolution of information recovery delivers a feature mapping for constructing a Mercer kernel. In some ways kernel integrates a similarity measure between instances; It is advised to follow specific application domain experts, who have already identified effective actions, chiefly in areas,such as, information retrieval and generative models.

Traditional classification methods are not applicable when working in a straight line because of high dimensionality of data, but SVMs can handle high dimensional representations. An analogous approach to techniques designated for text categorization can also be used for image classification. In such cases, linear hard-margin machines are often able to oversimplify. The first real-world job on which SVMs were used was hand-written character recognition. In addition, multiclass SVMs have been established on these data. It is fascinating not only to compare SVMs with other classifiers, but also to compare different SVMs; They turn out to have similar performance, that is, recording their support vectors, independent of the selected kernel. SVM can accomplish most things as well as these systems without any of prior knowledge (Jakkula, 2011).

5.2.5 Strength and Weakness of SVM

The greatest strength of SVM is that, unlike NNs training is comparatively easy with no local optimal. It balances with high-dimensional data comparatively well and balance between classifier complexity and error can be well ordered. The weakness is its inability to perform a decent kernel function (Burges, 1998; Cristianini and Shawe-Taylor, 2000; Jakkula, 2011; Moore, 2003).

SVM is one of the best methodology for data modeling. They use generalization regulator as a technique to control dimensionality. Kernel mapping delivers a collective base for most working model architectures, and supporting assessments, to be executed. In classification problems, generalization control is achieved by maximizing margin that matches the minimization of weight vector in an acknowledged framework. The solution is attained as a set of support vectors that can be scarce. The minimization of weight vector can be used as a measure in regression problems, with a revised loss function. Current research is investigating a modus operandi for cherry-picking the kernel function and, developing kernels with invariance and further capacity control . To conclude, new guidelines are cited in novel SVM-related learning originations recommended by Vapnik (2006).

5.3 Bayesian Networks–Based Approaches

Bayesian decision theory is a vital statistical approach to pattern classification problem. This approach is built on enumerating quid pro quo between various classification decisions by means of probability, and costs of such decisions. It assumes decision problem is understood in probabilistic positions and all pertinent probability values are known.

Let us pretend to design a classifier for two kinds of fish: catfish and batfish. Suppose an observer watching the fish being caught by a fisherman finds it hard to predict the kind that will arrive next because of random sequence of types of fish appearance. Decision-theory vocabulary says that as each fish arrives, its nature has only two possibilities: the fish is either a catfish or a batfish. Let ω denote the state of nature, with $\omega = \omega_1$ for catfish and $\omega = \omega_2$ for batfish. Because the state of nature is random, we consider ω to be a variable that must be designated probabilistically.

If the hook catches as much catfish as batfish, the next fish is equally likely to be catfish or batfish. It is presumed that there is a certain a priori probability (or simply prior) $P(\omega_1)$ that the next fish is catfish or that it is batfish. If we assume there are no new types of pertinent fish, then $P(\omega_1)$ and $P(\omega_2)$ are equal to one. These probabilities imitate our prior information about how likely we are going to get a catfish or batfish before looking at the fish; it might, for example, rest on fishing period of a year or selection of a fishing area (Duda et al., 2001).

Let us suppose for a moment, that we are obliged to make a decision about the kind of fish that will appear next without being allowed to see it. For now, we must assume that any misclassification carries same cost or consequence, and that only information we are allowed to use is the value of the prior probabilities. If a decision must be made with so little information, it seems logical to use the following decision rule: decide ω_1 if $P(\omega_1) > P(\omega_2)$; otherwise decide ω_2.

This seem logical for reviewing just one fish, but, using this rule recurrently to review numerous fish, may seem a little strange. Because, we would make a similar verdict, continually, even though we know that both types of fish will be caught eventually. How well it works is contingent on values of prior probabilities. If $P(\omega_1)$ is much greater than $P(\omega_2)$, our conclusion will be right most of the time, that is, in favor of ω_1. If $P(\omega_1) = P(\omega_2)$, we have only a 50–50 chance of being correct. In general, probability of an error is smaller of $P(\omega_1)$ and $P(\omega_2)$:and under this condition, no other decision rule can yield a higher probability of being right.

In more complex conditions, we cannot with so little information. Extending the example above, we might use a nimbleness measurement x to advance a classifier. Different fish will produce different lightness, and senses, Let's consider this inconsistency in probabilistic terms; we consider x to be a continuous random variable whose distribution is to be determined by the state of nature, expressed as $p(x/\omega_1)$. This is a class-conditional probability density function. The probability density function $p(x|\omega_1)$ should be written as $pX(x/\omega_1)$ to specify that we are talking about a specific density function for a random variable X. This more intricate subscripted representation makes it clear that $pX(\cdot)$ and $pY(\cdot)$ symbolize two different functions, a point (dot) hidden when in writing $p(x)$ and $p(y)$. Since this possible misperception seldom arises in an exercise, we have chosen to implement easier symbolization.

This is the probability density function $p(x/\omega_1)$ for x given that the state of nature is ω_1. Then the difference between $p(x/\omega_1)$ and $p(x/\omega_2)$ describes the difference in nimbleness between populations of catfish and batfish (Figure 5.10). Now, assume the prior probabilities $P(\omega)$ and the conditional densities $p(x/\omega)$ are identified together. Further suppose that we measure nimbleness of a fish and learn that its value is x. How does this measurement affect our assertiveness about factual state of nature—that is, the class of the fish?

The (joint) probability density of an outcome of a pattern that is in category ω_j with feature value x can be written in two ways: $p(\omega_j, x) = P(\omega_j/x)p(x) = p(x/\omega_j)*P(\omega_j)$. Reorganizing these, helps us to answer our question and produces Bayes' formula:

$$P(\omega_j/x) = \frac{p(\omega_j/x) * P(\omega_j)}{p(x)}, \qquad (5.33)$$

where,

$$p(x) = \sum_{j=1}^{2} p\left(\frac{\omega_j}{x}\right) * P(\omega_j), \qquad (5.34)$$

Bayes' formula can be expressed as

$$\text{Posterior} = \frac{\text{likelihood} \times \text{prior}}{\text{evidence}}, \qquad (5.35)$$

Bayes' formula demonstrates that by perceiving value of x we can adapt prior probability $P(\omega_j)$ to *a posteriori* probability (or posterior) probability $P(\omega_j/x)$; the probability of the state of nature being ω_j, given that feature value x has been measured. This is called $p(x/\omega j)$, in the likelihood of ω_j with respect to x. Note that the product of the likelihood and the prior probability is most significant in influencing the posterior probability; the evidence factor, $p(x)$, can be observed as simply a scale factor that assures the posterior probabilities summation to one, as all good probabilities must. The variation of $P(\omega j/x)$ with x is illustrated in Figure 5.18 for the case $P(\omega_1) = 2/3$ and $P(\omega_2) = 1/3$.

If there is a reflection x, where $P(\omega/x)$ is greater than $P(\omega_2/x)$, the resolution is likely to be that the true state of nature is ω_1. Similarly, if $P(\omega_2/x)$ is greater than $P(\omega_1/x)$, the probable choice is ω_2. To justify this procedure, the probability of error is planned whenever we have to make a judgment. On every occasion that we perceive a particular x,

$$P(\text{error}/x) = \begin{cases} P(\omega_1 \, / \, x) & \text{if is decided } \omega_2 \\ P(\omega_2 \, / \, x) & \text{if is decided } \omega_1 \end{cases}, \qquad (5.36)$$

Clearly, for a given x we can minimize the probability of an error by determining ω_1 if $P(\omega_1/x) > P(\omega_2/x)$ and ω_2 (Figure 5.19). Surely, we may never have

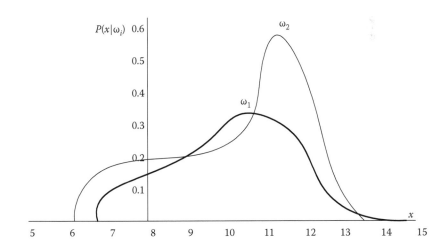

FIGURE 5.18
Hypothetical class-conditional probability density functions show the probability density of measuring a particular feature value x if the pattern is in category ω_i. If x represents the length of a fish, the two curves might describe the difference in length of populations of two types of fish. Density functions are normalized, and thus the area under each curve is 1.0. (From Duda, R. O., Hart, P. E. and Stork, G. D., 2001. *Pattern Classification*. 2nd edition. Canada: John Wiley & Sons.)

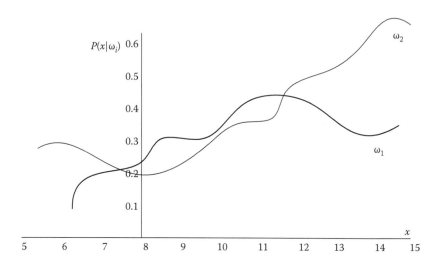

FIGURE 5.19
Posterior probabilities for the particular priors $P(\omega_1) = 2/3$ and $P(\omega_2) = 1/3$ for the class-conditional probability densities shown in Figure 5.18. Thus, in this case, given that a pattern is measured to have feature value $x = 14$, the probability in category ω_2 is roughly 0.08, and in ω_1 is 0.92. At every x, the posteriors sum to 1.0. (From Duda, R. O., Hart, P. E. and Stork, G. D., 2001. *Pattern Classification*. 2nd edition. Canada: John Wiley & Sons.)

identical value of x twice. Will this rule diminish average probability of error? Yes, since average probability of an error is specified by

$$P(\text{error}) = \int_{-\infty}^{\infty} P(\text{error}, x)dx = \int_{-\infty}^{\infty} P(\text{error}/x)p(x)dx, \tag{5.37}$$

and if, for every x we ensure $P(error|x)$ is as small as possible, then the integral must be as low as possible. Thus, we have vindicated the following Bayes' decision rule for minimizing the probability of error:

$$\text{Decide } \omega_1 \quad \text{if } P(\omega_1/x) > P(\omega_2/x); \quad \text{otherwise decide } \omega_2, \tag{5.38}$$

and under this rule Equation 5.43 becomes
$P(\text{error}) = \min [P(\omega_1/x), P(\omega_2/x)]$.

$$P(\text{error}) = \min\left[P(\omega_1/x), P(\omega_2/x)\right], \tag{5.39}$$

This procedure of decision law stresses on the role of posterior probabilities. By using Equation 5.40, this can, as an alternative, prompt the standing rule of conditional and prior evidence probabilities. First, a reminder that the indication, $p(x)$, in Equation 5.40 is insignificant for constructing a verdict; It is essentially just a scale factor that states how often it will actually measure a pattern with feature value x. Its presence in Equation 5.33 ensures that $P(\omega_1/x) + P(\omega_2/x) = 1$. We can derive corresponding decision rule by removing this scale factor:

$$\text{Decide } \omega_1 \text{ if } p(x/\omega_1) * P(\omega_1) > p(x/\omega_2) * P(\omega_2), \text{otherwise decide } \omega_2. \quad (5.40)$$

Some additional insights can be obtained by considering a few special cases. If for some xs $p(x/\omega_1) = p(x/\omega_2)$, then that specific reflection provides no evidence about nature; in this case, conclusion centers entirely on prior probabilities. On the other hand, if $P(\omega_1) = P(\omega_2)$, the states of nature are likely similar; in this case, the decision is founded completely on the possibilities $p(x/\omega_j)$. In a broad sense, both these factors are significant in constructing a decision and the Bayes decision rule uses them to create the minimum probability of error (Duda et al., 2001).

5.3.1 Bayesian Decision Theory—Continuous Features

Simply stated, the above concepts offer following four features:

- Permit the use of more than one feature
- Allow more than two states of nature
- Agree to actions, other than merely determining the state of nature
- Introduce a loss function more general than the probability of error.

Permitting the use of more than one feature merely necessitates substituting the scalar x by the feature vector x, where x is in a d-dimensional Euclidean space Rd, called the feature space. Allowing more than two states of nature provides a good generalization for a slight notational overhead. Agreeing to actions other than classification mainly precludes the possibility of rejection; if not too costly, this could be a convenient choice. The loss function states accurately, cost of each action, and is used to change a probability into a decision. Cost functions give us circumstances in which some types of classification errors are more expensive than others, though, frequently, we consider a situation, where all errors are overpriced.

Let $\omega_1, \ldots, \omega_c$ be the finite set of c states of nature ("categories") and $\alpha_1, \ldots, \alpha_a$ be the finite set of a probable actions. The loss function $\lambda(\alpha_i/\omega_j)$ designates the loss sustained for taking action α_i when the state of nature is ω_j. Let the feature vector x be a d-component vector-valued random variable, and let $p(x/\omega_j)$ be the state conditional probability density function for x—the probability density function for x conditioned on ω_j being the accurate state of nature. As before, $P(\omega_j)$ defines the prior probability that nature is in state

ω_j. Then, the posterior probability $P(\omega_j/x)$ can be calculated from $p(x/\omega_j)$ by Bayes' formula:

$$P(\omega_j/x) = \frac{p(x/\omega_j) * P(\omega_j)}{p(x)}, \qquad (5.41)$$

where the evidence is now,

$$p(x) = \sum_{j=1}^{c} p(x/\omega_j) * P(\omega_j), \qquad (5.42)$$

Supposing that a specific x is pragmatic and it takes action α_i. If the true state of nature is ω_j, by characterization it will incur the loss $\lambda(\alpha_i/\omega_j)$. Since $P(\omega_j/x)$ is the probability that the true state of nature is ω_j, the predictable cost linked with taking action α_i is

$$R(\alpha_i/x) = \sum_{j=1}^{c} \lambda(\alpha_i/\omega_j) * P(\omega_j/x), \qquad (5.43)$$

In decision-theory vocabulary, a predictable cost is called a risk, and $R(\alpha_i/x)$ is called the conditional risk. On every occasion with a specific reflection x, the predictable cost can be minimized by choosing an action that reduces the conditional risk. Bayes decision procedure essentially offers optimal performance to determine global risk.

The difficulty is to find a decision rule for $P(\omega_j)$ that minimizes global risk. A broad decision rule is a function $\alpha(x)$ that expresses which action to take for every probable reflection. More specifically, for every x, the decision function $\alpha(x)$ assumes one of the "a" values, $\alpha_1,...,\alpha_a$. The overall risk R is the predictable cost connected to a particular decision rule. Since $R(\alpha_i/x)$ is the conditional risk accompanying action α_i, and since the decision rule postulates the action, the global risk is given by,

$$R = \int R(\alpha(x) / x) * p(x)dx, \qquad (5.44)$$

where, dx denotes a d-space volume element, and integral spreads over whole feature space. Clearly, if $\alpha(x)$ is chosen so that $R(\alpha_i(x))$ is as small as possible for every x, then global risk will be minimized. This validates the expression, "to minimize the global risk, calculate the conditional risk" of the Bayes decision rule:

$$R(\alpha_i/x) = \sum_{j=1}^{c} \lambda(\alpha_i/\omega_j) * P(\omega_j/x), \qquad (5.45)$$

for $i = 1, \ldots, a$ and choose the action α_i for which $R(\alpha_i/x)$ is minimum. The subsequent minimum global risk is called the Bayes risk, denoted by R^*. It is the best determination of global risk (Duda et al., 2001).

5.3.1.1 Two-Category Classification

In a singular instance of two-category classification problems, action α_1 expresses determination that the true state of nature is ω_1, and action α_2 expresses it as, ω_2. For notational ease, let $\lambda_{ij} = \lambda(\alpha_i/\omega_j)$ be the loss suffered for determining ω_i when the true state of nature is ω_j. If we writ the conditional risk assumed by Equation 5.51, we get

$$R(\alpha_1/x) = \lambda_{11}P(\omega_1/x) + \lambda_{12}P(\omega_2/x) \quad \text{and}$$
$$R(\alpha_2/x) = \lambda_{21}P(\omega_1/x) + \lambda_{22}P(\omega_2/x) \tag{5.46}$$

There are multiple methods of articulating minimum-risk decision rule, each with its own benefits. The central rule is to choose ω_1 if $R(\alpha_1/x) < R(\alpha_2/x)$. In terms of the posterior probabilities, ω_1 is certain if,

$$(\lambda_{21} - \lambda_{11}) * P(\omega_1/x) > (\lambda_{12} - \lambda_{22}) * P(\omega_2/x), \tag{5.47}$$

In general, loss suffered for creating an error is bigger than that of loss suffered for being right, and both factors $\lambda_{21} - \lambda_{11}$ and $\lambda_{12} - \lambda_{22}$ are positive. Thus, in an exercise, resolution is mostly affected by other probable state of nature, even if we balance posterior probabilities against loss differences. By employing Bayes' formula, we can substitute posterior probabilities for prior probabilities and conditional densities.

We get ω_1 if,

$$(\lambda_{21} - \lambda_{11}) * p(x/\omega_1) * P(\omega_1) > (\lambda_{12} - \lambda_{22}) * p(x/\omega_2) * P(\omega_2), \tag{5.48}$$

and ω_2 otherwise.

Another margin, which follows sensible assumption that, $\lambda_{21} > \lambda_{11}$, is to decide on ω_1 if

$$\frac{p(x/\omega_1)}{p(x/\omega_2)} > \frac{(\lambda_{12} - \lambda_{22}) * P(\omega_2)}{(\lambda_{21} - \lambda_{11}) * P(\omega_1)}, \tag{5.49}$$

This form of decision rule emphasizes x-dependence of probability densities. It can consider $p(x/\omega_j)$ a function of ω_j (i.e., likelihood function), and then form likelihood ratio $p(x/\omega_1)/p(x|\omega_2)$. Thus, Bayes decision rule can be understood as calling for ω_1 if likelihood ratio exceeds a threshold value that is independent of the observation x (Duda et al., 2001).

5.3.2 Minimum-Error-Rate Classification

In classification problems, each state of nature is usually accompanied with one of c classes, and action α_i is usually construed as decision tha true state of nature is ω_i. If action α_i is taken and true state of nature is ω_j, then conclusion is correct, if $i = j$, and an error, if $i \neq j$. If errors are to be avoided, it is best to find a decision rule that minimizes probability of error, that is, error rate.

This results in so-called symmetrical or *zero-one* loss function,

$$\lambda(\alpha_i/\omega_j) = \begin{cases} 0 & i = j \\ 1 & i \neq j \end{cases} \qquad i, j = 1, \ldots, c, \tag{5.50}$$

This loss function allocates no loss to a correct decision, and allocates a unit loss to any error; thus, errors are expensive. Risk equivalent to this loss function is average probability of error, and, conditional risk is

$$R(\alpha_i/x) = \sum_{j=1}^{c} \lambda(\alpha_i/\omega_j) * P(\omega_j/x)$$

$$= \sum_{j \neq 1} P(\omega_j/x) \tag{5.51}$$

$$= 1 - P(\omega_i/x),$$

and $P(\omega_i/x)$ is conditional probability that action α_i is correct. Bayes decision rule to minimize risk requires selecting action that minimizes conditional risk. Thus, to minimize average probability of error, we should choose i that maximizes posterior probability $P(\omega_i/x)$. In other words, for minimum error rate,

$$\text{Decide } \omega_i \quad \text{if } P(\omega_i/x) > P(\omega_j/x) \quad \text{for all } j \neq i. \tag{5.52}$$

This is an equivalent rule to Equation 5.40.

Figure 5.11 shows some class-conditional probability densities and posterior probabilities, Figure 5.12 shows likelihood ratio $p(x/\omega_1)/p(x/\omega_2)$ for identical instance. In a broad spectrum, this ratio can range between zero and infinity. Threshold value θ_a is noticeable from same prior probabilities, but with *zero-one* loss function. Notice that this indicates same judgment boundaries as in Figure 5.11. If there are faults in classifying ω_1 patterns as ω_2 more than converse (i.e., $\lambda_{21} > \lambda_{12}$), then Equation 5.55 obviously tips towards θ_b. Note that, as shown in Figure 5.20, range of x values for which we classify a pattern as ω_1 lessens (Duda et al., 2001).

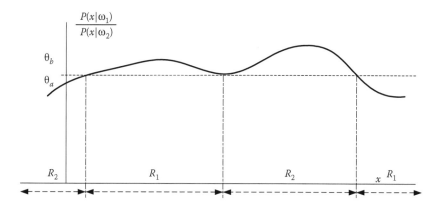

FIGURE 5.20
If the likelihood ratio $p(x/\omega_1)/p(x/\omega_2)$ for the distributions shown in Figure 5.18 employs a zero-one or classification loss, the decision boundaries are determined by the threshold θ_a. If the loss function penalizes miscategorizing ω_2 as ω_1 patterns more than the converse (i.e., $\lambda_{12} > \lambda_{21}$), the larger threshold θ_b, is obtained; hence, R1 becomes smaller. (From Duda, R. O., Hart, P. E. and Stork, G. D., 2001. *Pattern Classification*. 2nd edition. Canada: John Wiley & Sons.)

5.3.3 Bayesian Classifiers

A Bayesian classifier is constructed on the idea that character of a (natural) class is to forecast values of features for associates of that class. Instances are congregated in classes since they have shared values for features. Such classes are frequently termed natural kinds. In this section, target feature recognizes a discrete class as binary.

The notion behind a Bayesian classifier is that if an agent recognizes a class, it can forecast values of the other features. If it does not recognize a class, Bayes' rule can be used to calculate (some) feature values of the specified class. In a Bayesian classifier, learning agent constructs a probabilistic model of features and uses that model to predict classification of a new example.

A dormant variable is a probabilistic variable that is not detected. A Bayesian classifier is a probabilistic model where classification is a latent variable that is probabilistically connected to detected variables. Classification then becomes an implication in probabilistic model.

A simple example is a naive Bayesian classifier used when input features are conditionally independent of each other, particularly classification. Independence of a naive Bayesian classifier is indicated in a confidence network where features are nodes, target variable (classification) has no parents, and classification is only parent of each input feature. This confidence network involves probability distributions $P(Y)$ for target feature Y and $P(X_i/Y)$ for each input feature X_i. For each example, prediction can be calculated by conditioning detected values for input features and by questioning classification.

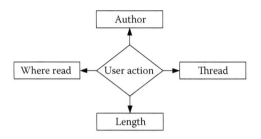

FIGURE 5.21
Belief network corresponding to a naive Bayesian classifier.

Following is an example with inputs $X_1 = v_1, \ldots, X_k = v_k$, Bayes' rule is used in the following to calculate posterior probability distribution of the example's classification, Y:

$$P(Y/X_1) = v_1, \ldots, X_k = v_k$$
$$= (P(X_1 = v_1, \ldots, X_k = v_k/Y)x\ P(Y))/(P(X_1 = v_1, \ldots, X_k = v_k))$$
$$= (P(X_1 = v_1/Y)x\ P(X_k = v_k/Y)x\ P(Y))/$$
$$\left(\sum\nolimits_Y P(X_1 = v_1/Y)x\ P(X_k = v_k/Y)x\ P(Y) \right)$$

where denominator is a normalizing constant to make sure probabilities sum to 1. The denominator does not rest on class and, for that reason, it is not desirable to conclude prospective class (Figure 5.21).

5.3.4 Bayesian Estimation

The Bayesian estimation or Bayesian learning approach works on pattern classification problems. Granted, answers we acquire using this method will generally be indistinguishable from those acquired by maximum likelihood. However, there is a theoretical difference, in maximum likelihood methods true parameter vector is expressed as θ, whereas in Bayesian learning we consider θ a random variable, and training data permit us to change a distribution on this variable into a posterior probability density (Duda et al., 2001).

5.3.4.1 Class-Conditional Densities

Calculation of the posterior probabilities $P(\omega_i/x)$ lies at the heart of Bayesian classification. Bayes' formula lets us determine these probabilities from prior probabilities $P(\omega_i)$ and the class-conditional densities $p(x/\omega_i)$. But what happens when these quantities are unidentified? Simply stated, we must calculate $P(\omega_i/x)$ by means of all information at our disposal. Part of this information might be prior knowledge, such as knowledge of functional arrangements for unidentified densities and ranges for values of unidentified parameters. Portions of this information might exist in a set of training samples.

If we again let D mean set of samples, we can highlight role of samples by saying that the goal is to determine the posterior probabilities $P(\omega_i/x, D)$. From these probabilities we can acquire Bayes classifier.

Given the sample D, Bayes' formula then becomes,

$$P(\omega_i/x, D) = \frac{p(x/\omega_i, D) * P(\omega_i/D)}{\sum_{j=1}^{c} p(x/\omega_j, D) * P(\omega_i/D)}, \tag{5.53}$$

This equation can use information provided by training samples to help determine both class-conditional densities and a priori probabilities.

From this point on, we will assume true values of a priori probabilities are identified or attainable from a trivial calculation; thus, we specify $P(\omega_i) = P(\omega/D)$.

Moreover, since we are discussing supervised instance, we can distinguish training samples by class into c subsets $D_1, \ldots, D_c$, with samples in D_i belonging to ω_i. As mentioned, when addressing maximum likelihood methods, in most cases, samples in D_i have no impact on $p(x/\omega_j, D)$ if $i \neq j$. This has two resulting penalties; First, it permits working with each class using only samples in D_i to define $p(x|\omega_i, D)$, used in combination with supposition that prior probabilities are recognized. This combines with Equation 5.59 to create,

$$P(\omega_i/x, D) = \frac{p(x/\omega_i, D) * P(\omega_i)}{\sum_{j=1}^{c} p(x/\omega_j, D) * P(\omega_i)}, \tag{5.54}$$

Second, since each class can be picked autonomously, it can distribute unnecessary class differences and shorten symbolization. In essence, we have c isolated problems of next form if we use a set D of samples, drawn independently of stable but unidentified probability distribution $p(x)$, to decide $p(x/D)$. This is the central problem of Bayesian learning (Duda et al., 2001).

5.4 Liquid State Machines and Other Reservoir Computing Methods

Liquid state machines (LSMs) represent a new direction in machine learning. They explain time-series problems in an entirely different way from most recurrent neural network (RNN) systems and avoid the difficulties RNNs regularly have learning them. LSMs use a vibrant reservoir or liquid (LM) to grip time-series data, as shown in Figure 5.14 (Kok, 2007).

After a definite time-period, state of liquid $x^M(t)$ is read out to be used as input for a readout network f^M (e.g., an FNN). This readout network learns to map states of liquid to target outputs. This means there is no necessity to train weights of RNN, which decreases calculation time along with complexity of learning time-series data. A liquid can be characterized in different forms. For example, it can be a real liquid, where waves can be understood as short-term memory. But a spiking neural network (SNN) can add more supplementary data than a real liquid. For that reason, in principle, a real liquid can have information in three spatial dimensions. But, a SNN can have distant additional neurons greater than three, thus being able to store more information.

LSMs are a good tool for classification problems. In classification problems, the LSM should not distinguish between same inputs and classify them. A sample of this is, classifying composer of music. The musical piece is input for a liquid, and readout network should classify which composer wrote musical piece. For classification, the readout network wants to distinguish dissimilar states from the liquid. Assuming there are sufficient units in the liquid, it can generate diverse patterns for each time-series pattern.

LSMs have a separation property (SP) and an approximation property (AP). SP addresses ability to distinguish between two different input sequences. This is significant, as readout network needs to be capable of isolating two input patterns to have a good presentation. If two patterns appear identical, even if they should not, readout network cannot distinguish between them and, thus, cannot state which pattern belongs to which class. AP concerns ability of readout network to differentiate two dissimilar patterns and convert states of liquid into certain target output.

References

Bledsoe, W. W., 1961. Further results on the n-tuple pattern recognition method. *IRE Transactions in Computers*, EC-10, 96.

Bledsoe, W. W. and Browning, I., 1959. *Pattern Recognition and Reading by Machine*. Boston, pp. 232–255.

Bousquet, O., Boucheron, S. and Lugosi, G., 2004. *Introduction to Statistical Learning Theory*. In: *Advanced Lectures on Machine Learning*. Heidelberg, Berlin: Springer, pp. 169–207.

Burges, C., 1998. A tutorial on support vector machines for pattern recognition. *Data Mining and Knowledge Discovery*, 2(2), 121–167.

Cover, T. M., 1965. Geometrical and statistical properties of systems of linear inequalities with applications in pattern recognition. *IEEE Transactions on Electronic Computers*, 14, 326–334.

Cristianini, N. and Shawe-Taylor, J., 2000. *An Introduction to Support Vector Machines and Other Kernel-Based Learning Methods*. Cambridge, UK: Cambridge University Press.

Duda, R. O., Hart, P. E. and Stork, G. D., 2001. *Pattern Classification*, 2nd edition. Canada: John Wiley & Sons.

Evgeniou, T., Pontil, M., and Poggio, T. (2000). Statistical learning theory: A primer. *International Journal of Computer Vision*, 38(1), 9–13.

Fahlman, S., 1988. *An Empirical Study of Learning Speed in Back-Propagation*, USA.

Fahlman, S. E. and Lebiere, C., 1990. The cascade correlation learning architecture. In: *Advances in Neural Information Processing Systems 2*. Morgan Kaufmann, pp. 524–532.

Fisher, R., 1936. The use of multiple measurements in taxonomic problems. *Annals of Eugenics*, 7, 179–188.

Hebb, D., 1949. *The Organisation of Behaviour: A Neuralpsychological Theory*. New York: John Wiley and Sons.

Hoehfeld, M. and Fahlman, S. E. 1992. Learning with limited numerical precision using the cascade-correlation algorithm. *Neural Networks, IEEE Transactions on*, 3(4), 602–611.

Jakkula, V., 2011. *Tutorial on Support Vector Machine (SVM)*. [Online] Available at: eecs. wsu.edu/~vjakkula/SVMTutorial.doc.

Kim, M., College, P. and Hardin, J., 2010. *Statistical Classification*. [Online] Available at: http://pages.pomona.edu/~jsh04747/Student%20Theses/MinsooKim10.pdf.

Kohonen, T., 1984. *Self-Organization and Associative Memory*. Berlin: Springer-Verlag.

Kohonen, T., Barna, G. and Chrisley, R., 1988. *Statistical Pattern Recognition with Neural Networks: Benchmarking Studies*. IEEE, New York, San Diego, pp. 61–68.

Kok, S., 2007. Liquid state machine optimization (Doctoral dissertation, Master Thesis, Utrecht University).

Krishnaiah, P. and Kanal, L., 1982. Classification, pattern recognition, and reduction of dimensionality. In: *Handbook of Statistics*. North Holland, Amsterdam.

Lewis, J., 2004. *Tutorial on SVM*. [Online] Available at: http://www.scribblethink. org/Work/Notes/svmtutorial.pdf.

McCulloch, D., 2005. *An Investigation into Novelty Detection*.

McCulloch, W. and Pitts, W., 1943. A logical calculus of the ideas immanentin nervous activity forms. *Bulletin of Methematical Biophysics*, 7, 127–147.

Michie, D., Spiegelhalter, D. and Taylor, C., 1994. *Machine Learning, Neural and Statistical Classification*. [Online] Available at: http://www1.maths.leeds.ac.uk/~charles/ statlog/whole.pdf.

Mitchell, T., 1997. *Machine Learning*. Boston, MA: McGraw-Hill Computer Science Series.

Møller, M., 1993. A scaled conjugate gradient algorithm for fast supervised learning. *Neural Networks*, 4, 525–534.

Moore, A., 2003. Andrew W. *Moore's Home Page*. [Online] Available at: http://www. cs.cmu.edu/~awm.

Scalero, R. and Tepedelenlioglu, N., 1992. A fast new algorithm for training feedforward neural networks. *IEEE Transactions on Signal Processing*, 40, 202–210.

Schölkopf, B., Burges, C. J., and Smola, A. J. (Eds.), 1999. *Advances in Kernel Methods: Support Vector Learning*. MIT Press.

Skapura, D. M., 1996. *Building Neural Networks*. Addison-Wesley Professional. New York: ACM Press.

Tato, R., Santos, R., Kompe, R. and Pardo, J. M., 2002. Classifiers. In: *Emotion Recognition in Speech Signal*. Sony.

Vapnik, V., 2006. Estimation of dependencies based on empirical data. In: *Empirical Inference Science: Afterword of 2006*. New York: Springer.

Vapnik, V., Golowich, S. and Smola, A., 1997. Support vector method for function approximation, regression estimation, and signal processing. In: *Advances in Neural Information Processing Systems 9*. Cambridge, MA: MIT Press, pp. 281–287.

Wynne-Jones, M., 1993. Node splitting: A constructive algorithm for feed-forward-neural. *Neural Computing and Applications*, 1(1), 17–22.

6

Nearest Neighbor–Based Techniques

The nearest-neighbor (NN) problem is important in many areas of computer science, spanning pattern recognition, searching multimedia data, vector compression, computational statistics, data mining, and so on, and has been given a number of names, for example, the post office problem, the best match problem, and so on.

Large amounts of data are available for many of these applications, including some described here (Shakhnarovich et al., 2006), thus making NN approaches appealing. At the same time, the wealth of data increases the computational complexity of an NN search. Algorithms must be designed for these searches, not to mention the various related classification, regression, and retrieval tasks, which must be efficient even if the number of points or the dimensionality of the data expand. This research area finds itself on the boundary of several disciplines, including computational geometry, algorithmic theory, and various application fields such as machine learning.

Below, we define the exact and approximate NN search problems, and briefly survey a number of popular data structures and algorithms developed for these problems. We also discuss the relationship between the NN search and machine learning. Finally, we summarize the contents of the chapters that follow.

Maintenance engineering must deal with poorly defined problems that cannot be defined by any mathematical theory developed to date. Traditionally, engineers design ad hoc strategies based on a body of heuristics collected over years of research, but today there are more systematic ways to solve difficult real-world problems (e.g., machines performing handwriting recognition). If we can extract behavioral measures (or examples) from a problem, with the help of a learning machine, we can build a model or a device that, under certain conditions, will reflect the problem's computational structure. A statistical model is typically inferred from these data, one able to deal with the inherent uncertainty or imprecise nature of the examples. A set of adjustable parameters is estimated in the learning phase using a set of examples (the training set), but the learning machine may not be able to handle all possible parameters. A good estimation of parameters will lead to good generalizations (e.g., correct responses to unseen examples). Therefore, we must make a trade-off between capacity and the information about the problem available in the training set. *Note*: This problem can be formulated as a bias–variance trade-off or a balance between the approximation and the estimation error of the learning machine, but all formulations are similar qualitatively.

Recent work in machine learning addresses the problem of increasing generalization through capacity control, as for example, ensemble learning. The idea behind ensemble learning is combining an uncorrelated collection of learning systems (or predictors) trained in the same task. Typically, the combination is done through the use of the techniques of finding the majority in classification or averaging in regression. These techniques generally control capacity and are able to stabilize the solution by reducing dependence on the training set and using the relevant optimization algorithms.

Large margin classifiers can also be used to control capacity. In classification problems, the learning machine must assign input patterns to one of the predefined categories. These systems are usually designed to minimize the number of misclassifications in the training set, but researchers have recently found that to ensure a small generalization error, the validity of the classifications must also be considered. At the same time, classifiers must be designed to have a large margin distribution for the training samples; in other words, we must be confident that the training samples are being assigned to the correct class A large margin distribution stabilizes the solution and helps to control capacity. Support vector-learning machines (SVMs) and boosting classifiers are examples of large margin classifiers.

Other recent work considers scaling up the learning algorithms to cope with difficult high-dimensional real-world problems with large databases. Recent developments include modular and hierarchical networks, along with other types of cooperative learning machines. Many existing approaches use gradient-based learning as the unifying principle for training the whole system globally; they then back propagate errors through their complex architectures to compute updated equations for global training algorithms.

This chapter considers learning-pattern recognition in the light of current developments in machine learning. In pattern recognition, a number of machines group complex input data (or patterns) into categories (or classes) with the help of a (supervised) learning device that uses a set of labeled patterns. The core of a pattern recognizer generally includes a feature extractor and a classifier. The former reduces the input by measuring certain invariant "features" or "properties." *Note*: This reduces the complexity of the original problem and the design of the classifier, thus, making feature extraction a mechanism to control the capacity of the recognizer. Once the features are measured, the classifier uses them to assign the input pattern to a class.

In the past, the feature extractor was often done manually, as it is usually specific to the problem, but the current trend is to use learning devices able to automatically extract features and to rely less on manual feature extraction. *Note*: While unsupervised-learning algorithms are commonly used to build feature extractors from training data, their use can lead to losing important discriminatory information because they do not consider class labels.

A powerful alternative is to integrate the feature extractor into the classifier and globally train both systems to alleviate separate and uncoupled training. This novel method, termed oriented principal component analysis (OPCA), performs a global gradient-based training of a feature extractor using several lineal combinations of input variables and any classifier whose architecture allows the back propagation of an error signal (e.g., feedforward networks).

Among the many classification methods, NN is one of the most famous in machine learning. These methods are well-recognized in many fields, such as statistics, machine learning, and pattern recognition, because in spite of their simplicity, these so-called lazy learning or memory-based methods turn out to be powerful nonparametric classification systems for real-world problems. When an input pattern is presented to them, these classifiers compute the k-closest prototypes (using a distance metric specified by the user). At this point, the classifier assigns the class label using a majority vote among the labels of the k-nearest prototypes. For example, if $k = 1$, the classifier simply assigns the label of the nearest prototype to the input pattern. The parameters of these classifiers, as determined in the learning phase, are often the set of prototypes and (sometimes) the distance metric (e.g., in the context of NN classifiers, OPCA becomes a problem of learning the distance metric.) Storing the whole training database as the set of prototypes is perhaps the most direct way to compute prototypes, but storage and computational requirements and the belief that simpler solutions achieve better generalization (Occam's razor) suggest, rather, the use of more condensed sets (Shakhnarovich et al., 2006).

6.1 Concept of Neighborhood

Machine-learning systems often do not have enough training examples of a task to develop an accurate hypothesis. For example, an organization may have records on only 100 machines with a particular type of problem, and this may not be enough to formulate a hypothesis about that problem. One way to overcome a lack of training examples is to use knowledge acquired during the learning of previous related tasks. If we have learned a model to identify machines with high vibration, for example, we can use its knowledge to identify assets with an imbalance or misalignment. The process requires transferring the previously acquired knowledge (high vibration) to a new but related learning task (diagnosis). Due to "neighboring" symptoms and/or features extracted from condition indicators and collected by users, the concept can be applied to assets (Su, 2005).

Some work has been done on the fundamental theory of knowledge transfer and a method of selective knowledge transfer in the context of k-nearest neighbors (k-NNs) has been proposed (Silver, 2000; Caruana, 1993; Thrun, 1995). But these methods use the similarity between the distance metric (a structural measure) for each task, without considering the functional relationship between the tasks' output values.

Other previous work has explored methods of determining to what extent two behaviors are functionally related; these methods include linear coefficient of correlation, coefficient of determination, and Hammer distance (Silver, 2000). All methods measure functional relatedness at the task level: thus, the relationship between behaviors is based on all target values. Yet it may be useful to measure the relationship at the classification level. For example, if the output value of a previously learned behavior T1 is the same as that of the new behavior T0 for a particular class value, this relationship should not be dismissed out of hand. The two tasks are similar at some subregion of the input attribute space, and the transfer could be useful.

Few researchers have considered knowledge transfer in the context of the *k-NN* algorithm; all methods propose a transfer based on structural measures at the task level. In this chapter, we develop a functional measure of relatedness at the classification level and use it for knowledge transfer between *k-NN* tasks.

6.2 Distance-Based Methods

The best way to describe distance-based methods is to use related outlier definitions. There are currently three outstanding definitions associated with distance-based techniques. Distance-based outlier detection techniques, in general, exploit distances of data points to their corresponding neighborhood to flag outliers. The distance, also called outlier score, can be computed using only one neighbor or *k-NN*s. It can simply be used to count the total number of r-neighbors, that is, the number of data points within distance **r**, of each data point. Normally, distance-based techniques do not assume any specific distribution of the data. However, they suffer expensive computational costs when searching the nearest neighborhood. This limitation has recently motivated researchers to develop more efficient techniques with lower time complexity. These have excellent applicability for large and multidimensional data sets.

In the first distance-based detection technique, outliers comprise points from which there are fewer than **P** other points within distance **r**. To detect such outliers, researchers have introduced a nested loop and a cell-based algorithm; here, the nested-loop algorithm has time complexity $O(N^2)$ and, hence, is usually not suitable for applications on large data sets. Meanwhile, a cell-based algorithm has time complexity linear with **N**, but is exponential with the number of dimensions, or dim. In practice, this can only work efficiently when dim ≤ 4, making it inapplicable to high-dimensional data sets.

In the second distance-based detection technique, instead of counting the r-neighborhood of a data point, only the data point's distance to its kth NN is taken into account. This definition of an outlier is not intuitive enough since information of other neighbors is simply ignored when computing the outlier score.

Currently, three algorithms are proposed: nested loop with $O(N^2)$ time complexity, index-based, and partition-based algorithms. The general idea of an index-based algorithm is that by maintaining a list of top n outliers, we can prune data points whose outlier score computed so far is less than the minimum score in the list. Usually, this idea can be used in techniques where an outlier score computed so far is always an upper bound true score. An index-based algorithm is illustrated in Algorithm 1. *Note*: In this algorithm, OutHeap is the top **n** outlier based on the defined outlier score while Min(OutHeap) returns the minimum outlier score of the heap. **NN (p; k)** contains the set of k-NNs of a data point **p**. PointHeap is a data structure for maintaining the set of data points utilized in the iterations of k-NNs computation. For each data point, OutScore is its outlier score computed so far. The computation process of a data point terminates whenever its OutScore falls below the Min(OutHeap), causing the time complexity to be reduced.

Partition-based algorithm goes even further in pruning the searching space. The underlying data set is first grouped into clusters. Each cluster is assessed as to whether it contains some candidate outliers, or else it will be eliminated. With the remaining clusters, index-based or nested-loop algorithm can be used to detect outliers, while the outlier definition only considers the distance from a data point to its k-**th** NN as the outlier score. The increase in the number of distances used for computing the outlier score does not lead to any increase in time complexity, since the number of NNs that must be found for each data point in each definition is still the same, which is **k**.

Briefly stated, the notion of outliers is better and more intuitive. As mentioned, distance-based techniques usually involve computing points' neighbors, which is very time consuming. Accordingly, more recent techniques in distance-based outliers aim to introduce algorithms with less time complexity. Among the methods for reducing the computational cost, pruning outlier-searching space and computation reduction dominate. Computation-reduction techniques usually try to limit the number of detected outliers (e.g., top n outliers), and use data structures similar to those used in Ramaswamy's index-based algorithm. More specifically, a list of top n outliers found and the minimum outlier score found so far are employed to reduce the computational cost, but their proposed technique, ORCA, depends on certain assumptions, such as (a) the data are in random order and (b) the data points' values are independent. The analysis also depends on the cutoff threshold c, which is identical to Min(OutHeap). As can be observed in Algorithm 1, Min(OutHeap) usually starts at **0**. However, domain knowledge or a training phase can help to achieve a better pruning value. In particular, by training a subset of the original data set, an initial cutoff threshold can be obtained. The training phase continues if the obtained threshold at the first attempt is not as expected. During the testing phase, the final training set is placed at the top of the data set so that the cutoff threshold calculated during the training phase can be retrieved quickly; hence, the pruning occurs at the very first stage of the detecting process.

Domain knowledge can also help in choosing a suitable value for Min(OutHeap). The linear time complexity presented can only be obtained if the cutoff threshold c converges to **O(pN)** quickly, but that only happens when the data set contains many outliers, making its asymptotic time complexity **O(N. lgN)**. Instead of finding the exact NNs for each data point, this algorithm searches for the approximate ones. The approximate NNs of a data point **p** are **k** points within distance **c** from **p**. A clustering algorithm is employed (e.g., k-means clustering) to partition points into bins such that points close to each other in space are likely to be assigned to the same bin. Data point p's approximate NNs are searched in p's bin and the consecutive bins. For all normal points, the searching time is linear with respect to the size of data set, that is, **O(N)**. However, we need to perform a full scan on the entire data set for each outlier; and that searching strategy leads to a reduction in execution time. A detection technique using the Hilbert space-filling curve has been proposed to map a multidimensional space to the interval $I = [0; 1]$ to reduce the computational cost for finding k-NNs. This is done in two steps. First, map the data set **DS** to $D = [0; 1]$ dim, where dim is the number of dimensions of DS. Second, use the Hilbert space-filling curve to map **D** to **I**. Two data points that are close in I will be close in **D**, but the reverse is not always true. Searching for a data point p's NNs becomes searching for p's approximate NNs in **I** by assessing p's predecessors and successors in **I**. The proposed technique consists of two phases. During the first phase, the approximate outliers (based on approximate outlier score) are extracted from the data set using the mentioned mapping. The approximate score is always an upper bound true score. In the second phase, true outliers are extracted from the set of approximate ones. The time complexity of the first phase is reported to be O(dim². N. k), where k is the number of NNs taken into account. The second phase has time complexity to be O (N'. N. dim), where N' is the number of candidate outliers left after the end of the first phase (Vu, 2010).

6.2.1 Cell-Based Methods

Because of the computationally expensive distance function, the naive approach is unfeasible. Hence, two pruning techniques are proposed in this section. These techniques are used to detect distance-based outliers without the need of an actual distance function (Salman and Kitagawa, 2013).

6.2.1.1 Cell-Based Pruning

The cell-based pruning technique is proposed to quickly identify and prune cells containing only inliers. Similarly, it can detect cells containing outliers such as the cell-based approach of Knorr. Since the cell-based approach by Knorr deals with only deterministic data, it considers two cell layers that lie within certain distances from a target cell for its pruning. However, the objects are infinitely uncertain; hence, all the cell layers in the Grid need to be considered for pruning of the target cell.

6.2.1.1.1 Grid Structure

To identify distance-based outliers using the cell-based technique, each object $O_i \in GDB$ is mapped to a k-dimensional space that is partitioned into cells of length l. Let $C_{x,y}$ be any cell in the Grid G, where positive integers x and y denote the cell indices. The layers $(L_1,...,L_n)$ of $C_{x,y} \in G$ are the neighboring cells of $C_{x,y}$ as shown in Figure 6.1, and are defined as follows:

$$L_1(C_{x,y}) = \{C_{u,v} | u = x \pm 1, v = y \pm 1, C_{x,y} \neq C_{u,v}\}$$

$$L_2(C_{x,y}) = \{C_{u,v} | u = x \pm 2, v = y \pm 2, C_{u,v} \neq L_1(C_{x,y}), C_{x,y} \neq C_{u,v}\}$$

$L_3(C_{x,y}),...,L_n(C_{x,y})$ are defined in a similar way. The considerable maximum number of layers depends on the position of the target cell in the Grid. A cell $C_{x,y}$ in G can have the maximum number of layers if it exists at the corner of the Grid and the minimum number of layers if it exists at the center of the

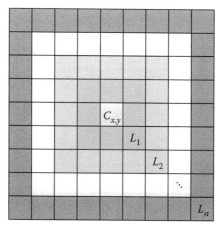

FIGURE 6.1
Cell layers. (Redrawn from Shaikh, A. S. and Kitagawa, H., 2013. Efficient distance-based outlier detection on uncertain datasets of Gaussian distribution. *World Wide Web* 1–28.)

Grid. Let **n** denote the maximum number of layers; then the minimum number of layers is given by $[n/2]$.

6.2.1.1.2 Cell Bounds

Like the cell-based approach by Knorr, the goal of the proposed cell-based technique is to identify and prune cells guaranteed to contain only inliers or outliers. A cell $\mathbf{C}_{x,y}$ can be pruned as an "outlier cell" if the expected number of D-neighbors for any object in $\mathbf{C}_{x,y}$ is less than or equal to the threshold θ. Similarly, a cell can be pruned as an "inlier cell" if the expected number of D-neighbors for any object in cell $\mathbf{C}_{x,y}$ is greater than θ. Hence, bounds on the expected number of D-neighbors of $\mathbf{C}_{x,y} \in$ are defined so as to prune them. The upper and lower bounds bind the possible expected number of D-neighbors without expensive object-wise distance computation.

Upper bound: The upper bound of a cell $\mathbf{C}_{x,y}$, U B($\mathbf{C}_{x,y}$), binds the maximum expected number of D-neighbors in Grid **G** for any object in cell $\mathbf{C}_{x,y}$. Since the Gaussian distribution is infinite, two objects in the same cell may reside at the same coordinate. Hence, the maximum expected number of D-neighbors in $\mathbf{C}_{x,y}$ for any object in cell $\mathbf{C}_{x,y}$ itself is equal to the number of objects in $\mathbf{C}_{x,y}$ denoted by $N(\mathbf{C}_{x,y})$. Similarly, the maximum expected number of D-neighbors in cells in layer $L_m(\mathbf{C}_{x,y})$ $(1 \le m \le n)$ for any object in $\mathbf{C}_{x,y}$ can be obtained as follows:

$$\sum_{m=1}^{n} N(L_m(\mathbf{C}_{x,y})) * P_r((m-1)l, D)$$

where $N(L_m(\mathbf{C}_{x,y}))$ denotes the number of objects in layer $L_m(\mathbf{C}_{x,y})$. Figure 6.2 shows how the $\alpha = (m-1)$ values are obtained for the upper bounds. Hence, $UB(\mathbf{C}_{x,y})$ of $\mathbf{C}_{x,y} \in G$ is derived as follows:

$$UB(\mathbf{C}_{x,y}) = N(\mathbf{C}_{x,y}) + \sum_{m=1}^{n} N(L_m(\mathbf{C}_{x,y})) * P_r((m-1)l, D).$$

Lower bound: The lower bound of a cell $\mathbf{C}_{x,y}$, LB($\mathbf{C}_{x,y}$), binds the minimum expected number of D-neighbors in the Grid for any object in cell $\mathbf{C}_{x,y}$. When two objects in the same cell reside at opposite corners, the probability that they are D-neighbors takes the minimum value. Hence, the minimum expected number of D-neighbors in $\mathbf{C}_{x,y}$ for any object in cell $\mathbf{C}_{x,y}$ itself is equivalent to $1 + (N(\mathbf{C}_{x,y}) - 1) * P_r(\sqrt{2}.l, D)$. Similarly, the minimum expected number of D-neighbors in cells in layer L_m $(\mathbf{C}_{x,y})$ $(1 \le m \le n)$ for any object in $\mathbf{C}_{x,y}$ can be obtained as follows:

$$\sum_{m=1}^{n} N(L_m(\mathbf{C}_{x,y})) * P_r((m+1)\sqrt{2}.l, D).$$

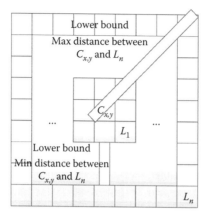

FIGURE 6.2
Cell and layers bounds. (Redrawn from Shaikh, A. S. and Kitagawa, H., 2013. Efficient distance-based outlier detection on uncertain datasets of Gaussian distribution. *World Wide Web* 1–28.)

Figure 6.2 shows how the $\alpha = (m + 1) \sqrt{2l}$ values are obtained for the lower bounds. Hence, $L_B(\mathbf{C}_{x,y})$ of $\mathbf{C}_{x,y} \in$ is derived as follows:

$$LB(C_{x,y}) = 1 + N(C_{x,y}) - 1) * P_r(\sqrt{2l}, D) + \sum_{m=1}^{n} N(L_m(C_{x,y})) * P_r(m + 1)\sqrt{2l}, D)$$

Look-up table: The bounds discussed above are required by each $\mathbf{C}_{x,y} \in G$ for pruning. Each bound computation requires evaluation of the costly distance function $P_r(\alpha, D)$ and the object counts of the respective cell $\mathbf{C}_{x,y}$ and its layers $L_m(\mathbf{C}_{x,y})$. The number of distance function computations for the bounds calculation can be reduced by precomputing $P_r(\alpha, D)$ values for $\mathbf{C}_{x,y}$ bounds. Since the $P_r(\alpha, D)$ values are decided only by the α-values and are independent of the locations of $\mathbf{C}_{x,y}$, $P_r(\alpha, D)$ values need to be computed only for $\alpha = m \sqrt{2l}$ $(1 \le m \le n + 1)$ and $\alpha = ml$ $(0 \le m \le n - 1)$. The precomputed values are stored in a look-up table to be used by the cell-based pruning technique.

6.2.1.1.3 Cell Pruning

Having defined bounds and the look-up table, a cell $\mathbf{C}_{x,y} \in G$ can be pruned as an inlier cell or identified as an outlier cell as follows. If $LB(\mathbf{C}_{x,y})$ is greater than θ, $\mathbf{C}_{x,y}$ cannot contain outliers. Hence, it can be pruned as an inlier cell.

Alternatively, if $UB(\mathbf{C}_{x,y})$ is less than or equal to θ, $\mathbf{C}_{x,y}$ is identified as an outlier cell. Lines 1–12 in Algorithm 1 show the cell-based pruning technique (Salman and Kitagawa, 2013).

Algorithm 6.1 Cell-based outlier detection

Input: GDB, D, p, l
Output: Set of distance-based outliers O
  1: Create cell grid G depending upon dataset GDB values and cell length l;
  2: Initialize $Count_k$ of each cell $C_k \in G$;
  3: Map each object o in GDB to an appropriate C_k, and increment $Count_k$
     by 1;
  4: $\theta \leftarrow |GDB|(1 - p)$, $O = \{\}$; (θ correspond to the threshold)
     /*Bounds computation*/
  5: Compute $P_r(\alpha, D)$ values for the computation of bounds as discussed
     in Section 4.1.2;
     /*Pruning cells using bounds*/
  6: **for each** non-empty C_k in G **do**
  7:    **if** $LB(C_k) > \theta$ **then**
  8:        C_k is an inlier cell, mark C_k green. GOTO Next C_k;
  9:    **else if** $UB(C_k) \leq \theta$ **then**
  10:       C_k is an outlier cell, add objects of C_k to o, mark C_k black. GOTO
            Next C_k;
  11:   **end if**
  12: **end for**
     /*Object-wise pruning*/
  13: $O = O \cup ObjectWisePruning(G, D, \theta)$;
     /*Unpruned objects processing*/
  14: **for each** object o_i in non-empty, uncoloured $C_k \in G$ **do**
  15:   **if** o_i is uncoloured **then** compute EN_{oi} (expected number of
         D-neighbours of o_i) using objects in C_k and higher layers of $C_k \in G$;
  16:   **if** $EN_{oi} \leq \theta$ **then** o_i is outlier. Add o_i to o;
  17: **end for**
  18: **return** o;

6.2.2 Index-Based Methods

Let **N** be the number of objects in dataset **T**, and let **F** be the underlying distance function that gives the distance between any pair of objects in **T**.

For an object **0**, the D-neighborhood of **0** contains the set of objects $Q \varepsilon T$ that are within distance D of **0** (i.e., $\{Q \varepsilon T] F(O,Q) \leq D\}$). The fraction **p** is the minimum fraction of objects in **T** that must be outside the D-neighborhood of an outlier. For simplicity of discussion, let **M** be the maximum number of objects within the D-neighborhood of an outlier; that is, $M = N(1 - P)$.

From the formulation above, it is obvious that, given **p** and **D**, the problem of finding all DB(p, D)-outliers can be solved by answering a NN or range query centered at each object **0**. More specifically, based on a standard multidimensional indexing structure, we execute a range search with radius **D** for each object **0**. As soon as $(M + 1)$ neighbors are found in the D-neighborhood, the search stops, and 0 is declared a nonoutlier; otherwise, 0 is an outlier. Analyses of multidimensional indexing schemes reveal that for variants of R-trees and K-d trees, the lower bound complexity for a range search is $R(N^{1-1/k})$, where **k** is the number of dimensions or attributes and **N** is the number of data objects. As **k** increases, a range search quickly reduces to $O(N)$, giving, at best, a constant time improvement reflecting sequential search. Thus, the above procedure for finding all DB(p, D)-outliers has a worst-case complexity of $O(kN^2)$. Two points are worth noting:

- Compared to the depth-based approaches, which have a lower bound complexity of $\Omega(N^{\lceil k/2 \rceil})$, DB-outliers scale much better with dimensionality. The framework of DB-outliers is applicable and computationally feasible for datasets with many attributes, that is, $k \geq 5$.

  This is a significant improvement on the current state of the art, where existing methods can only realistically deal with two attributes.

- The above analysis only considers search time. When it comes to using an index-based algorithm, most often, for the kinds of datamining applications under consideration, it is a very strong assumption that the right index exists.

6.2.3 Reverse NN Approach

The most popular variant of NN query is reverse nearest neighbor (RNN), a query that focuses on the inverse relation among points. An RNN query variant of an NN query is RNN, a query that focuses on the inverse relation among points. An RNN query **q** is to find all the objects for which **q** is their NN. It is formally defined below.

Definition 6.1

Given a set of objects **P** and a query object **q**, an RNN query finds a set of objects RNN such that for any object **p** ε **P** and *r* ε RNN, dist $(r, q) \leq$ dist (r, p) (Muhammad, 2007). ∎

The RNN set of a query **q** may either be empty or have one or more elements. Korn and Muthukrishnan (2000) have defined RNN queries and suggested numerous applications. A two-dimensional RNN query, for example, may ask for the set of customers most likely to be affected by the opening of a new store so they may be informed about the opening. Alternatively, the query can be used to identify the location likely to maximize the number of potential customers. In a related example, an RNN query may be asked

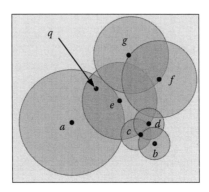

FIGURE 6.3
The objects *a* and *e* are the RNNs of *q*.

to find other stores affected by the opening of a new store at some specific location. *Note*: The first example has two different sets (customers and stores) involved in the query whereas the second example has only one (stores) (Korn and Muthukrishnan, 2000) defining two variants of RNN queries. A bichromatic query (the first example) seeks find RNNs when the underlying data set comprises two different types of objects. In contrast, a monochromatic RNN query (the second example) is asked to find RNNs when the data set has only one type.

The problem of RNNs has been extensively studied in the past few years. Korn and Muthukrishnan (2000) answer the RNN query by precalculating a circle of each object **p** such that the NN of **p** lies on the perimeter of the circle, as shown in Figure 6.3. The minimum bounding rectangles (MBRs) of all these circles are indexed by an R-tree called RNN-tree. The problem of an RNN query is reduced to a point location query on the RNN-tree that returns all the circles containing **q**. For example, the circle of **a** and **e** contains **q**; so both are the RNNs of **q**. The intermediate nodes contain the MBRs of underlying points, along with the maximum distance from every point in the subtree to its NN. The problem with the above-mentioned techniques is that they rely on precomputation and cannot deal with efficient updates. To alleviate this problem, many suggest using density-based methods.

6.3 Density-Based Methods

Density-based methods (local outlier factor [LOF], local correlation integral [LOCI]), in general, assign to each data point a factor describing the relative density of that data point's neighborhood. Similar to the distance-based

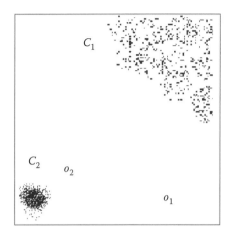

FIGURE 6.4
Example showing the case of distance-based outlier definitions. (Redrawn from Vu, N. H., 2010. *Outlier Detection Based on Neighborhood Proximity.* Singapore: Nanyang Technological University.)

approach, density-based detection also involves the computation of data point NNs. However, the measurement of a data point p to its NNs is then compared to its neighbors' same measurement. The purpose of doing so is to overcome different effects of dense and sparse clusters on points' neighborhood in detecting outliers, but this comes with a trade-off, as the computational cost becomes even more expensive than that of distance-based techniques. In spite of this, once again, because of their applicability for large and high-dimensional data, such kinds of methods still attract considerable attention from the research community (Vu, 2010).

There is a popular example that is often used to highlight the advantage of the density-based approach (Figure 6.4). Assume the distance from every object p_3 in C_1 to its NN is greater than the distance from p_2 to its NN in C_2. If it is a distance-based definition, there will be no values of P and r such that p_2 will be an outlier while every object in C_1 is not.

The outlier score used, called LOF, is a measure of difference in neighborhood density of a point p and with the same measurement as other points in its local neighborhood, LOF is able to capture local outliers. For data points that belong to a cluster, the LOFs are approximately equal to 1, while for each outlier, the corresponding value should be much higher. All the computations of LOF depend on MinPts, which is used to compute the neighborhood density for each data point. The choice of MinPts, however, is not simple.

6.3.1 Local Outlier Factor

Between formal definitions, we have the following LOF:

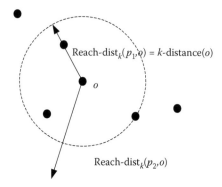

FIGURE 6.5
Reach-dist(p_1,o) and reach-dist(p_2,o), for $k = 4$.

Definition 6.2

k-Distance of p: The k-distance of p, denoted as k-distance (p) is defined as the distance $d(p; o)$ between p and o such that

- For at least k data points $0' \varepsilon \, DS \backslash p$, it holds that $d(p; 0') \leq d(p; o)$
- For at most (k–1) data points $0' \varepsilon \, DS \backslash p$, it holds that $d(p; 0') < d(p; o)$

Figure 6.5 shows an example of such a definition.   ∎

Definition 6.3

k-Distance of p's neighborhood: The k-distance of p's neighborhood contains every object whose distance from p is not greater than the k-distance, and is denoted as (Wen et al., 2001)

$$N_k(p) = \{q \ \varepsilon \ DS \backslash p \backslash D(p, q) \leq k\text{-distance}(p)\}$$

These objects q are called the k-NNs of p. Whenever no confusion arises, we simplify our notation to use $N_k(p)$ as a shorthand for $N_{k\text{-distance}}(p)(p)$. Note: In Definition 6.2, the k-distance(p) is well-defined for any positive integer k, although the object o may not be unique. In this case, the cardinality of $N_k(p)$ is greater than k. For example, suppose there is (i) one object with distance 1 unit from p; (ii) two objects with distance 2 units from p; and (iii) three objects with distance 3 units from p. Then, 2-distance(p) is identical to 3-distance(p), and there are three objects of 4-distance(p) from p. Thus, the cardinality of $N_4(p)$ can be greater than 4, in this case 6.   ∎

Definition 6.4

Reachability distance of p w.r.t. o: The reachability distance of data point p with respect to o is defined as (Ankerst et al., 1999)

$$\text{Reach-dist}_k (p, o) = \max\{k\text{-distance}(o), d(p, o)\}$$

Figure 6.5 illustrates the idea of reachability distance with $k = 4$. Intuitively, if object p is far away from o (e.g., p_2 in Figure 6.5), then the reachability distance between the two is simply their actual distance. However, if they are "sufficiently" close (e.g., p_1 in Figure 6.5),the actual distance is replaced by the k-distance of o. The reason is that in doing so, the statistical fluctuations of $d(p, o)$ for all ps close to o can be significantly reduced. The strength of this smoothing effect can be controlled by the parameter k. The higher the value of k, the more similar the reachability distances for objects within the same neighborhood.

So far, we have defined k-distance(p) and reach-distk(p) for any positive integer k. But for the purpose of defining outliers, we focus on a specific instantiation of k that links us back to density-based clustering. In a typical density-based clustering algorithm, two parameters define the notion of density: (i) a parameter MinPts specifying a minimum number of objects; (ii) a parameter specifying a volume. These two parameters determine a density threshold for the clustering algorithms. That is, objects or regions are connected if their neighborhood densities exceed the given density threshold. To detect density-based outliers, however, it is necessary to compare the densities of different sets of objects, which means we must determine the density of sets of objects dynamically. Therefore, we keep MinPts as the only parameter and use the values reach-distMinPts(p, o), for o Î NMinPts(p), as a measure of the volume to determine the density in the neighborhood of an object p (Breunig et al., 2000). ■

Definition 6.5

Local reachability density of p: The local reachability density of a data point p is the inverse of the average reachability distance from the k-NNs of p (Wen et al., 2001) and is shown as

$$Lrdk(p) = 1 \Bigg/ \left[\frac{\sum_{o \varepsilon Nk(p)} \text{reach-dist}_k(p, o)}{|N_k(p)|} \right]$$

Intuitively, the local reachability density of an object p is the inverse of the average reachability distance based on the MinPts-NNs of p. *Note:* The local density can be ¥ if all the reachability distances in the summation are 0. This may occur for an object p if there are at least MinPts objects, different from p, but sharing the same spatial coordinates; that is, if there are at least MinPts duplicates of p in the data set. For simplicity, we will not handle this case explicitly but simply assume there are no duplicates. (To deal with duplicates, we can base our notion of neighborhood on a k-distinct distance, defined analogously to k-distance in Definition 6.2, with the additional requirement that there must be at least k objects with different spatial coordinates.) ∎

6.3.1.1 Properties of Local Outliers

In this section, we conduct a detailed analysis of the properties of LOF. The goal is to show that our definition of LOF captures the spirit of local outliers, and has many desirable properties. Specifically, we show that for most objects p in a cluster, the LOF of p is approximately equal to 1. As for other objects, including those outside a cluster, we provide a general theorem giving a lower and upper bound on the LOF. Furthermore, we analyze the tightness of our bounds and show that the bounds are tight for important classes of objects. However, for other classes of objects, the bounds may not be as tight. For the latter, we give another theorem specifying better bounds (Breunig et al., 2000).

- LOF for objects deep in a cluster.

Below, we show that for most objects in C1, the LOF is approximately 1, indicating that they cannot be labeled as outlying.

Lemma 6.1

Let C be a collection of objects. Let reach-dist-min denote the minimum reachability distance of objects in C; that is, reach-dist-min = min {reach-dist (p, q) $|p, q \varepsilon C$}.
   Similarly, let reach-dist-max denote the maximum reachability distance of objects in C.
   Let ε be defined as (reach-dist-max/reach-dist-min–1). Then, for all objects $p\varepsilon C$, such that:

   i. All the MinPts-NNs q of p are in C
   ii. All the MinPts-NNs o of q are also in C

Therefore, it holds that $1/(1 + \varepsilon) \leq LOF(p) \leq (1 + \varepsilon)$. ∎

Proof (Sketch)

For all MinPts-NNs q of p, reach-dist$(p, q) \geq$ reach-dist-min. Then, the local reachability density of p, as per Definition 6.5, is ≤1/reach-dist-min.

However, reach-dist$(p, q) \leq$ reach-dist-max. Thus, the local reachability density of p is ≥1/reach-dist-max.

Let q be a MinPts-NN of p. By an argument identical to the one for p above, the local reachability density of q is also between 1/reach-dist-max and 1/reach-dist-min. Thus, we have reach-dist-min/reach-dist-max ≤ LOF(p) ≤ reach-dist-max/reach-dist-min and can establish $1/(1 + \varepsilon) \leq$ LOF$(p) \leq (1 + \varepsilon)$.

The interpretation of Lemma 6.1 is as follows. Intuitively, C corresponds to a "cluster." Let us consider the objects p that are "deep" inside the cluster; this means all the MinPts-NNs q of p are in C, and, in turn, all the MinPts-NNs of q are also in C. For such deep objects p, the LOF of p is bounded. If C is a "tight" cluster, the e value in Lemma 6.1 can be quite small, forcing the LOF of p to be quite close to 1. To return to the example in Figure 6.4, we can apply Lemma 6.1 to conclude that the

LOFs of most objects in cluster C_1 are close to 1.

- A general upper and lower bound on LOF.

Lemma 6.1 above shows a basic property of LOF, namely that for objects deep inside a cluster, the LOFs are close to 1, and should not be labeled as local outliers. A few immediate questions come to mind. What about those objects near the periphery of the cluster? And what about those objects outside the cluster, such as o_2 in Figure 6.4? Can we get an upper and lower bound on the LOF of these objects?

Theorem 6.1 below shows a general upper and lower bound on LOF (p) for any object. As such, Theorem 6.1 generalizes Lemma 6.1 along two dimensions. First, it applies to any object p, and is not restricted to objects deep inside a cluster. Second, even for objects deep inside a cluster, the bound given by Theorem 6.1 can be tighter than the bound given by Lemma 6.1, implying that the epsilon defined in Lemma 6.1 can be made closer to zero because in Lemma 6.1, the values of reach-dist-min and reach-dist-max are obtained based on a larger set of reachability distances. In contrast, in Theorem 6.1, this minimum and maximum are based on just the MinPts-nearest neighborhoods of the objects under consideration giving rise to tighter bounds.

Before we present Theorem 6.1, we define the following terms. For any object p, let directmin(p) denote the minimum reachability distance between p and a MinPts-NN of p, that is,

$$\text{direct}_{min}(p) = \min\{\text{reach} - \text{dist}(p,q) | q \; \varepsilon \; N_{MinPts}(p)\}$$

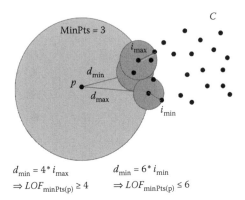

$$d_{min} = 4^* \, i_{max}$$
$$\Rightarrow LOF_{minPts(p)} \geq 4$$

$$d_{min} = 6^* \, i_{min}$$
$$\Rightarrow LOF_{minPts(p)} \leq 6$$

FIGURE 6.6
An example illustrating the general upper and lower bound on LOF with MinPts = 3.

Similarly, let $direct_{max}(p)$ denote the corresponding maximum, that is,

$$direct_{max}(p) = max\{reach - dist(p,q) | q \, \varepsilon \, N_{MinPts}(p)\}$$

To generalize these definitions to the MinPts-NN q of p, let $indirect_{min}(p)$ denote the minimum reachability distance between q and a MinPts-NN of q; that is,

$$indirect_{min}(p) = min\{reach - dist(q,o) | q \, \varepsilon \, N_{Minpts}(p) \text{ and } o \, \varepsilon \, N_{Minpts}(q)\}$$

Similarly, let $indirect_{max}(p)$ denote the corresponding maximum. In the sequel, we refer to p's MinPts-nearest neighborhood as p's direct neighborhood, and refer to q's MinPts-NNs as p's indirect neighbors, whenever q is a MinPts-NN of p. Figure 6.6 gives a simple example to illustrate these definitions. In this example, object p lies some distance away from a cluster of objects C. For ease of understanding, let MinPts = 3. The $direct_{min}(p)$ value is marked as dmin in the figure; the $direct_{max}(p)$ value is marked as d_{max}. Since p is relatively far away from C, the 3-distance of every object q in C is much smaller than the actual distance between p and q. Thus, from Definition 6.4, the reachability distance of p with respect to q is given by the actual distance between p and q. Now, among the three-NNs of p, we, in turn, find their minimum and maximum reachability distances to their three-NNs. In the figure, the $indirect_{min}(p)$ and $indirect_{max}(p)$ values are marked as i_{min} and i_{max}, respectively.

Theorem 6.1

Let p be an object from the database D, and $1 \leq MinPts \leq |D|$. Then, it is the case that

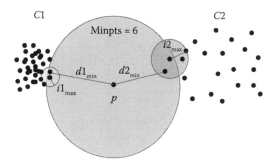

FIGURE 6.7
Relative span for LOF depending on the percentage of fluctuation for *d* and *w*.

$$\frac{\text{direct}_{min}(p)}{\text{indirect}_{max}(p)} \leq LOF(p) \leq \frac{\text{direct}_{max}(p)}{\text{indirect}_{min}(p)}$$

To illustrate the theorem using the example in Figure 6.7, suppose d_{min} is 4 times that of i_{max}, and d_{max} is 6 times that of i_{min}. Then, by Theorem 6.1, the LOF of p is between 4 and 6. It should also be clear from Theorem 6.1 that LOF(p) has an easy-to-understand interpretation: it is simply a function of the reachability distances in p's direct neighborhood relative to those in p's indirect neighborhood. The figure below gives an example of Theorem 6.1 (Ankerst et al., 1999).

- The tightness of the bounds.

As discussed before, Theorem 6.1 is a general result with the specified upper and lower bounds for LOF applicable to any object p. An immediate question comes to mind. How good or tight are these bounds? In other words, if we use LOF_{max} to denote the upper bound $\text{direct}_{max}/\text{indirect}_{min}$, and use LOF_{min} to denote the lower bound $\text{direct}_{min}/\text{indirect}_{max}$, how large is the spread or difference between LOF_{max} and LOF_{min}? In the following, we study this issue. A key part of the following analysis is to show that the spread $\text{LOF}_{max} - \text{LOF}_{min}$ is dependent on the ratio of direct/indirect. It turns out that the spread is small under some conditions, but not so small under other conditions.

Given $\text{direct}_{min}(p)$ and $\text{direct}_{max}(p)$ as defined above, we use direct(p) to denote the mean value of $\text{direct}_{min}(p)$ and $\text{direct}_{max}(p)$. Similarly, we use indirect(p) to denote the mean value of $\text{indirect}_{min}(p)$ and $\text{indirect}_{max}(p)$. In the sequel, whenever no confusion arises, we drop the parameter p, for example, direct as a shorthand of direct(p).

To summarize, if the fluctuation of the average reachability distances in the direct and indirect neighborhoods is small (i.e., pct is low) Figure 6.8, Theorem 6.1 estimates the LOF very well, as the minimum and maximum

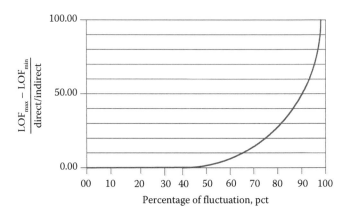

FIGURE 6.8
Illustration of Theorem 6.2.

LOF bounds are close to each other. There are two important instances when this is true (Breunig et al., 2000):

1. The percentage pct is very low for an object p, if the fluctuation of the reachability distances is rather homogeneous, that is, if the MinPts-NNs of p belong to the same cluster. In this case, the values $direct_{min}$, $direct_{max}$, $indirect_{min}$, and $indirect_{max}$ are almost identical, resulting in LOF being close to 1. This is consistent with the result established in Lemma 6.1.
2. The argument above can be generalized to an object p that is not located deep inside a cluster, but whose MinPts-NNs all belong to the same cluster (as depicted in Figure 6.7). In this case, even though LOF may not be close to 1, the bounds on LOF as predicted by Theorem 6.1 are tight.
3. Bounds for objects whose direct neighborhoods overlap multiple clusters.

To this point, we have analyzed the tightness of the bounds given in Theorem 6.1, and provided two conditions under which the bounds are tight. An immediate question that comes to mind is under what condition are the bounds not tight? On the basis of Figure 6.8, if the MinPts-NNs of an object p belong to different clusters having different densities, the value for pct may be very large. Then, based on Figure 6.8, the spread between LOF_{max} and LOF_{min} value can be large. In this case, the bounds given in Theorem 6.1 do not work well.

Theorem 6.2 intends to give better bounds on the LOF of object p when p's *MinPts*-nearest neighborhood overlaps with more than one cluster. The intuitive meaning of Theorem 6.2 is that when we partition the *MinPts*-NNs of p into several groups, each group proportionally contributes to the LOF of p.

An example is MinPts = 6. In this case, three of object p's six-NNs come from cluster C_1, and the other three come from cluster C_2. Then, as shown in Figure 6.7, according to Theorem 6.2, LOF_{min} is given by $(0.5 * d1_{min} + 0.5 * d2_{min})/(0.5/i1_{max} + 0.5/i2_{max})$, where $d1_{min}$ and $d2_{min}$ give the minimum reachability distances between p and the six-NNs of p in C_1 and C_2, respectively, and $i1_{max}$ and $i2_{max}$ give the maximum reachability distances between q and q's six-NNs, where q is the six-NN of p from C_1 and C_2, respectively. For simplicity, Figure 6.8 does not show the case for the upper bound LOF_{max}. ∎

Theorem 6.2

Let p be an object from the database D, $1 \leq$ MinPts $\leq |D|$, and $C_1, C_2,...,C_n$ be a partition of $N_{MinPts}(p)$, that is, $N_{MinPts}(p) = C_1 \cup C_2... \cup C_n \cup \{p\}$ with $C_i \cup C_j = \emptyset$, $C_i = \emptyset$ for $1 \leq ij \leq n$, $i \neq j$.

Furthermore, let $\xi_i = |C_i|/|N_{MinPts}(p)|$ be the percentage of objects in p's neighborhood, which are also in C_i. Let the notions direct $i_{min}(p)$, direct $i_{max}(p)$, indirect $i_{min}(p)$, and direct $i_{max}(p)$ be defined analogously to direct$_{min}(p)$, direct$_{max}(p)$, indirect$_{min}(p)$, and indirect$_{max}(p)$ but restricted to the set C_i (e.g., direct $i_{min}(p)$, denoting the minimum reachability distance between p and a MinPts-NN of p in the set C_i). Then, it holds that (a)

$$LOF(p) \geq \left(\sum_{i=1}^{n} \xi_i . \text{direct } i_{min}(p) \right) . \left(\sum_{i=1}^{n} \frac{\xi_i}{\text{indirect } i_{min}(p)} \right)$$

and

$$LOF(p) \geq \left(\sum_{i=1}^{n} \xi_i . \text{direct } i_{max}(p) \right) . \left(\sum_{i=1}^{n} \frac{\xi_i}{\text{indirect } i_{min}(p)} \right)$$

Theorem 6.2 generalizes Theorem 6.1, taking into consideration the ratios of the *MinPts*-NNs coming from multiple clusters. As such, the following corollary can be formulated. Corollary 1: If the number of partitions in Theorem 6.2 is 1, then LOF_{min} and LOF_{max} given in Theorem 6.2 are exactly the same corresponding bounds as those given in Theorem 6.1. ∎

6.3.1.2 Connectivity Outlier Factor

We can compute the connectivity-based outlier factor (COF) at data record p with respect to its k-neighborhood with (Pokrajac et al., 2008)

$$COF(p) = \frac{ac - dist_{N_k(p) \cup p}(p)}{\frac{1}{k} \sum_{o \varepsilon N_K(P)} ac - dist_{N_k(o) \cup o}(o)} \quad (6.1)$$

COF is computed as the ratio of the average chaining distance from data record p to $Nk(p)$ and the averaged average chaining distances at the record's neighborhood.

Applying static COF outlier detection algorithms to data streams would be extremely computationally inefficient, thus making incremental outlier techniques essential. The static COF algorithm can be applied to data streams in an "iterated" way by reapplying it every time a new data record pc is inserted into the data set. However, each time a new record is inserted, the algorithm recomputes COF values for all the data records from the data set. Given the time complexity of the COF algorithm of $O(N\log_n)$ (for moderate data record dimensions), where n is the current number of data records in the data set, after the insertion of N records, the total time complexity for the "iterated" approach is

$$O\left(\sum_{n=1}^{N} n\log n\right) = O(N^2 \cdot \log N)$$

Alternatively, we could perform static COF periodically after inserting particular data blocks, but this "periodic" method cannot identify the exact time when an outlier appears in the database; nor can it detect outliers that will later be classified as normal records, because the data set is not stationary.

Our proposed incremental COF algorithm will provide the same results detecting outliers as the "iterated" COF described above. All existing records in the database retain the same COF values as the "iterated" COF algorithm. Because the proposed algorithm has time complexity $O(N \cdot \log N)$, it clearly outperforms the static "iterated" COF approach. After the N data records are all inserted into the data set, the result of the incremental COF algorithm on these N data records is independent of the order of insertion and equivalent to the static COF that is performed after all data records are inserted.

6.3.1.2.1 Methodology

We have two aims in our design of an incremental COF algorithm. First, we want the result of the incremental algorithm to be equivalent to the result of the "iterated" static algorithm every time a new record is inserted into a data set. Nor should there be a difference between the application of incremental COF and static COF when all data records up to a certain time are available. Second, the asymptotic time complexity of an incremental COF algorithm must be comparable to the static COF algorithm. To have a feasible incremental algorithm, at any moment in time, the insertion/deletion of the data record must result in small (and preferably limited) numbers of updates of the algorithm parameters. To be clear, the number of updates per insertion/ deletion must be independent of the current number of records in the data set, or the time complexity of the incremental COF algorithm will be $\Omega(N2)$ (*note*: N is the size of the final dataset). In this section, we suggest ways to insert and delete records for the incremental COF algorithm and discuss the time complexities of each.

- Incremental COF algorithm.

Our proposed incremental COF algorithm computes the COF value for each data record that is inserted into the data set and immediately determines whether it is an outlier. COF values for the existing data records are also updated if needed (Pokrajac et al., 2008).

- Insertion.

The following two tasks are performed: (a) insertion of a new record into the database and computation of its *ac-dist* and COF; (b) maintenance, if the ac-dist and COF values require updating for the records already in the database.

For example, consider the insertion of a new record p into a database of two-dimensional records; ac-dist may change for a certain record q if the set of its k-NNs $Nk(q)$ changes due to the insertion of p. Put otherwise, ac-dist(q) can change if a data record q is among the reverse k-NN of p. Since p is among k-NNs of q (see Figure 6.9), the *ac-dist(q)* should be updated. *Note:* The set of records where *ac-dist* needs to be updated after the insertion of p is denoted by $S_{update\_ac\_dist}(p)$ in the rest of the chapter. Now, denote the set of records where COF should be updated as $S_{update\_COF}(p)$. Therefore, $COF(o)$ for an existing record o needs to be updated if: (a) ac-dist(o) is updated; (b) the insertion of p changes the neighborhood of o (in other words, p is among k-NNs of o); and (c) ac-dist is updated for some of k-NNs of o.

Since ac-dist(o) is updated only if o is among RNNs of p, conditions (a) and (b) above imply that $S_{update\_ac\_dist}(p) \subset S_{update\_COF}(p)$. Condition (c) indicates COF must be updated for all RNNs of points from the set $S_{update\_ac\_dist}(p)$. In Figure 6.9, for example, COF will be updated on data record r, since its

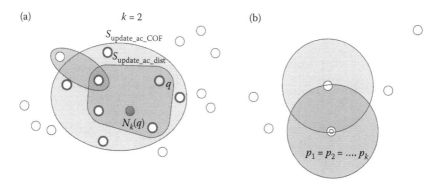

FIGURE 6.9
(a) Illustration of updates for COF insertion and (b) degenerate case of the incremental COF algorithm.

k-NN contains a record from $S_{\text{update\_ac\_dist}}(p)$. As in the LOF approach, we define *k*-th-NN of a record *p* as a record *q* from the data set *S*, such that for at least *k* records $o' \in S\text{-}\{p\}$, it holds that $d(p, o') \leq d(p, q)$, and for at most $k-1$ records $o' \in S\text{-}\{p\}$, it holds that $d(p, o') < d(p, q)$. In this case, $d(p, q)$ denotes the Euclidean distance between data records *p* and *q*, and *k*-NNs $(Nk(p))$ include all data records $r \in S\text{-}\{p\}$ such that $d(p, r) \leq d(p, q)$. We also define *k* reverse nearest neighbors of *p* (called *kRNN* (*p*)) as all data records *q* for which *p* is among their *k*-NNs. For a given data record *p*, $N_k(p)$ and $kRNN(p)$ may be, respectively, retrieved by executing NN and RNN (a.k.a. inverse) queries on a data set *S*. The general framework for the insertion of new data record for the COF algorithm is given in Figure 6.10a (Pokrajac et al., 2008).

- Deletion.

In data stream applications, it is frequently necessary to delete irrelevant data records because of obsolescence or changes in regime. Figure 6.10b shows the general framework for deleting a block of data records S_d from data set *S*. Note that it is very similar to the insertion scheme. At the start of the process, the record that must be deleted from the set is marked accordingly. Of course, the removal of a record *p* may affect the *k*-RNN of the other records. Because of the update of *ac-dist*, the COF value must be updated. In addition to data records where *ac-dist* values change, this update includes RNNs. At the end of the procedure, the data record can be deleted from the database (Pokrajac et al., 2008).

Two technical comments are relevant at this point. First, in the algorithms shown above, $S_{\text{update\_ac\_dist}}$ (*p*) denotes the set where *ac-dist* should be recomputed, not where the *ac-dist* will actually change. Yet our experimental evidence suggests *ac-dist* will always change for all points from $S_{\text{update\_ac\_dist}}$ (*p*).

(a)

```
IncCOF_insertion(Dataset S)
Given: Set S {p₁, ... ,pₙ}pᵢ∈R^D to be inserted
into the database
For ∀pₑ∈S
        insert(pₑ);
        compute Nₖ(pₑ)=kNN(pₑ)
        compute ac-dist_{Nk(pe)∪pe}(pₑ) using Eq.
(1)
        S_{update_ac_dist} =reverse k-NN(pₑ);
        //candidates for update of ac_dist;
        (∀pⱼ ∈ S_{update_ac_dist})
                update ac-dist_{Nk(pi)∪pi} (pⱼ);
//Eq. (1)
        S_{update_cof} = S_{update_ac_dist};
        (∀pₘ ∈ S_{update_ac_dist})
                S_{update_cof} = S_{update_cof}∪reverse k-
NN(pₘ);
        (∀p₁ ∈ S_{update_cof})
                update COF(p₁) using Eq.
```

(b)

```
IncCOF_deletion(Database S, Dataset Sₔ)
Given: Set S_D={p₁,...,pₙ} pᵢ∈R^D to be
deleted from the Database S
For ∀pₑ∈S_D
        compute reverse-kNN(pₑ);
        compute ac-dist_{Nk(pe)∪pe} (pₑ) using
Eq.(1);
        S_{update_ac_dist} =reverse k-NN(pₑ);
        (∀pⱼ ∈ S_{update_ac_dist})
        update ac-dist_{Nk(pj)∪pj} (pⱼ);
        S_{update_cof} = S_{update_ac_dist};
        (∀pₘ ∈ S_{update_ac_dist})
                S_{update_cof} = S_{update_cof} ∪ reverse k-
NN(pₘ);
        (∀p₁ ∈ S_{update_cof})
                update COF(p₁) Eq.(2);
        delete (pc);
```

FIGURE 6.10
The pseudo code for (a) insertion and (b) deletion of incremental COF algorithm.

Second, should the reverse k-NN of points from $S_{update\_ac\_dist}$ be computed with or without considering the deleted record p? In our view, the data record p can be deleted from a database before S_{update_COF} is computed. We justify this as follows: let r be an existing data record such that $r \notin S_{update\_ac\_dist}(p)$. Then, $p \notin Nk(r)$. Now assume $r \in S_{update\_COF}(p)$. Then, there is an $o \in S_{update\_ac\_dist}(p)$ such that $o \in Nk(r)$. Since $p \notin Nk(r)$, the deletion of p will not affect the neighborhood of r and S_{update_COF} can safely be determined after p is deleted.

The static COF algorithm does not cover a degenerate case where the denominator in Equation 6.1 equals to zero, but this scenario is possible, as a database of dynamic data could contain identical records (corresponding to different time instants). For example, consider k identical records $p_1,\dots,p_k$ and record q such that records $p_1,\dots,p_{k+1}$ are k-NNs of q and q is among k-NNs of each of the records $p_1,\dots,p_{k+1}$ (see Figure 6.1). In this case, clearly $ac\text{-}dist_{Nk(pi)} \cup_{pi}(pi) = 0$, $i = 1,\dots,k+1$ and $ac\text{-}dist_{Nk(qi)} \cup_{qi}(q) > 0$.

$COF(p_i)$ will involve division by 0 and therefore be undefined. But COF is defined as the ratio of $ac\text{-}dist$ at the data record and the average $ac\text{-}dist$ at its neighborhood. If the $ac\text{-}dist$ at the record equals the average $ac\text{-}dist$, COF should be equal to 1. Therefore, for data records where the numerator and denominator of Equation 6.1 are both zeros, as for example, data records pi, we stipulate $COF(p_i) \equiv 1$. But when only the denominator of Equation 6.1 is equal to 0, as in record q, the neighborhood of the record will have an infinitely larger density than the record itself will have. Accordingly, for these data records, we stipulate $COF(q) \equiv \infty$ (*note*: We assign $COF(q)$ a large number, in practice).

6.4 Multigranularity Deviation Factor

In this section, we introduce the multigranularity deviation factor (MDEF), which satisfies the properties listed above. Let the r-neighborhood of an object p_i be the set of objects within distance r of p_i.

Intuitively, the MDEF at radius r for a point p_i equals the relative deviation of its local neighborhood density from the average local neighborhood density in its specific r-neighborhood. Thus, an object whose neighborhood density matches the average local neighborhood density will have an MDEF of 0. In contrast, outliers will have MDEFs far from 0.

To be more precise, we define the following terms. Let $n(p_i, \alpha_r)$ be the number of objects in the α_r-neighborhood of p_i. Let $\hat{n}(p_i, r, \alpha)$ be the average, over all objects p in the r-neighborhood of p_i, of $n(p, \alpha r)$ (see Figure 6.11). The use of two radii serves to decouple the neighbor size radius r from the radius r over which we are averaging. We denote the function $\hat{n}(p_i, \alpha, r)$ over all r as the LOCI.

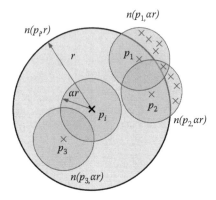

FIGURE 6.11
Definitions for n and $\hat{n}$—for instance, $n(pi, r) = 4$, $n(pr, \_r) = 1$, $n(p1, \_r) = 6$ and $\hat{n}$ $(pi, r, \_) = (1 + 6 + 5 + 1)/4 = 3.25$.

Definition 6.6

MDEF: For any pi, r, we define the MDEF at radius (or scale) r as (Papadimitriou et al., 2003).

$$MDEF(p_i, r, \alpha) = \frac{\hat{n}(p_i, r, \alpha) - n(p_i, \alpha r)}{\hat{n}(p_i, r, \alpha)} = 1 - \frac{n(p_i, \alpha r)}{\hat{n}(p_i, r, \alpha)} \qquad (6.2)$$

Note: The r-neighborhood for an object p_i always contains p_i. This implies $\hat{n}$ $(p_i, \alpha, r) > 0$ and the above quantity is always defined. For faster computation of MDEF, we will sometimes estimate both $n(p_i, \alpha r)$ and $\hat{n}(pi, r, \alpha)$. This leads to the following definition.

Definition 6.7

Counting and sampling neighborhood: The counting neighborhood (or αr-neighborhood) is the neighborhood of radius αr, over which each $n(p, \alpha r)$ is estimated. The sampling neighborhood (or r-neighborhood) is the neighborhood of radius r, over which we collect samples of $n(p, \alpha r)$ to estimate $\hat{n}(pi, r, \alpha)$ (Papadimitriou et al., 2003).

In Figure 6.11, for example, the large circle bounds the sampling neighborhood for p_i, while the smaller circles bound counting neighborhoods for various p. The main outlier detection scheme we propose relies on the standard deviation of the r-neighbor count over the sampling neighborhood of p_i. We thus define the following quantity:

$$\sigma_{MDEF}(p_i, r, \alpha) = \frac{\sigma\hat{n}(p_i, r, \alpha)}{\hat{n}(p_i, r, \alpha)}$$

Advantages of our definitions: Among several alternatives for an outlier score (such as max($\hat{n}/n$, $n/\hat{n}$), to give one example), our choice allows us to use probabilistic arguments for flagging outliers. This is a very important point.

The above definitions and concepts make minimal assumptions. The only general requirement is that a distance be defined. Arbitrary distance functions are allowed, and these may incorporate domain-specific, expert knowledge, if desired. Furthermore, the standard-deviation scheme assumes pairwise distances at a sufficiently small scale are drawn from a single distribution, which is reasonable. For the fast-approximation algorithms, we make the following additional assumptions (the exact algorithms do not depend on these):

- Objects belong to a k-dimensional vector space, that is, $p_i = (p_i^1, p_i^2,..., p_i^k)$. This assumption holds in most situations. However, if the objects belong to an arbitrary metric space, it is possible to embed them into a vector space. There are several techniques for this using the L1 norm on the embedding vector space (Ankerst et al., 1999).

- We use the L_1 norm, defined as $\|p_i - p_j\| \infty \equiv \max_1 \le m \le k |p_i^m - p_j^m|$. This is not a restrictive hypothesis, since it is well known that, in practice, there are no clear advantages of one particular norm over another.

Finally, see Figure 6.11.

6.5 Use of Neural Network Based in Semisupervised and Unsupervised Learning

6.5.1 Semisupervised Learning with Neural Networks

It is possible to apply the principles of graph-based semisupervised learning to neural networks to provide more efficient ways to exploit a higher number of either labeled or unlabeled pixels. Because neural networks can be trained by the stochastic gradient descent (SGD), they can easily scale to millions of samples, something unfeasible with SVM-based methods. This obviously increases the attraction of neural networks for large-scale remote-sensing problems, such as those encountered in semisupervised image classification (Ratle et al., 2010).

Following the semisupervised regularization framework, we propose to minimize the following function:

dsdsds

$$L = \frac{1}{l} \sum_{i=1}^{l} V(X_i, y_i \cdot f) + \gamma m \frac{1}{(l+u)^2} \sum_{i,j=1}^{l+u} L(f_i, f_j, W_{ij}),$$

where the edge weights W_{ij} define pairwise similarity relationships between unlabeled examples.

Note: This problem, like the transductive or Laplacian support vector machines (TSVMs) objective, is nonconvex in the nonlinear case. It cannot be solved by a simple optimization scheme; even linear models such as kernel machines cannot solve it. However, we propose to minimize this function in the primal by SGD, and use a multilayer perceptron for solving the nonlinear case.

The general scheme of semisupervised neural networks goes as follows. Essentially, the algorithm is given both labeled and unlabeled samples. For each iteration, it takes gradient steps to optimize the loss function of errors (labeled information), V, and the regularizer (unlabeled information), L. *Note*: Both supervised and unsupervised neural networks use this approach. Because it allows the weighting of V and L in the same neural net, the model offers a general learning framework for classification problems.

The implementation of the proposed algorithm is to define a loss function for both labeled and unlabeled samples, a neural network topology, an optimization algorithm, and balancing constraints. We analyze these issues in more detail in the following subsections. For the most part, in this framework, neighbors are defined according to a k-NNs algorithm.

- The loss function for supervised classification.

We use the hinge loss function in Equation 6.4 for V, as do SVM, LapSVM, or TSVM. Traditionally, neural networks use a squared loss function that is appropriate for Gaussian-noise distributions. When this assumption does not hold, using the hinge loss function may be more appropriate. In fact, for the classification setting, the use of entropic or hinge loss functions may be more suitable (Vapnik, 1998).

- The loss function for unsupervised classification.

LapSVM implements a functional version of Laplacian Eigenmaps. Instead of learning a one-to-one mapping between the input and the embedding space, it learns a function (an SVM) preserving neighborhood relations. But optimizing the Laplacian Eigenmaps loss term $Wij \, \|f(xi) - f(xj)\|^2$ can be difficult, especially as we must enforce the following constraints: $f^tDf = I$ and f $^tD1 = 0$. These ensure the new features have zero mean and unit variance. Because these constraints make the optimization difficult, we seek a loss function that will permit unconstrained optimization. Accordingly,

$$L(f_i, f_j, W_{ij}) = \begin{cases} \sum_{ij} \|f_i - f_j\|^2 & \text{if } W_{ij} \quad \text{if } W_{ij} = 1 \\ \sum_{ij} \max(0, m - \|f_i - f_j\|^2 & \text{if } W_{ij} = 0 \end{cases} \tag{6.3}$$

where $W_{ij} = 1$ if i and j are deemed similar and 0 otherwise; m is a margin implemented in this loss in a neural network and is termed "DrLIM" (dimensionality reduction by learning an invariant mapping). DrLIM permits similar examples to be mapped closely, with dissimilar ones separated by at least the distance m, thereby preventing the embedding from collapsing and making the use of constraints unnecessary, where the LapSVM approach is generalized to networks of several layers. When we adapt this type of objective function directly to the task of classifying the unlabeled data (Karlen et al., 2008), we get an approach interestingly related to TSVM. Rather than performing an optimization over the coordinates of the samples in the embedding, we propose the direct optimization of the labels of unlabeled data by directly encoding into the algorithm that neighbors with $W_{ij} > 0$ should have the same class assignment. To this end, we can optimize a general objective function such as

$$L(f_i, f_j, W_{ij}) = \left\{ \begin{array}{ll} \sum_{ij} \eta^{(+)} V(x_i, f(x_i), c) & with\ c = sign(f_i + f_j) \\ \sum_{ij} -\eta^{(-)} V(x_i, f(x_i), c) & with\ c = sign(f_i) \end{array} \right\} \qquad (6.4)$$

where
   W is a pairwise similarity matrix defined *a priori*, as previously shown.
   $V(\cdot)$ is the hinge loss function. For the multiclass case, we can sum over the classes Ω, so that $V(x, f(x), y) = \sum_{c=1}^{\Omega} \max(0, 1 - y(c)f(x))$, where $y(c) = 1$ if $y = c$ and -1 otherwise.

$\eta(+)$ and $-\eta(-)$ are learning rates. The classifier should be trained to classify x_i and x_j in the same class if $W_{ij} = 1$, with learning rate $\eta(+)$, and also trained to push them in different classes if $W_{ij} = 0$, with learning rate $-\eta(-)$.

Intuitively, Equation 6.4 assigns a pair of neighbors to the cluster with the most confident label from the pair. Examples x_j that are not neighbors of x_i, in other words, when $W_{ij} = 0$, are encouraged to fall into different clusters. Figure 6.12 illustrates this principle.

Note: If only the L cost is used (Equation 6.4), the method only works with unlabeled samples, thereby performing unsupervised learning. We can do this by setting M arbitrarily large. As a result, the proposed method constitutes a generalization of both supervised and unsupervised approaches. The NCutEmb approach provides algorithms for the binary and the multiclass case in an unsupervised setting. In the binary case, Equation 6.4 describes the minimized loss using a neural network $f(x)$ with one output $y \in \{\pm1\}$ and trained online via SGD. For the multiclass case, two algorithmical variants are provided, NCutEmbmax and NCutEmball; these; these differ in the way the "winning" class is chosen. They are also evaluated in this chapter. NCutEmbmax is very similar to the binary case:

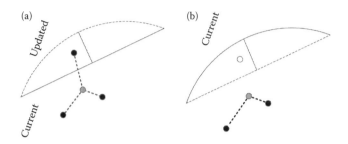

FIGURE 6.12
Unsupervised learning using NCutEmb. A hyperplane that separates neighbors is pushed to classify them in the same class ((a) the classifier cuts through an edge of the graph and is pushed upward), while the hyperplane that classifies nonneighbors in the same class is modified to separate them ((b) the hyperplane is pushed to separate the unconnected points).

we push neighbors toward the most confident label. With NCutEmball, we push toward all clusters simultaneously, with one learning rate η per class. Each learning rate is weighted by the outputs of the neural network (Ratle et al., 2008).

- Neural network architecture.

We propose using a multilayer neural network model, whose neuron j in layer $l + 1$ yields

$$x_j^{l+1} = g\left(\sum_i w_{i,j}^l x_i^l + w_{bj}^l \right)$$

where $w_{i,j}^l$ are the weights connecting neuron i in layer $l - 1$ to neuron j in layer l, w_{bj}^l are the bias term of neuron j in layer l (we fix it to +1), and g is a nonlinear activation function (we use the hyperbolical tangent). As in the LapSVM case, we denote the model's output (prediction) for the sample xi as $f(x_i)$, and the vector of all predictions as $f = [f(x_1),\dots, f(x_l + u)]^\tau$. For the multiclass case, the network has an output node for each class, but all outputs share the same hidden neurons, as is often the case in neural networks.

The advantages of neural networks for semisupervised learning include the following:

- Kernel methods, by way of contrast are computationally demanding as the number of samples increases.
- In neural networks, regularization can be easily controlled by limiting the number of hidden neurons in the network.

- The model is readily applicable to new pixels that become available.
- It is easy to encode prior knowledge using the neural network's architecture.
- Neural networks permit stochastic gradient optimization.

To clarify the final point, gradient descent works by moving toward the minimum of the loss function by taking a step proportional to the negative of the loss function's gradient. This technique, commonly called batch gradient descent, generally requires computing the average value of the gradient over the whole data set. This can be costly when working with a very large number of pixels. To avoid this problem, we propose training using SGD (Bottou and LeCun, 2005). Unlike batch gradient descent, SGD processes one example at a time, requiring fewer calculations and often yielding better generalization properties, as it finds an approximate minimum rather than an exact one. With SGD, parameters w_{ij} are updated using

$$W^{t+1} = w^t - \eta \nabla l$$

where w^t is the weight vector at training epoch t, and l refers to any differentiable loss function. If we train with SGD, we can handle very large databases, as every update involves one (or a pair) of examples, and grows linearly in time with the size of the data set. In addition, the algorithm will converge for low-enough values of η.

6.5.2 Unsupervised Learning with Neural Networks

The artificial neural networks with unsupervised learning (or self-supervised) do not require any external element to adjust the weight of the communication links to their neurons. They do not receive the information from the environment that indicates if the generated output, in response to a determined input, is correct or incorrect: hence the general understanding that unsupervised artificial neural networks are capable of self-organization (Masters, 1993).

The main problem in the unsupervised classification is to how divide the space where the objects are (space of characteristics) into groups or categories. Intuitively, closeness criteria are used for this: in other words, an object belongs to a group if it is similar to the elements that integrate that group.

When there is no supervisor, the network must determine for itself the characteristics, regularities, correlations, or categories of input data. There are many different interpretations of the output of the unsupervised networks, depending on their structure and the learning algorithm used. In some cases, the output represents a degree of familiarity with or similarity between the signal being introduced into the network and the information gathered to that point. Under other circumstances, information can be grouped (clustered) to generate a category structure, but in this case, the network detects the categories from the correlations between the presented information. In such a situation, the output of the network codifies the input data, keeping the relevant information. Finally, some networks with unsupervised learning can map characteristics (i.e., feature mapping), generating a geometric disposition in the output neurons that represents a topographic map of the characteristics of the input data. This gives the network similar information that will always affect output neurons close to each other, that is, in the same mapping zone.

In general, there only two kinds of unsupervised learning (Eduardo, 2009):

- Hebbian learning
- Competitive learning

6.5.2.1 Hebbian Learning

Hebbian learning applies to a group of neurons strongly attached by a complex structure. Efficiency is identified by the intensity or the magnitude of a connection, that is, the weight. Thus, Hebbian learning basically consists of the adjustment of the weights of the communication links according to the correlation of the values of activation (outputs) of two connected neurons; when the two neurons are activated at the same time, the weight is reinforced through the following expression:

$$\Delta w_{ki}(t) = \eta * y_k(t) * x_i(t) \tag{6.5}$$

where η is a positive constant-denominated learning rate, y_k is the state of activation of the kth computational neuron in observation, and x_i corresponds to the state of activation of the ith node of the preceding layer. This expression is Hebbian, because if the two units are activated (positive), the connection is reinforced. But if one unit is activated and the other is not (negative), the value of the connection weakens. A basic rule of unsupervised learning is that the weight is modified as a function of the states (outputs) of the obtained nodes without asking if these states of activation will be attained or not.

A Hebbian process is characterized by four key properties:

1. It is time dependant: The communication link occurs when the two computational neurons are activated.
2. It is local: For the effect described by Hebb to take place, the nodes must be continuous in the space, but the modification only has a local effect.

3. It is interactive: Modification occurs only when both units are activated. However, it is not possible to predict the activation.

4. It is correlated. Given the co-occurrence of the activations to be produced in very short periods of time, the effect described by Hebb is also termed a compound synapse. For one of the computational neurons to be activated, however, the activation must be related to the activation of one or many previous nodes. This is why it is called a correlated synapse.

A disadvantage of the first point (time dependency) is that the exponential growth leads to a weight saturation. To avoid this, Kohonen proposes the incorporation of a term to modulate growing, thereby obtaining the following expression:

$$\Delta w_{ki}(t) = \eta * y_k(t) * x_i(t) - \alpha * y_k(t) * w_{ki}(t)$$

Another version of the learning rule is the so-called Hebbian differentail which uses the correlation of the derivative of the functions of the computational neurons.

6.5.2.2 Competitive Learning

In networks with competitive and cooperative learning, neurons compete and cooperate with other neurons. In Hebbian learning, many output nodes can be activated simultaneously, but in competitive learning, only one output node can be activated at a time. In other words, competitive learning tries to find a winner to take over all units, or to find a winner in a group of nodes, for example, a node that introduces information. The rest of the nodes are forced to take minimum-value answer keys.

Competition occurs in all layers of the network. It appears in neighboring nodes (of the same layer) as recurrent connections of excitation or inhibition. If the learning is cooperative, these connections with the neighbors will be of excitation.

Since we are not talking about supervised learning, the objective of competitive learning is to group the data that are introduced into the network. Similar neurons are organized into the same category and activate the same exit neuron. The network creates groups by detecting correlations between the input data. Consequently, the individual units of the network learn to specialize in a set of similar elements, and for that reason, they become detectors of characteristics.

The simplest artificial neural network using a competitive learning rule is formed by a totally connected input and exit layer; the exit nodes also include lateral connections. The connections between layers can be ones of excitation, for example, with the lateral connections between nodes of the exit layer inhibiting. In this type of network, each neuron in the exit layer is assigned a gross weight, the sum of all the weights of its connections. Learning affects only the connections with the winning neuron, however. Therefore, it redistributes its gross weight between its various connections, removing a portion of the weights of all connections that feature the winning neuron and distributing this amount between all the connections coming from active units. Therefore, if j is the winning node, the variation of the weight between unit i and j is null if neuron j does not receive excitation from neuron i. In other words, it does not win in the presence of a stimulus on the part of i, and it will be modified (i.e., it will be reinforced) if it is excited by this neuron i. In the end, each unit in the winning exit has discovered a group.

References

Ankerst, M., Breunig, M. M., Kriegel, H.-P., and Sander, J., 1999. *OPTICS: Ordering Points to Identify the Clustering Structure*. Philadelphia, PA: ACM.

Bottou, L. and LeCun, Y., 2005. On-line learning for very large datasets. *Journal of Applied Stochastic Models in Business and Industry*, 21(2), 137–151.

Breunig, M., Kriegel, H.-P., Raymond, T. N., and Sander, J., 2000. *LOF: Identifying Density-Based Local Outliers*. Dallas, Texas, USA: ACM Press, pp. 93–104.

Caruana, R. A., 1993. Multitask connectionist learning. In: *Proceedings of the 1993 Connectionist Models Summer School*. Carnegie Mellon University.

Eduardo, G. A., 2009. *Artificial Neural Networks*. [Online] Available at: http://edugi.uni-muenster.de/eduGI.LA2/downloads/02/ArtificialNeuralNetworks240506.pdf.

Karlen, M., Weston, J., Erken, A., and Collobert, R., 2008. Large scale manifold transduction. In: *Proceedings of the 25th International Conference on Machine Learning*. Helsinki, Finland: ACM, pp. 448–455.

Korn, F. and Muthukrishnan, S., 2000. *Influence Sets Based on Reverse Nearest Neighbor Queries*. New York: ACM, pp. 201–212.

Masters, T., 1993. *Practical Neural Networks Receipes in C++*. London: Academic Press.

Muhammad, A. C., 2007. *Reverse Nearest Neighbor Queries*. [Online] Available at: http://users.monash.edu.au/~aamirc/thesis/node22.html.

Papadimitriou, S., Kitagawa, H., Gibbons, P., and Faloutsos, C., 2003. *LOCI: Fast Outlier Detection Using the Local Correlation Integral*. Pittsburgh, PA, USA: Carnegie Mellon University, pp. 315–326.

Pokrajac, D., Reljin, N., Pejcic, N., and Lazarevic, A., 2008. *Incremental Connectivity-Based Outlier Factor Algorithm*. London, UK: British Computer Society, pp. 211–223.

Ratle, F., Camps-Valls, G., and Weston, J., 2010. Semi-supervised neural networks for efficient hyperspectral image classification. *IEEE Transactions on Geoscience and Remote Sensing*, 48(5), 2271–2282.

Ratle, F., Weston, J., and Miller, M., 2008. Large-scale clustering through functional embedding. In: *Machine Learning and Knowledge Discovery in Databases.* Heidelberg, Berlin: Springer, pp. 266–281.

Shaikh, A. S. and Kitagawa, H., 2013. Efficient distance-based outlier detection on uncertain datasets of Gaussian distribution. *World Wide Web* 1–28.

Shakhnarovich, G., Darrell, T., and Indyk, P., 2006. *Nearest-Neighbor Methods in Learning and Vision: Theory and Practice.* Cambridge, MA, USA: The MIT Press.

Silver, D. L., 2000. *Selective Transfer of Neural Network Task Knowledge.* London: Faculty of Graduate Studies, University of Western Ontario.

Su, Y., 2005. *Selective Knowledge Transfer from K-Nearest Neighbour Tasks Using Functional Similarity at the Classification Level.* Nova Scotia, Canada: Acadia University.

Thrun, S., 1995. *Lifelong Learning: A Case Study (No. CMU-CS-95-208).* Pittsburgh, PA: Carnegie-Mellon University.

Vapnik, V., 1998. *Statistical Learning Theory.* New York: John Wiley and Sons.

Vu, N. H., 2010. *Outlier Detection Based on Neighborhood Proximity.* Singapore: Nanyang Technological University.

Wen, J., Anthony, K. H. T., and Han, J., 2001. *Mining Top-n Local Outliers in Large Databases.* San Francisco, California, USA: ACM Press, pp. 293–298.

7

Cluster-Based Techniques

Our world is filled with data. Every day, whether we realize it or not, we measure and observe data. Data can be as mundane as a grocery list or more complex, a description of the characteristics of a living species or a 'natural phenomenon, a summary of the results of a scientific experiment, or a record of the events in the life cycle of a mechanical system. Data provide a basis for understanding all kinds of objects and phenomena, allowing us to analyze them and make decisions about them. Grouping data into categories or clusters based on certain similar properties is one of the most important analytic activities, and is the topic of this chapter.

Grouping or classifying data is a basic human activity (Anderberg, 1973; Everitt, 2001), indispensable to human development (Xu and Wunsch, 2009). When faced with a new object or a new phenomenon, people look for features that are found in other, better-known objects or phenomena; in effect, they compare the new object to another, looking for both similarities and differences. When people are presented with a new object in nature, for example, they will classify that object into one of three groups: animal, plant, or mineral. If it can be classified as an animal, it can be further classified into kingdom, phylum, class, order, family, genus, and species, working from general categories to specific ones. Thus, we have animals named tigers, lions, wolves, dogs, horses, sheep, cats, mice, and so on, and within those categories, we would have the names of specific types of sheep, horses, cats, and so on. Naming and classifying are essentially synonymous, according to Everitt (2001). With such classification information at hand, we can infer the properties of a specific object based on the category to which it belongs. For instance, when we see a seal lying on the ground, we know it is a good swimmer without seeing it swim.

This need for classification is endemic in all activities of our lives. Especially when dealing with the malfunction of a system, human beings need to identify the kind of failure before fixing it so they can schedule the proper remedial actions and avoid possible disasters. Clearly, getting a proper diagnosis depends on the quality of available data and what we know about the system. When this knowledge covers all possible types of failure in the system, we can take a model-based approach, but this is not always the case.

The diagnosis of a complex process, in the absence of precise knowledge and without a mathematical model, can be developed from measures recorded during previous normal and abnormal situations. These data can be mined and used to define the operational states through training mechanisms and expertise. For example, classification techniques allow us to establish a model of a system's states (behavioral model) by extracting its attributes (raw or statistical characteristics, including average and standard deviation, as well as qualitative information) related to a particular behavior without this behavior being represented by a set of analytical relations. Changes in these characteristics enable the detection of abnormal operations (Claudia Isazaa, 2008; Sylviane, 2007).

Among the many classification techniques, those using fuzzy logic have advantages when we want to express the degree of an observation's (data) membership in several classes, as this can model knowledge uncertainty and imprecision. In general, these methods work well if their initial parameters are carefully selected. Several different approaches have been proposed for the selection of data (Anon, 2001; Claudia Isazaa, 2008; Wang, 2005).

7.1 Categorization versus Classification

At this point, it is relevant to mention two key words, well-known to condition monitoring experts: classification and categorization. Novices often think they are synonyms, but they are not at all. When we talk about classification, we are basically talking about the projection of collected data into a closed catalog of known failures or states of malfunction. In this scenario, we simply decide what category the failure belongs to.

In categorization, however, things are entirely different. The maintainer collects data but the states of malfunction are unknown; therefore, these data are automatically grouped in a finite number of states not known *a priori*. These data can be updated and refreshed during a process of ongoing categorization.

Process monitoring uses the classification method as well. Here, the current classification and its association with a previously determined functional state of the process must be determined at each sample time. There are two phases: training and recognition. In training, the objective is to find the process behavior characteristics, as these will allow us to differentiate the process states (each associated with a class). The initial algorithm parameters are selected by a process expert who validates the obtained behavioral model. In recognition, the operator recognizes and identifies the current process state. More specifically, at each sampling time, a vector collects the accessible information (raw data or pretreated data: filtered, FFT, etc.) provided for monitoring; the class recognition procedure tells the operator the current functional state of the process. To optimize the procedure, we propose including in the training phase a step to automatically validate and adjust the clusters. The proposed approach automatically improves a nonoptimal initial partition in terms of compactness and class separation, thus facilitating the ability to discriminate between classes, that is, between operation modes (Claudia Isazaa, 2008).

7.2 Complex Data in Maintenance: Challenges or Problems?

Modern maintenance storage and condition monitoring technology makes the accumulation of data increasingly easy. However, it is not just a matter of data size; complexity is also an issue. Data in maintenance departments, as described in previous chapters, comprise many different data sets: condition indicators (online and off-line); original equipment manufacturers' (OEMs) data with images, documents, and so on; work orders reported by maintainers; operator expertise; and so on. The combination of maintenance information sources creates a complex scenario where faults may happen in a number of different contexts. Today, failure diagnosis must identify faulty states in many contexts and conditions, something traditional approaches are unable to tackle.

Given the dramatic and unprecedented growth in data, our prediction and data analysis tools must be continually updated to handle the volume. Data mining extracts information from large databases, but for data mining to work well, the storage structure must be considered. A recently developed mining technique for large databases is Association Rule Mining; more traditional techniques from statistics and machine learning, such as classification and clustering, have also been adapted and used. Classification is used to predict a data point's membership in one of a finite number of classes on the basis of a certain attribute. The attribute indicating class membership is the class label attribute. Clustering is used to identify classes in data without a predefined class label. This chapter looks at techniques of both classification and clustering (Denton, 2003).

7.3 Introduction

Three major techniques in machine learning are clustering, classification, and feature reduction, all widely used in condition monitoring. Classification and clustering are also broadly known as unsupervised and supervised learning. In supervised learning, the object is to learn predetermined class assignments from other data attributes. For example, given a set of condition monitoring data for samples with known faults, a supervised learning algorithm might learn to classify fault states based on patterns of condition. In unsupervised learning, there are no predetermined classes, or class assignments are ignored. Cluster analysis is the process by which data objects are grouped together based on some relationship defined between objects. In both classification and clustering, an explicit or implicit model is created from the data to help predict future data instances or understand the physical process behind the data. Creating these models can be a very intensive task, for example, training a neural network (NN). Feature reduction or selection reduces the data attributes used to create a data model, thereby reducing analysis time and creating simpler and (sometimes) more accurate models.

Which is better, unsupervised or supervised algorithms? The debate is ongoing in condition monitoring. Academia likes supervised ones as these data sets are clean and have good quality. However, the ugly truth is data sets are seldom complete or clean. In fact, we cannot really train an expert system to handle all potential malfunctions, especially when we add in the various combinations. Accordingly, the use of unsupervised algorithms, even if not yet mature, seems to have promise if academicians want to transfer knowledge about clustering to industry (Denton, 2003).

7.3.1 What Is Clustering?

The main idea of cluster analysis is very simple (Bacher, 1996, p. 1–4): "Find K clusters (or a classification that consists of K clusters) so that the objects of one cluster are similar to each other whereas objects of different clusters are dissimilar."

Clustering can be a problem in *unsupervised learning*, as the goal is to find a *structure* in a set of unlabeled data (Matteucci, 2003). However, clustering is a tremendously powerful technique for identifying an unknown number of faulty states, including different combinations of failure modes that eventually may happen, as well as those that remain unknown to the user *a priori*.

A broad definition of clustering is "the process of organizing objects into groups whose members are similar in some way." Thus, a cluster is a collection of "similar" items that are very different from objects belonging to other clusters. A simple graphic example appears in Figure 7.1 (Matteucci, 2003).

In the figure, we can easily identify four clusters of data, with distance being used as the similarity criterion: two or more objects belong to the same group if they are close according to a certain distance (i.e., geometric distance). This is known as *distance-based clustering*.

Conceptual clustering is another type of clustering: two or more objects belong to the same cluster if they share a common *concept*. In other words, objects are grouped, not based on simple similarity measures, but on how well they fit the descriptive concepts. In maintenance, the technique can be used to identify faulty states: a variable represents a fault only when the environment or operation displays certain aspects; otherwise, a fault is not occurring.

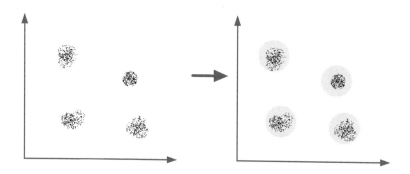

FIGURE 7.1
Example of four clusters.

7.3.2 Goal of Clustering

The goal of clustering is to determine the intrinsic grouping in a set of unlabeled data. But how do we decide what constitutes good clustering? How do we decide when a failure is not a failure or is a different kind of failure? There is no absolute "best" criterion that will be independent of the final clustering. In other words, the users must supply their own criterion, one that will adapt the clustering results to their specific needs. For example, we may want to find "natural clusters" and describe their unknown properties (*"natural" types of data*); alternatively, we may be looking for representatives of homogeneous groups (*reduction of data*), or groupings of useful and suitable data ("useful" classes of data), or even unusual data objects (*outlier detection*) (Matteucci, 2003).

This Information Technology (IT) vocabulary may sound strange to maintainers, but it suits their purposes. When we look for natural clusters, for example, we may be interested in identifying the original failure modes rather than the different combinations present in the system. After identification, we may want to know the commonalities of things, findings, or states clustered in the same group as such knowledge may be helpful in decision making. Finally, identifying abnormal behavior is always relevant in finding NFF (no fault found) situations or isolating infrequent failure modes that do not follow normal clustering techniques due to their low frequency or intensity (Matteucci, 2003).

7.3.3 Clustering as an Unsupervised Classification

Classification systems are either supervised or unsupervised; in other words, they may assign new data objects to one of a finite number of discrete supervised or unsupervised categories (Bishop, 1995; Cherkassky and Mulier, 1998; Duda et al., 2001; Xu and Wunsch, 2009).

In supervised classification, the mapping from a set of input data vectors, denoted as $x \in \mathcal{R}d$, where d is the input space dimensionality, to a finite set of discrete class labels, represented as $y \in 1,\ldots,$ where C is the total number of class types, is modeled in terms of some mathematical function $y = y(x,w)$, where w is a vector of adjustable parameters. The parameters' values are determined (optimized) by an inductive learning algorithm (an inducer), whose goal is to minimize an empirical risk functional (related to an inductive principle) on a finite dataset of input–outputs (x_i, y_i), $i = 1,\ldots,$ N, where N is the finite cardinality of the available representative data-set (Bishop, 1995; Cherkassky and Mulier, 1998; Kohavi, 1995). An induced classifier is generated when the inducer reaches convergence or terminates (Kohavi, 1995).

Supervised classification is only possible when deep knowledge of the system is available or the system has been observed over a long period. In both situations, the maintainer may have information about all faulty states (or at least the most frequent) with certain characteristics that allow their fast identification from within the collected data. That is why classification with supervised diagnosis is more pattern recognition than anything else.

In unsupervised classification, also called clustering or exploratory data analysis, no labeled data are available (Everitt, 2001; Jain and Dubes, 1988). The goal of clustering is to separate a finite, unlabeled data set into a finite and discrete set of "natural" hidden data structures, not to provide an accurate characterization of unobserved samples generated from the same probability distribution (Baraldi and Alpaydin, 2002; Cherkassky and Mulier, 1998). This can put the task of clustering outside the framework of unsupervised predictive learning problems, such as vector quantization (Cherkassky and Mulier, 1998), probability density function estimation (Bishop, 1995; Fritzke, 1997), and entropy maximization (Fritzke, 1997). *Note*: Clustering differs from multidimensional scaling (perceptual maps), in that the latter seeks to depict all evaluated objects in such a way as to minimize topographical distortion while using as few dimensions as possible. Also note: Many (predictive) vector quantizers are used for (nonpredictive) clustering analysis in practice (Cherkassky and Mulier, 1998).

It is clear from the above discussion that a direct reason for unsupervised clustering is the need to explore the unknown nature of data that are integrated with little or no prior information. For example, consider the diagnosis of failure and maintenance actions. For a particular type of failure, there may be several unknown subtypes with a similar morphological appearance that respond differently to the same maintenance actions. In this context, cluster analysis with contextual data that measure the activities of assets in certain scenarios is a promising method to uncover the subtypes and thereby determine the corresponding maintenance actions for real scenarios. The process of labeling data samples can become expensive and time consuming, making clustering a good option, as it can save both money and time. Finally, cluster analysis provides a compressed representation of the data and is useful in large-scale data analysis. Aldenderfer and Blashfield (1984) summarize the goals of cluster analysis:

1. Developing a classification
2. Investigating useful conceptual schemes for grouping entities
3. Generating hypotheses through data exploration
4. Testing hypotheses or determining if a dataset contains types defined in other procedures

Nonpredictive clustering is a subjective process in nature that precludes the necessity to pass absolute judgments (Baraldi and Alpaydin, 2002; Jain et al., 1999). Backer and Jain (1981, p. 4, 129) say, "In cluster analysis a group of objects is split up into a number of more or less homogeneous subgroups on the basis of an often subjectively chosen measure of similarity (i.e., chosen subjectively based on its ability to create 'interesting' clusters), such that the similarity between objects within a subgroup is larger than the similarity between objects belonging to different subgroups."

In fact, a different clustering criterion or clustering algorithm, even for the same algorithm but with different parameters, may cause completely different clustering results. For instance, human beings may be classified based on a wide variety of factors: ethnicity, region, age, gender, socioeconomic status, education, job, hobbies/interests, weight/height, favorite food, fashion sense, and so on.

Many different clusters can be created in any situation, depending on the criteria used for clustering. In our work-related scenario, the proper selection of variables sensitive to certain operating conditions will obviously perform better clustering, thus more clearly identifying faults and preventing confusion.

Depending on the clustering criterion, an individual may be assigned to a number of very different groups. There is no "best" criterion. Each has an appropriate use corresponding to a particular occasion, although some criteria have broader application than others. A coarse partition divides regions into four major clusters, while a finer one uses nine clusters. Whether we adopt a coarse or fine scheme depends on the requirement of the specific problem. Hence, we cannot say which clustering results are better, in general.

Classification rests on a set of multivariate data containing a set of cases; each case has multiple attributes/variables. Set up in chart form, each attribute/variable has an associated column, and each case is represented as a row. The totality constitutes the set of cases. Attributes can be of any kind: ordinal, nominal, continuous, and so on, but in cluster analysis, none of these attributes is used as a variable of classification. Rather, the goal is to derive a rule that puts all cases in groups or clusters. We can then define a nominal classification variable and assign a distinct label to each cluster.

Obviously, we do not know the final classification variable before we do the clustering, as our goal is to define this variable. Thus, clustering is a method of unsupervised classification; before we do the clustering we do not know the classes of any of the cases. We do not have the class values to guide (or "supervise") the training of our algorithm to obtain the rules of classification.

Note: One or more of the attributes of the case set can be nominal classification variables, but they are treated as nominal variables, not as variables of classification in clustering analysis. Also note: The term "unsupervised" occasionally causes confusion. We use the term as defined above.

Some computer procedures are completely automated and require no controlling input (or supervision) from the user. They can, therefore, be termed "unsupervised." Other procedures are not completely automatic and require some input from the user at various points and are called "supervised." Since such "supervision" requires human controlling input, we term this supervision "h-supervised" (alternatively, h-unsupervised). Hence, if a clustering algorithm (an unsupervised classification process) requires human intervention, we might say we are doing an "h-supervised unsupervised classification."

h-Supervised unsupervised classification is very popular in condition monitoring since the maintainer can add variables to the clustering to speed up the process and contextualize the results, thus obtaining a meaningful outcome. In the operator's dynamic introduction of environmental variables, operating conditions, or experience from work orders, clustering for diagnosis becomes an h-supervised unsupervised classification. However, in this process, the interface between the user and the need to be understood to achieve good results is as important as the performance of the algorithm itself.

7.3.3.1 Distance Metrics

To perform a diagnosis, users collect several different signals and extract certain features from them (time and frequency domain). These n-dimensional features can be displayed in an n-dimensional space where health conditions can be compared. The simplest representation is a plane with two features on the x- and y-axis. If we have two cases, that is, C_1 and C_2, with continuous variables x and y, taking values (x_1, y_1) and (x_2, y_2), respectively, we can graph the cases in x–y space as in Figure 7.2.

Using the Pythagorean Theorem, we may write

$$d_{12} = \sqrt{(x_2 - x_1)^2 + (y_2 - y_1)^2}$$

representing the Euclidean distance between the two cases, that is, "in the x–y state space." When we define the formula for the distance between two cases, we frequently say we have defined a "distance metric." This obvious distance may represent the difference between healthy and unhealthy conditions or different failure modes. This is a key factor in the clustering process: the longer the distance the distance between, the more different the conditions between them (Rennols, 2002).

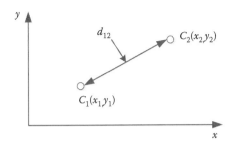

FIGURE 7.2
Euclidean distance.

If the two cases are characterized by p continuous variables (say, $x_1, x_2, \ldots x_{iy} \ldots x_p$) rather than two (i.e., x, y) [*Note:* $x \rightarrow x_1$, so $x(\text{case } 2) \rightarrow x_1(\text{case } 2) \equiv x_1(2)$; similarly, $y_2 \rightarrow x_2(2)$], we may generalize the Euclidean distance to

$$d_{12} = \sqrt{\sum_{i=1}^{p} (x_i(2) - x_i(1))^2}$$

This can be generalized further to the Minkowski metric

$$d_{12} = \sqrt[m]{\sum_{i=1}^{p} |(x_i(2) - x_i(1))|^m}$$

where $|x|$ denotes the absolute value of x (i.e., the size, without the sign).

If $m = 2$, we have the Euclidean metric. If $m = 1$, we have what is called the "city-block" metric, sometimes known as the "Manhattan" metric; it is the sum of the distances along each axis (variable) between the two cases.

7.3.3.2 Standardization

If the values of the variables are in different units, some variables will likely have large values and vary greatly between units; hence, the "distance" between cases (health conditions in the transformed space) can be large. Other variables may be small in value or vary little between cases, in which case, the difference in the variable between cases will be small. The distance metrics considered above are dependent on the choice of units for the variables involved. The metric will be dominated by those with high variability, but this can be avoided by standardizing the variables.

If $\bar{x}$ and s_x^2 are the mean and variance of the x-variables over all cases in the case set, then $z = x - \bar{x}/s_x$ or $z_i = x_i - \bar{x}_i/s_{x_i}$ is the standardized x-value for the case. Hence, the Minkowski metric for standardized variables will be

$$d_{12} = \sqrt[m]{\sum_{i=1}^{p} |(z_i(2) - z_i(1))|^m} = \sqrt[m]{\sum_{i=1}^{p} \left| \left(\frac{x_i(2) - x_i(1)}{s_{x_i}} \right) \right|^m}$$

The standardization prepares the process for proper visualization and tracking of the condition. As mentioned, the n-dimensional space requires a similar range of values to have similar values in all dimensions and be meaningful for both users and algorithms. In addition, the selection of the variables of this n-dimensional space is key to identifying the most sensitive variables in terms of degradation. An incorrect selection may produce dimensions where the variables do not change at all, making the segregation and identification of the health condition impossible.

7.3.3.3 Similarity Measures

A measure of the similarity (or closeness) between two cases takes its highest value when the cases have identical values of all variables (i.e., when the cases in the multivariable space are coincident). The measure of similarity (s_{12}) should decrease monotonically as the case variable increases, that is, as the distance increases between cases. This means any monotonically decreasing function of distance is a possible measure of similarity. If we want the similarity measure to have a value of 1 when cases are coincident, we could consider the following:

$$s_{12} = \frac{1}{1 + d_{12}} \quad \text{or} \quad s_{12} = \exp(-d_{12})$$

or similar expressions with d_{12} (distance) replaced by $(d_{12})2$.
Note: All measures defined above are suitable for continuous variables.

7.3.4 Clustering Algorithms

Clustering analysis can identify clusters embedded in data, where a cluster is a collection of "similar" data objects. Similarity is expressed by distance functions, as specified by users or experts. A good clustering method produces high-quality clusters, with low intercluster similarity and high intracluster similarity.

For example, we may cluster the centrifugal pumps in a plant according to vibration, rotation speed, flow meter reading, temperature, manufacturer, and so on.

As has been noted, unlike classification, clustering and unsupervised learning do not rely on predefined classes or class-labeled training examples. Clustering is a way of learning by observation, not by examples. Conceptual clustering groups objects to form a class, described by a concept. This differs from conventional clustering, which measures similarity by geometric distance. Conceptual clustering has two functions: (1) discovering the appropriate classes and (2) forming descriptions for each class, as in classification. The guideline of striving for high intraclass and low interclass similarity still applies.

This conceptual clustering is especially relevant in maintenance where condition indicators extracted from physical variables are small pieces of available information that must be fused with experience, OEM information, work orders, and so on. Numerical variables may be fused with additional information, creating both metadata and a class that can be clustered. This differs from traditional condition monitoring where only numerical distance is considered.

The contextual approach requires processing many databases which "speak" different languages to create the classes to be clustered. Data mining research has been focused on high-quality and scalable clustering methods for large databases and multidimensional data warehouses like the ones generated today by maintenance departments and asset managers.

An example of clustering is what most people perform when they group their clothes at the laundry: permanent press, whites, and colors. These clusters have important common attributes in the way they behave when washed.

Clustering is straightforward but often difficult because it can be dynamic. The typical requirements of clustering in data mining are:

1. Highly scalable clustering algorithms are needed in large data sets to prevent biased results.

2. Many algorithms are designed to cluster interval-based (numerical) data, but applications may require clustering of other types of data, such as binary, categorical (nominal), and ordinal data, or a mixture of data types.

3. Clusters can take any shape; algorithms should be able to detect clusters with arbitrary shapes.

4. Clustering results can be sensitive to input parameters, so the minimum requirements must be met.

5. Some clustering algorithms are sensitive to missing, unknown, outlier, or erroneous data, leading to clusters of poor quality; therefore, they must be able to deal with noise.

6. Clustering must be insensitive to the order of input data; some clustering algorithms are sensitive to order.

7. High dimensionality must be achieved. It is challenging to cluster data objects in high-dimensional space, especially as such data can be very sparse and highly skewed.

8. Applications may need to perform clustering under various kinds of constraints. It may be challenging to find groups of data with good clustering behavior under specified constraints.

9. Clustering needs to be tied into specific semantic interpretations and applications. It is important to study how an application goal may influence the selection of clustering methods.

There are many clustering techniques, organized into the following categories: partitioning, hierarchical, density-based, grid-based, and model-based methods. Clustering can also be used for outlier detection. In the following sections, we consider three popular and easy-to-deploy methods; nearest-neighbor methods, hierarchical methods, and mixed model methods.

7.3.4.1 Nearest-Neighbor Methods

At first glance, the people in a given neighborhood seem to have similar incomes. The nearest-neighbor prediction algorithm (or the k-means method) works in much the same way except that nearness in a database may consist of a variety of factors. It performs quite well for our present needs because many of the algorithms are robust with respect to dirty and missing data.

The nearest-neighbor prediction algorithm, simply stated, is as follows: "Objects that are 'near' each other will also have similar prediction values. Thus, if you know the prediction value of one of the objects, you can predict it from its nearest neighbours" (Berson et al., 2008, p. 9).

The k-means method seems suitable for diagnosis since similar behavior will generate similar data and, therefore, clustering will work.

Mathematically, it is very simple and can be expressed as follows: the multidimensional mean of a set of cases is the point (centroid) with coordinates $(\bar{x}_1, \bar{x}_2, \bar{x}_3, ..., \bar{x}_p)$, called the mean point of a set of cases (Figure 7.3).

Determining the k-means algorithm follows this procedure:

1. Decide on the value of k.

2. Start with k arbitrary centers; these may be chosen randomly, or as the centroids of arbitrary starting partitions of the case set.

3. Consider each case in sequence, and find the center to which the case is closest. Assign the case to that cluster. Recalculate the center of the new and old clusters as the centroids of the points in the cluster.

4. Repeat until the clusters are stable.

5. Repeat for different initial centers. Choose the best clustering, in terms of minimum within cluster sum of squares.

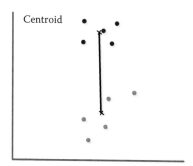

Centroid

FIGURE 7.3
Centroid of two clusters. (From Anon., 2008. *Clustering and Classification Methods for Biologists.*
[Online] Available at: http://www.alanfielding.co.uk/multivar/ca_alg.htm.)

7.3.4.2 Hierarchical Clustering Methods

As the name suggests, a hierarchical method hierarchically decomposes
the given set of data objects. A hierarchical method can be classified as
either agglomerative or divisive, based on the formation of the hierarchical
decomposition (Figure 7.4). The agglomerative approach, or the bottom-up
approach, starts with each object constituting a separate cluster. In successive iterations IT continues to merge the objects until eventually there is one
cluster, or a termination condition holds. The divisive or top-down approach
moves in the opposite direction.

Once a step (merge or split) is done, it can never be undone. This rigidity is
useful as its lack of choices leads to smaller computation costs. Yet erroneous
decisions cannot be corrected. To improve the quality of hierarchical clustering, we need to: (1) carefully analyze object "linkages" at each hierarchical
partitioning; (2) integrate hierarchical agglomeration with iterative relocation by using a hierarchical agglomerative algorithm and refining the result
with iterative relocation (Han and Kamber 2001).

Figure 7.5 shows agglomerative and divisive methods.

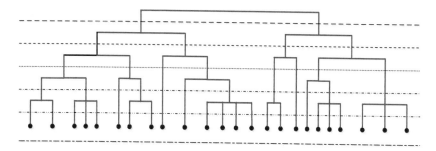

FIGURE 7.4
Hierarchical clustering methods.

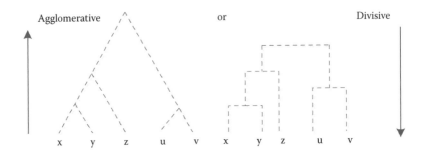

FIGURE 7.5
Agglomerative and divisive methods.

7.3.4.2.1 Agglomerative

In this method, find the two cases that are closest and form them into a cluster. Continue agglomerating cases and clusters on the same basis (single nearest neighbor; single linkage method) until one cluster is obtained.

Single linkage clustering can lead to chaining. This can be avoided by defining the "working distance" between two cases at any iterative stage to be the distance between the most distant members of the two clusters to which they belong. This is called the complete linkage method.

A cluster, once initiated by bringing together two cases, may be represented by its centroid. Clusters can be merged based on the distance between their centroids, creating an agglomerative k-means method (Al-Akwaa, 2012).

7.3.4.2.2 Divisive

In this method, start with the whole case set. Divide the set into two subsets in some optimal way. Then subdivide each subset, again optimally.

7.3.4.3 Mixed Models

Consider one continuous variable, x, say, to start with. A normal distribution over the x-variable is given by the following formula for the probability density function:

$$f(x \mid \mu, \sigma^2) = \frac{1}{\sqrt{2\pi\sigma^2}} e^{-\frac{1}{2\sigma^2}(x-\mu)^2}$$

This is a model with two parameters, the distribution mean μ, and the distribution variance σ^2.

A mixed model of k normal distributions would be given by

$$f(x \mid \mu, \sigma^2) = \sum_{i=1}^{k} p_i \frac{1}{\sqrt{2\pi\sigma_i^2}} e^{-\frac{1}{2\sigma_i^2}(x-\mu_i)^2} \; ; \quad \sum_{i=1}^{k} p_i = 1$$

where the proportions, p_i, are often called weights. The value of k might be taken to indicate the underlying number of groups, or clusters. Such models can be generalized to many dimensions/variables, and the parameters estimated by the use of maximum likelihood methods, implementing the E–M algorithm. However, if correlation/covariance parameters are included in the general mixed model, the number of parameters increases rapidly with the number of variables involved. In general, if we are working in an n-dimensional variable space, there will be $n(n + 3)/2$ parameters for each centroid.

7.4 Categorization: Semisupervised and Unsupervised

Clustering involves grouping data points together according to some measure of similarity. One goal of clustering is to extract trends and information from raw data sets. An alternative goal is to develop a compact representation of a data set by creating a set of models representing it (Buddemeier et al., 2002).

As noted previously, there are two general types of clustering: supervised and unsupervised. Supervised clustering uses a set of sample data to classify the rest of the data set. This can be considered classification, and the task is to learn to assign instances to predefined classes (Keller and Crocker, 2003). For example, consider a set of colored balls (all colors) that we want to classify into three groups: red, green, and blue. A logical way to do this is to pick out one example of each class—a red ball, a green ball, and a blue ball—and set each next to a bucket. Then we go through the remaining balls, compare each ball to the three examples, and put each ball in the bucket whose example it best matches.

This example of supervised clustering is illustrative because it shows two potential problems. First, the result will depend on the balls selected as examples. If we select a red, an orange, and a blue ball, it may be difficult to classify a green ball. Second, unless we are careful about selecting examples, we may select ones that do not represent the distribution of data. For example, we might select red, green, and blue balls, only to discover that most of the colored balls are cyan, purple, and magenta (located between the other three primary colors). It is extremely important to select representative samples in supervised clustering.

Unsupervised clustering tries to discover the natural groupings inside a data set with no input from a trainer. The main input for a typical unsupervised clustering algorithm is the number of classes it should find. In the case of the colored balls, this would be like dumping the balls into an automatic sorting machine and telling it to create three piles. The goal of unsupervised clustering is to create three piles where the balls within each pile are very similar, but the piles are different. Here, no predefined classification is required. The task is to learn a classification from the data.

One of the most important characteristics of any supervised or unsupervised clustering process is how to measure the similarity of two data points. Clustering algorithms divide a data set into natural groups (clusters). Instances in the same cluster are similar to each other, as they share certain properties.

Clustering algorithms can be one of the following (Keller and Crocker, 2003):

1. *Hierarchical*: These include techniques where the input data are not partitioned into the desired number of classes in a single step. Instead, a series of successive fusions of data are performed until the final number of clusters is obtained (Center for the New Engineer, 1995).

2. *Nonhierarchical or iterative*: These include techniques where a desired number of clusters is assumed at the start. Instances are reassigned to clusters to improve them.

3. *Hard and soft*: Hard clustering assigns each instance to exactly one cluster. Soft clustering assigns each instance a probability of belonging to a cluster.

4. *Disjunctive*: Instances can be part of more than one cluster. Figure 7.6 shows an illustration of the properties of clustering.

7.4.1 Unsupervised Clustering

Unsupervised clustering tries to discover the natural groupings inside a data set without any input from a trainer. A typical unsupervised clustering algorithm needs the number of classes it should find as input. Unsupervised data clustering (Buddemeier et al., 2002; Keller and Crocker, 2003) is an intelligent tool for delving into unknown and unexplored data, one that brings out the hidden patterns and associations between variables in a multivariate data set. Although mostly used in large databases, the unsupervised clustering method has enjoyed success in a variety of industries. From finding patterns in customers' buying habits, to fraud detection, to discovering clusters in genetic microarray data, these data exploration models have proved beneficial when dealing with data without *a priori* information.

Attracted by the success of the models in a number of different fields, researchers are attempting to use them as a tool for fault diagnosis in machine health. One of the current limitations in machine health monitoring is dealing with novel data and determining whether data can be labeled using existing trained classifiers or if new ones must be created.

Supervised fault classification algorithms, such as NNs, decision trees, and support vector machines (SVM) work efficiently for trained signatures only and generate erroneous results when encountering novel data not found in

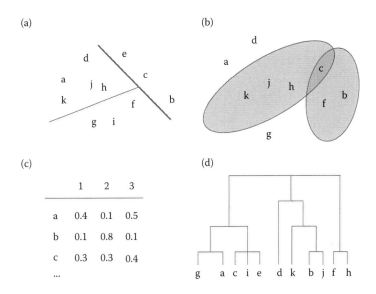

FIGURE 7.6
Properties of clustering. (a) Hard, nonhierarchical, (b) nonhierarchical, disjunctive, (c) soft, nonhierarchical, k disjunctive, (d) hierarchical, disjunctive. (From Pallavi, 2003. *Clustering.* [Online] Available at: ftp://www.ece.lsu.edu/pub/aravena/ee7000FDI/Presentations/ Clustering-Pallavi/CLUSTERING_report.doc.)

the domain of trained data. To correctly classify new data, models are often generalized (as opposed to fine-tuned). On the one hand, generalizing these supervised models to accommodate new data may have a bigger impact by increasing the classification error and, thus, reducing the overall performance of the model. On the other hand, fine-tuning the models to reduce error may lead to overfitting, restricting model performance (Rennols, 2002). To alleviate this rigidity in the trained classifier models and allow room to accommodate novel data, we turn to unsupervised clustering.

One of the simplest and most commonly used unsupervised clustering algorithms is the *k*-means algorithm:

1. Specify *k*, the number of clusters
2. Choose *k* points randomly as cluster centers
3. Assign each instance to its closest cluster center using Euclidian distance
4. Calculate the median (mean) for each cluster, and use it as its new cluster center
5. Reassign all instances to the closest cluster center
6. Iterate until the cluster centers no longer change Figure 7.7 explains the concept of *k*-means clustering.

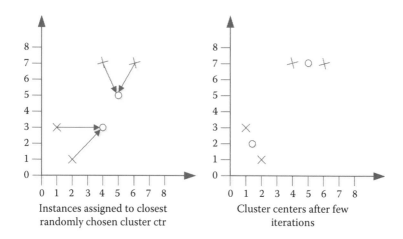

FIGURE 7.7
K-means algorithm. (From Bradley, P. and Fayyad, U. M., 1998. *15th International Conference on Machine Learning.* San Francisco: Morgan Kaufmann, pp. 91–99.)

7.4.2 Supervised Clustering Algorithms

In supervised clustering based on known health states, we have prior knowledge that the incoming data should be clustered in either faulty or nonfaulty classes. This is based on the premise that when a signal changes due to the occurrence of a fault, something in the signal is altered. These changes may be detected by using a filter bank and creating indicators. The indicators obtained are segregated using the clustering technique. Clustering is carried out in two phases, training and classification.

In the training phase for the supervised clustering case, the representatives of the predefined clusters are formed from the example data—indicator matrices are created after processing the simulated signal. The representatives of the clusters are called "signatures." In the classification phase, the distance of the indicators to the signatures is calculated and the indicators are grouped in the cluster with the closest signature.

There are two common techniques for signature creation. In the first technique, for the training phase, the signatures for faulty and nonfaulty cases are the mean vectors of the indicator vectors obtained for each window. The mean of prefault indicators is the signature for the nonfaulty class and the mean of postfault indicators is the signature for the faulty class. In the classification phase, to analyze the performance of this clustering technique, the same prefault and postfault indicators are classified into either faulty or nonfaulty classes, depending on their vector distance to the above predetermined signatures. The indicator is classified in the group with the minimum distance between the indicator and the group's signature. In the second clustering technique, clustering is based on principal component analysis (PCA) and the vector subspace signature concept. Here, the signatures are vector subspaces instead of the mean used in the previous case.

Four types of supervised clustering algorithms are vector quantization, fuzzy clustering, artificial neural net (ANN), and fuzzy–neural algorithms. Although fuzzy and neural nets initially go through unsupervised clustering to determine the cluster centers, only the supervised clustering algorithms are discussed here.

7.4.2.1 Vector Quantization

Shanon's source coding theory, used for the transmission and encoding of data, is the origin of this algorithm. In the algorithm, as shown in Figure 7.8, a vector quantizer maps k-dimensional vectors in the vector space R^k into a finite set of vectors $Y = \{y_i: i = 1, 2, \ldots, N\}$ (Qasem, 2009).

Each vector, y_i, is called a code vector or a code word, and the set of all the code words is called a code book. Associated with each code word, y_i, is a nearest-neighbor region called Voronoi region, defined by

$$V_i = \left\{ x \in R^k : \| x - y_i \| \leq \| x - y_j \|, \quad \text{for all } j \neq i \right\}$$

where the entire space R^k is partitioned by the set of Voronoi regions such that

$$\bigcup_{i=1}^{N} V_i = R^k$$

$$\bigcap_{i=1}^{N} V_i = \phi \quad \text{for all } i \neq j$$

As an example, we take vectors in the two-dimensional case. Figure 7.9 shows some vectors in space.

Associated with each cluster of vectors is a representative code word (cluster center or cluster representative obtained by k-means algorithm or similar algorithms). Each code word resides in its own Voronoi region. These regions are separated by imaginary lines in Figure 7.1 for illustration.

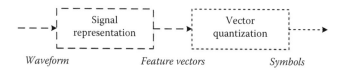

FIGURE 7.8
Vector quantization representation.

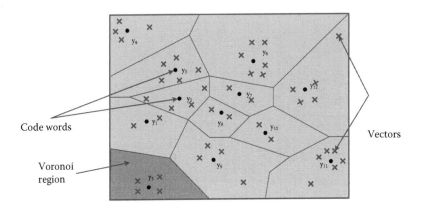

FIGURE 7.9
Code words in two-dimensional space. Input vectors are marked with an x, code words are marked with red circles, and the Voronoi regions are separated by boundary lines. (From Qasem, M., 2009. *Vector Quantization*. [Online] Available at: http://www.mqasem.net/vector quantization/vq.html.)

The representative code word (cluster center) is the closest in Euclidean distance from the input vector (instances). The Euclidean distance is defined by

$$d(x, y_i) = \sqrt{\sum_{j=1}^{k} (x_j - y_{ij})^2}$$

where x_j is the jth component of the input vector, and y_{ij} is the jth component of the code word y_i. The problem is that we transmit only the cluster index, not the entire data for each sample.

7.4.2.2 Fuzzy Supervised Clustering

Fuzzy logic first became popular in the field of automatic control, but was rapidly adapted to fields where no math expressions define the existing context or the rules of the process. Maintenance is one such field, where things are not black and white, and decisions are sometimes subjective due to the multiple factors to be considered. Fuzzy logic requires no analytical model of the system and offers the chance to combine heuristic knowledge with any available model knowledge (Dalton, 1999). Fuzzy logic can also deal with vague or imprecise data, relevant in maintenance where the amount of data is not an issue, but where quality of data and contextual engines to give meaning to data are often missing. In the field of fault diagnosis, fuzzy logic has been used successfully in many applications, both as a means of residual generation and to aid in the decision-making process of residual evaluation.

The idea behind fuzzy clustering is basically that of pattern recognition. Training data are used off-line to determine relevant cluster centers for each of the faults of interest. Online, the degree to which the current data belong to each of the predefined clusters is determined, resulting in a degree of membership to each of the predetermined faults. This method is useful in cases where there are many residuals, or when no expert knowledge of the system is available. Fuzzy clustering is different from fuzzy reasoning, which is also used in residual analysis. Fuzzy reasoning mainly consists of IF-THEN reasoning based on the sign of the residual. The following is an example of fuzzy reasoning:

1. IF residual 1 is positive and residual 2 is negative, THEN fault 1 is present.
2. IF residual 1 is zero and residual 2 is zero, THEN the system is fault free.
3. And so on.

Clustering is the allocation of data points to a certain number of classes. Each class is represented by a cluster center, or prototype, which can be considered as the point that best represents the data points in the cluster. The idea behind fuzzy clustering is that each data point belongs to all classes with a certain degree of membership. The degree to which a data point belongs to a certain class depends on the distance to all cluster centers. For fault diagnosis, each class could correspond to a particular fault. The general principle is shown for three inputs and three clusters in Figure 7.10.

The fuzzy clustering fault isolation procedure consists of the following two steps:

1. *Off-line phase*: This is a learning phase that consists of the determination of the characteristics (i.e., cluster centers) of the classes. A learning data set is necessary for the off-line phase, and it must contain residuals for all known faults. For more details on origin of the idea of fuzzy clustering, refer to Bezdsek (1991).
2. *Online phase*: This phase calculates the membership degree of the current residuals to each of the known classes. In this way, each data point does not belong to just one cluster, but its membership is distributed among all clusters according to the varying degree of resemblance of its features to those cluster centers (Marsili-Libelli, 1998).

It is important that the training data contain all faults of interest; otherwise, they cannot be isolated online—though unknown faults can sometimes be detected.

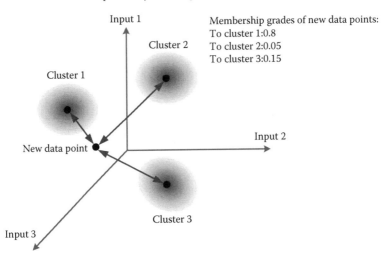

FIGURE 7.10

Fuzzy clustering concept showing cluster centers and the membership grade of a data point. (From Pallavi, 2003. *Clustering.* [Online] Available at: ftp://www.ece.lsu.edu/pub/aravena/ ee7000FDI/Presentations/Clustering-Pallavi/CLUSTERING_report.doc.)

The fuzzy membership matrix and the cluster centers are computed by minimizing the following partition formula:

$$J_f(C,m) = \sum_{i=1}^{C} \sum_{k=1}^{N} (u_{k,i})^m d_{k,i} \text{ subject to } \sum_{i=1}^{C} u_{i,k} = 1 \qquad (7.1)$$

where C denotes the number of clusters, N the number of data points, $u_{i,k}$ the fuzzy membership of the kth point to the ith cluster, $d_{k,i}$ the Euclidean distance between the data point and the cluster center, and $m \in (1,\infty)$ a fuzzy weighting factor which defines the degree of fuzziness of the results. The data class becomes fuzzier and less discriminating with increasing m. In general, $m = 2$ is chosen (*note*: this value of m does not produce an optimal solution for all problems).

The constraint in Equation 7.1 implies each point must distribute its entire membership among all the clusters. The cluster centers (centroids or prototypes) are defined as the fuzzy-weighted center of gravity of the data x, such that

$$v_i = \frac{\displaystyle\sum_{k=1}^{N} (u_{k,i})^m x_k}{\displaystyle\sum_{k=1}^{N} (u_{k,i})^m} \qquad i = 1,2,\ldots,C \qquad (7.2)$$

Since $u_{i,k}$ affects the computation of the cluster center v_i, the data points with high membership will influence the prototype location more than points with low membership. For the fuzzy C-means algorithm, distance $d_{k,i}$ is defined as follows:

$$(d_{k,i})^2 = \left\| x_k - v_i \right\|^2 \tag{7.3}$$

The cluster centers v_i represent the typical values of that cluster, whereas the $u_{i,k}$ component of the membership matrix denotes the extent to which the data point x_k is similar to its prototype. The minimization of the partition function (1) will give the following expression for membership:

$$u_{i,k} = \cfrac{1}{\displaystyle\sum_{j=1}^{C} \left(\cfrac{d_{k,i}}{d_{k,j}} \right)^{\frac{2}{m-1}}} \tag{7.4}$$

Equation 7.4 is determined in an iterative way since the distance $d_{k,i}$ depends on membership $u_{i,k}$.

The procedure to calculate the fuzzy C-means algorithm is as follows:

1. Choose the number of classes C, $2 \le C < n$.
2. Choose m, $1 \le m < \infty$.
3. Initialize $U^{(0)}$.
4. Calculate the cluster centers v_i using Equation 7.2.
5. Calculate new partition matrix $U^{(1)}$ using Equation 7.4.
6. Compare $U^{(j)}$ and $U^{(j+1)}$.
7. If the variation of the membership degree $u_{k,i}$, calculated with an appropriate norm, is smaller than a given threshold, stop the algorithm; otherwise go back to step 2.
8. The determination of the cluster centers is now complete.

Online, the U matrix is calculated for each data point. The elements of the U matrix give the degree to which the current data correspond to each of the fault classes.

Fuzzy reasoning and fuzzy clustering are chosen according to the system and availability of expert knowledge of the system. If expert knowledge of the system is available, fuzzy reasoning can be used; otherwise it is better to use the fuzzy clustering method.

7.4.2.3 ANN Clustering

The most basic components of NNs are modeled after the structure of the brain, and the most basic element of the human brain is a specific type of cell which gives us with the ability to remember, think, and apply previous experiences to all our actions. These cells are known as neurons; each neuron can connect with up to 200,000 other neurons. The power of the brain comes from the numbers of these basic components and the multiple connections between them.

All natural neurons have four components: dendrites, soma, axons, and synapses. Essentially, a biological neuron receives inputs from other sources, combines them in some way, performs a generally nonlinear operation on the result, and outputs the final result. A simplified biological neuron and its four components appear in Figure 7.11 (Rahman et al., 2011).

ANN clustering is a system loosely modeled on the human brain (Klerfors, 1998). It attempts to simulate the brain's multiple layers of simple processing elements or neurons using specialized hardware or sophisticated software.

The basic units of NNs, the artificial neurons, simulate the four basic functions of natural neurons. Artificial neurons are much simpler than biological neurons; Figure 7.12 shows the basics of an artificial neuron (Rahman et al., 2011). Each neuron is linked to certain of its neighbors with varying coefficients of connectivity that represent the strengths of these connections. Learning is accomplished by adjusting these strengths to cause the overall network to output appropriate results.

Hebb has postulated a principle for a learning process (Hebb, 1949) at the cellular level: if Neuron A is stimulated repeatedly by Neuron B at times when Neuron A is active, Neuron A will become more sensitive to stimuli

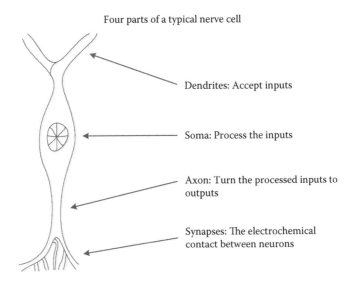

Four parts of a typical nerve cell

Dendrites: Accept inputs

Soma: Process the inputs

Axon: Turn the processed inputs to outputs

Synapses: The electrochemical contact between neurons

FIGURE 7.11
Four main parts of human nerve cells on which artificial neurons are designed. (From Klerfors, D., 1998. *Artificial Neural Networks*. [Online] Available at: http://osp.mans.edu.eg/rehan/ann/Artificial%20Neural%20Networks.htm.)

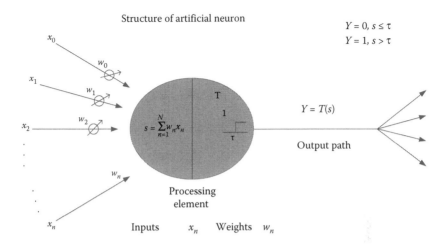

FIGURE 7.12
Structure of an artificial neuron with Hebbian learning ability (weights are adjustable). (From Klerfors, D., 1998. *Artificial Neural Networks.* [Online] Available at: http://osp.mans.edu.eg/rehan/ann/Artificial%20Neural%20Networks.htm.)

from Neuron B; this is the correlation principle (Carl, 2008). It implicitly involves adjusting the strengths of the synaptic inputs, leading to the incorporation of adjustable synaptic weights on the input lines to excite or inhibit incoming signals.

An input vector $x = (x_1...x_N)$, a column matrix vector, is linearly combined with the weight vector $w = (w_1...w_N)$ via the inner (dot) product to form the sum

$$s = \sum_{n=1}^{N} w_n x_n = w^T x$$

If the sum s is greater than the given threshold b, the output y is 1; otherwise, it is 0. The function that gives the output value is called the activation function. Figure 7.13 shows some activation functions.

Activation functions as in (a) and (b) give binary outputs (0 or 1/+1 or –1), whereas the functions in (c) and (d) give nonbinary outputs (output value varies between 0 and 1/+1 and –1). The functions are unipolar if the output range is between 0 and 1; they are bipolar if the output range is from +1 through –1.

The basic artificial neuron unit shown in Figure 7.12 is called a perceptron, and the architecture for a network consisting of a layer of M perceptrons is shown in Figure 7.14. An input feature vector $x = (x_1...x_N)$ is inputted into the network via the set of N branching nodes. The lines fan out at the branching nodes so that each perceptron receives an input from each component of x. At each neuron, the lines fan in from all of the input (branching) nodes.

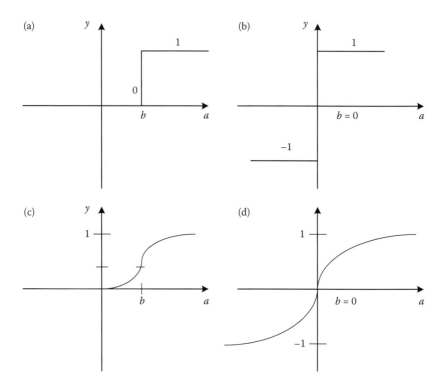

FIGURE 7.13
Some common activation functions. Activation functions as in (a) and (b) give binary output (0 or 1/+1 or −1) whereas the functions in (c) and (d) give non-binary output (output value varies anywhere between 0 and 1/+1 and −1). (From Pallavi, 2003. *Clustering*. [Online] Available at: ftp://www.ece.lsu.edu/pub/aravena/ee7000FDI/Presentations/Clustering-Pallavi/CLUSTERING_report.doc.)

Each incoming line is weighted with a synaptic coefficient (weight parameter) from the set $\{w_{nm}\}$, where w_{nm} weights the line from the nth component x_n coming into the m_{th} perceptron.

Because of their simplicity and ease of use, NNs have been widely used in academia to show the capabilities of artificial intelligence in the clustering of fault data. However, they have many limitations in diagnosis; practitioners must be aware that certain techniques may be unstable in certain conditions.

The main weakness of NNs is their need for data, especially good quality data; their lack of robustness with wrong data sets or missing data leads to incorrect diagnostics.

7.4.2.4 Integration of Fuzzy Systems and NNs

NNs process numerical information and exhibit learning capability. Fuzzy systems can process linguistic information and represent, say, expert

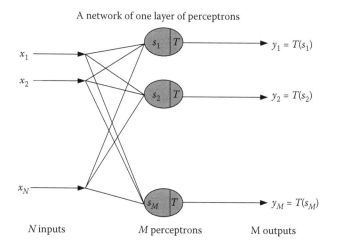

FIGURE 7.14
One layer of perceptron network with N inputs and M perceptrons. (From Pallavi, 2003. *Clustering*. [Online] Available at: ftp://www.ece.lsu.edu/pub/aravena/ee7000FDI/Presentations/Clustering-Pallavi/CLUSTERING_report.doc.)

knowledge using fuzzy rules. The fusion of these two technologies is the current research trend. The aim is to create machines with more intelligent behavior (Nguyen, 1994).

The following are some reasons for considering both fuzzy systems and NNs:

1. The knowledge base of a fuzzy system consists of a collection of "if... then..." rules in which linguistic labels are modeled by membership functions. NNs can be used to produce membership functions when available data are numerical.

2. We can take advantage of the learning capability of NNs to adjust membership functions, say, in control strategies, to enhance control precision.

3. NNs can be used to provide learning methods for fuzzy inference procedures.

4. In the opposite direction, we can use fuzzy reasoning architecture to construct new NNs.

5. We can also fuzzify the NN architecture to enlarge the domain of applications.

6. The fusion of NNs and fuzzy systems is based on the fact that NNs can learn expert knowledge (through numerical data) and fuzzy systems can represent expert knowledge (through the representation of in–out relationships by fuzzy reasoning).

The literature mentions two combinations:

1. *Neural–fuzzy system*: In this type of system, the learning ability of NNs is utilized to capitalize on the key components of a general fuzzy logic inference system. NNs are used to realize fuzzy membership functions.

2. *Fuzzy–NN system*: These models incorporate fuzzy principles into a NN to create a more flexible and robust system. The NNs model algorithm can be fuzzified with fuzzy neurons, fuzzified neural models, and NNs with fuzzy training.

These developments represent recent progress in this field, but a number of integrated systems and algorithms remain in the proposal stage. For a more detailed explanation of combinations and proposals, refer to Lin and George Lee (1996).

7.5 Issues Using Cluster Analysis

Accompanying the developments in computer technology is a dramatic increase in the number of numerical classification techniques and their applications. Various names have been applied to these methods by the disciplines using them, including numerical taxonomy (biology), Q-analysis (psychology), and unsupervised pattern recognition (AI), but the most common generic term is cluster analysis. Cluster analysis has been used to solve specific classification problems in psychiatry, medicine, social services, market research, education, and archaeology (Kural, 1999).

As mentioned above, classification is central to our understanding of any phenomenon: "All the real knowledge which we possess depends on methods by which we distinguish the similar from dissimilar. The greater number of natural distinctions this method comprehends, the clearer becomes our idea of things. The more numerous the objects which employ our attention, the more difficult it becomes to form such a method and the more necessary" (Linnaeus, 1737, p. 1).

In summary, a clustering algorithm should have the following characteristics if it is to be useful for maintainers:

1. Scalability
2. Ability to deal with different types of attributes
3. Ability to find clusters with arbitrary shapes
4. Minimum requirements of domain knowledge to determine the input parameters
5. Ability to handle noise and/or outliers

6. Insensitivity to input record order

7. High dimensionality

8. Good interpretability and usability

Unfortunately, even when classification might facilitate understanding, the process of measuring the similarities between the objects and identifying the classes based on these similaritiescan be difficult. Many numerical techniques have been created to assist in classification, "in an effort to rid taxonomy of its traditionally subjective nature and to provide objective and stable classifications" (Everitt, 1993, p. 20). Most have originated in the natural sciences, such as biology and zoology.

Even with these techniques, using cluster analysis is not straightforward. At every stage of analysis, investigators must make decisions based on their data and their purpose. This may even apply to the definitions of terms such as cluster, group, and class. Accordingly, Bonner (1964) suggests the meaning should reflect the value judgment of the user.

The decisions to be made by an investigator during cluster analysis center on the representation of the objects, measures of association, the method to be used, and the representation of the clusters. For example, Cormack (1971) and Gordon (1981) define clusters as having such properties as internal cohesion and external isolation. Unfortunately, most of these decisions often lack a sound theoretical basis. Everitt (1993, p. 256) elaborates:

> It is generally impossible a priori to anticipate what combination of variables, similarity measures and clustering techniques are likely to lead to interesting and informative classifications. Consequently the analysis proceeds through several stages with the researcher intervening if necessary to alter variables, choose a different similarity measure, concentrate on a particular subset of individuals etc. The final, extremely important stage, concerns the evaluation of the clustering solution(s) obtained. Are the clusters real or merely artefacts of the algorithms? Do other solutions exist which are better? Can the clusters be given a convincing interpretation?

Generally speaking, to ensure both the understanding of the data and the validity of the analysis, some graphical representation of the data should be designed before the application of any method. As Jain and Dubes (1988, p. 3) say: "Cluster analysis is a tool for exploring data and should be supplemented by techniques for visualising data."

Although there are several methods for such visualization (see Everitt, 1993), a common option is PCA. This transforms the original variables into a new set of independent variables; each accounts for decreasing portions of variance of the original variables and provides a two-dimensional view of the data. Another option, Andrew's plots, uses a trigonometric function to create a similar transformation by plotting multivariate data over the range $-\pi$ and π (Kural, 1999).

In the next stage, we must decide whether the raw data or data derived from them should be used as input for the cluster analysis. This decision depends on both the clustering technique and the type of raw data. Some reduction in the number of variables is often called for to make the clustering procedure feasible. Indeed, the interface of the clustering algorithm and the operator has systematically been ignored because academics expect to remove humans from the scene and make diagnosis and prognosis in a fully automated way. This is obviously not possible with present technology, especially in the maintenance field, due to the heterogeneity of the data, different data sources, and the need of human expertise as a context to create meaningful links between disparate information that cannot be connected using any current techniques.

7.5.1 Clustering Methods and Their Issues

There are many different clustering methods; these are not necessarily mutually exclusive, nor can they be neatly categorized into a few groups. Hierarchical methods are the broadest family to be categorized as a single group. Other categories are optimization methods, clumping methods (creating overlapping clusters), density search techniques, and mixed models.

No clustering method can be judged "best" in all circumstances. Certain methods will be better for particular types of data. In many applications, it may be reasonable to apply a number of clustering methods. If all methods produce very similar solutions, the investigator may be more confident that the results merit further investigation. Widely different solutions might be taken as evidence of a lack of a clear-cut cluster structure (Kural, 1999).

Therefore, the selection of a clustering technique should consider existing problems in the targeted field, available data, and quality of data, including the operators who may participate in the clustering.

There are a number of problems with clustering common to all fields and faced by all practitioners selecting a clustering technology. These include the following:

1. Current techniques do not address all requirements appropriately (and concurrently).
2. Many dimensions and many data items can cause problems if there are time constraints.
3. Effectiveness depends on definition of "distance" (for distance-based clustering).
4. Distance measures may not be obvious, thus requiring definition; this can be a problem in multidimensional spaces.
5. Results of the clustering algorithm are often arbitrary and may be interpreted in a number of ways.

A noncomprehensive outline of commonly used methods with their strengths and weaknesses is provided in the next subsections.

7.5.1.1 Hierarchical Methods

Hierarchical clustering, the most commonly used method, consists of a series of partitions that may run from a single cluster containing all individuals to n clusters, with each cluster containing a single individual. Hierarchical classifications are often represented by a two-dimensional diagram, called a dendrogram; this diagram shows the fusions or divisions occurring at each successive analytic stage.

For the most part, hierarchical clustering techniques are either agglomerative methods fusing individuals in successive steps; or divisive methods successively separating a group of *n* individuals into finer classes. As the investigator must decide at what point to stop dividing or fusing, the use of the techniques clearly requires the user to have a great deal of expert knowledge and a well-designed interface to carry on or stop the fusion/division (Kural, 1999).

A drawback is the possibility of influence from *a priori* expectations. Unfortunately, this happens frequently in maintenance; before starting the diagnosis process, the maintainer may be so confident about the diagnosis that the whole process is affected. Preconceptions alter the outcome, because once two objects are divided, they cannot be fused again and vice versa.

These methods impose a hierarchical structure on data; researchers, therefore, must consider whether this is merited or if it introduces unacceptable distortions of the original relationships of the individuals (Everitt, 1993).

Some commonly used agglomerative methods are outlined below, along with their strengths and weaknesses. As divisive methods are far less popular than agglomerative methods, they are not detailed here (Kural, 1999). *Note*: Studies comparing the performances of various hierarchical methods find performance varies according to input data types (Cunningham and Ogilvie, 1972; Milligan, 1980).

7.5.1.2 Single Link Clustering (Nearest-Neighbor Technique)

Single link clustering is one of the simplest agglomerative clustering techniques (Figure 7.15). Many consider it the archetypal clustering technique, as most other techniques are inspired by it. In this method, distance between groups is defined as that between the closest pair of individuals from different groups (Kural, 1999).

The method is deemed the best hierarchical method as far as theoretical soundness is concerned (Jardine and Sibson, 1971). The single linkage method can be applied efficiently to large data sets (Rohlf, 1973; Sibson, 1973). The hierarchy is progressively updated as new similarities become available, regardless of the order in which they are calculated; nor is there a need to

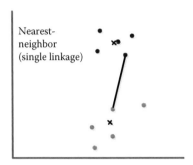

FIGURE 7.15
Nearest-neighbor single linkage for two clusters.

store an interdocument similarity matrix (Willet, 1988). Its invariance under monotonic transformations of the proximity matrix (Krzanowski, et al., 1995) is another plus, as it is susceptible to scaling and combining different variables.

That being said, in many applications, single linkage is the least successful method to produce useful cluster solutions. Notably, its tendency to cluster individuals linked by a series of individuals (chaining effect) leads to the creation of loosely bound clusters with little internal cohesion (El-Hamdouchi and Willett, 1989). This results in gray areas in diagnosis; if identification of the failure is unclear, maintenance decisions will be equally loose. While this is undesirable for many applications, it is not unreasonable however for subjects like taxonomy and its examination of evolutionary chain mechanisms (Krzanowski, 1988).

Another limitation is the method's inability to provide an immediate definition of the cluster center or representative (Murtagh, 1983).

Finally, in cluster-based retrieval (CBR), this method performs poorly, producing a small number of large, well-defined document clusters (Voorhees, 1985). Experiments by Griffiths et al. (1986) on document clustering reveal that complete linkage, group average, and Ward's method yield far superior results to the single linkage method (Kural, 1999).

7.5.1.3 Complete Linkage Clustering (Furthest-Neighbor Clustering)

Complete linkage clustering is the opposite of single linkage clustering in the sense that the distance between groups is now defined as the distance between the most distant pair of individuals from different groups (Figure 7.16). The definition of cluster membership is much stricter than single linkage, and the large straggly clusters are replaced by large numbers of small, tightly bound clusters (El-Hamdouchi and Willett, 1989) of equal diameter (Krzanowski, 1988). In theoretical graph terms, this method corresponds to the identification of maximally complete subgraphs at some threshold

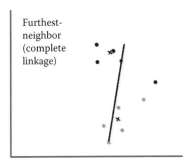

FIGURE 7.16
Furthest-neighbor complete linkage for two clusters.

similarity (Willet, 1988). This is especially relevant in diagnosis where a tightly bound cluster may identify single failures and combinations of failure modes, preparing the user for a diagnostic where both existing failures and degree of severity can be outputs of a system in which clusters are very well defined (Kural, 1999).

According to Willet (1988), complete linkage is an effective method for clustering, despite being the method that requires the greatest computational resources.

A disadvantage is that this method is sensitive to observational errors.

7.5.1.4 Group Average Clustering

Here, the distance between two clusters is defined as the average of the distances between all pairs of individuals composed of one individual from each group (Figure 7.17). It represents a midpoint between the two extreme types of linkage methods, that is, single linkage and complete linkage. The method is known to minimize the distortion imposed on the interobject similarity matrix when a hierarchic classification is generated (El-Hamdouchi and Willett, 1989).

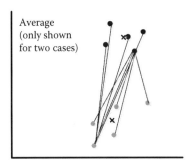

FIGURE 7.17
Average for two clusters.

Group average had the best overall performance in a comparative study involving seven hierarchical clustering methods (Cunningham and Ogilvie, 1972). But another study reports that the performance of this technique is affected by outliers (Milligan, 1980).

7.5.1.5 Centroid Clustering

Unlike the above three methods that operate directly on the proximity matrix and do not need access to the original variable values of the individuals, centroid clustering requires original data (Figure 7.18).

With this method, once groups are formed, they are represented by the mean values for each variable, in other words, the mean vector. Moreover, the intergroup distance is now defined in terms of the distance between two mean vectors. Although the use of a mean generally implies variables on an interval scale, the centroid clustering method is often used for other types of variables.

If the two groups to be fused have very different sizes, the centroid of the new group will be close to the centroid of the larger group and may remain within that group. If we assume the groups to be fused are of equal size, we can avoid this problem, as the apparent position of the new group will be somewhere between them (Kural, 1999).

Another disadvantage is that the method is biased toward finding "spherical" clusters.

7.5.1.6 Median Clustering

Although somewhat like centroid clustering, median clustering attempts to make the process independent of group size by assuming the groups to be fused are of equal size; as discussed above for centroid clustering, the apparent position of the new group will always be between the two groups to be fused.

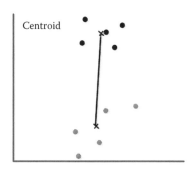

FIGURE 7.18
Centroid for two clusters.

Lance and Williams (1967) suggest this method is unsuitable for such measures as correlation coefficients, where geometrical interpretation is not possible. Finally, like the single link clustering method, median clustering suffers from the chaining effect (Kural, 1999).

7.5.1.7 Ward's Method

Ward (1963) has proposed a clustering procedure to form partitions in a manner that minimizes the loss associated with each grouping and to quantify that loss in a readily interpretable form. At each step in the analysis, the union of every possible pair of clusters is considered and the two clusters whose fusion results in the minimum increase in "information loss" are combined. Information loss is defined by Ward in terms of an error sum-of-squares criterion, or ESS (Kural, 1999).

Ward's method is biased toward finding "spherical," tightly bound clusters (El-Hamdouchi and Willett, 1989) and does not cope well with unequal sample sizes (Kuiper and Fisher, 1975). Moreover, it is only defined explicitly when the Euclidian distance is used to calculate inter-document similarities; using an association coefficient (e.g., the Dice's coefficient) will not yield an exact Ward classification (Willet, 1988).

That being said, in their CBR experiments with small document collections, Griffiths et al. (1986) find this method gives better results than single linkage, complete linkage, and group average methods.

7.5.1.8 Optimization Methods

Optimization methods numerically partition the individuals of specific groups, by minimizing or maximizing certain numerical criteria, not necessarily classifying data hierarchically. Essentially, an index, $f(n, g)$, the value of which indicates the "quality" of this particular clustering, is associated with each partition of n individuals into the required number of groups g. Associating a number with each partition allows their comparison.

Optimization methods have some limitations. Blashfield (1976) finds the result of optimization can be radically affected by the choice of the starting partition, although we can reasonably expect convergence if data are well-structured. In addition, these methods may impose a "spherical" structure on clusters even if the "natural" data clusters have different shapes. Finally, selecting the number of groups is a major issue; here, several methods have been suggested to aid users (Beale, 1969; Calinski and Harabasz, 1974; Marriott, 1971).

7.5.1.9 Mixed Models

Mixed models attempt to provide a way to derive inferences from sample to population. The most common approach is mixed distributions.

The mixed approach provides a better statistical basis than other methods and requires no decisions about what particular similarity or distance measure is appropriate for a data set. Nevertheless, mixed models have their own set of assumptions, such as normality and conditional independence, which may not be realistic in all applications (Kural, 1999).

7.5.1.10 Density Search Clustering Techniques

These techniques search for high-density regions in the data, with each such region taken to signify a different group. Although some density search clustering techniques originate in single link clustering, they attempt to overcome chaining as shown in Figure 7.19 (Kural, 1999).

7.5.1.11 Taxmap Method

The taxmap method detects clusters by comparing relative distances between points; it searches for continuous, relatively densely populated regions of the space, surrounded by continuous relatively empty spaces. Although the initial formation of clusters emulates the single linkage model, criteria are set to determine when to stop additions to the clusters.

7.5.1.12 Mode Analysis

Mode analysis is another derivative of single linkage clustering. It searches for natural subgroupings of the data by seeking disjoint density surfaces on the sample distribution. The search considers a sphere of a radius, R, surrounding each point and counts the number of points falling in the sphere. Individuals are labeled dense or nondense based on whether their spheres contain more or fewer points than the value of a linkage parameter, K; this parameter is preset at a value dependent on the number of individuals in the data set.

Unfortunately, mode analysis cannot simultaneously identify large and small clusters (Kural, 1999).

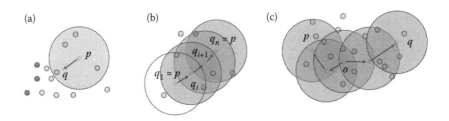

FIGURE 7.19
Types of density reachable techniques. (a) Directly density reachable, (b) density reachable, (c) density connected.

7.5.1.13 Clumping Techniques

Clumping techniques allow overlapping clusters. They often begin with the calculation of a similarity matrix, followed by the division of the data into two groups by minimizing the so-called cohesion function.

Minimizing algorithms perform successive reallocations of single individuals from an initially randomly chosen cluster center. Many divisions can be created by iterating from different starting points. In each case, members of the smaller group are noted; they constitute a class to be set aside for further examination (Kural, 1999).

7.5.2 Limitations of Associating Distances

As mentioned, all clustering methods must measure the association between objects in some fashion, for example, with similarity, dissimilarity, and distance measures.

However, when we look at the existing techniques, it is clear that distance and similarity are relative and dependent on many things, including available data, expertise from users, and so on.

Maintenance practitioners seek similarity and dissimilarity indicators to get an accurate diagnosis for maintenance actions. A similarity coefficient indicates the strength of the relationship between two objects, given the values of a set of p variables common to both. Most similarity measures are nonnegative and are scaled to have an upper limit of unity, although some are correlational so that the coefficient varies between −1 and 1.

Dissimilarity measures complement the similarity measures. There are also several dissimilarity coefficients representing a kind of "distance" function. These generally satisfy the following conditions (Kural, 1999):

$$D(X,Y) \geq 0 \qquad \text{for all } X, Y \in P$$
$$D(X,X) = 0 \qquad \text{for all } X \in P$$
$$D(X,Y) = D(Y,X) \qquad \text{for all } X,Y \in P$$
$$D(X,Y) \leq D(X,Z) + D(Y,Z) \qquad \text{for all } X,Y \in P$$

The fourth condition is triangle inequality, a theorem from Euclidian geometry, stating that the sum of the lengths of two sides of a triangle is always greater than the length of the third side.

The following is an example of the dissimilarity coefficient satisfying the above four conditions:

$$\frac{|X \nabla Y|}{|X| + |Y|}$$

where $(X \nabla Y) = (X \cup Y) - (X \cap Y)$ is the symmetric difference of sets X and Y.

It is simply related to Dice's coefficient by

$$1 - \frac{2|X \cap Y|}{|X| + |Y|} = \frac{|X \nabla Y|}{|X| + |Y|}$$

and it is monotone to Jaccard's coefficient subtracted from 1.

Some argue differences in output among different measures are insignificant, if these are appropriately normalized. Lerman (1970) finds many of the frequently used measures are monotone with respect to each other. Sneath and Sokal (1973) agree with Lerman (1970) and suggest the simplest type of coefficient that seems appropriate should be used. Yet some experimental results reveal coefficients may affect the outcome (Kirriemuir and Willett, 1975).

Because it organizes multivariate data into subgroups or clusters, clustering may assist investigators in determining the characteristics of any structure or pattern. Applying the methods in practice, however, requires considerable care to avoid overinterpreting the resulting solutions. Questions of cluster validity require considerable attention, although such questions are seldom straightforward and "full of traps for the unwary" (Dubes and Jain, 1979, p. 4).

Personal intuition and insight often dominate interpretations of the results of a clustering algorithm. Problematically, clustering algorithms generate clusters even when applied to random data; it is clearly necessary to avoid elaborate interpretation of the derived solutions in such cases. This is a major risk of clustering done by maintainers who can influence the process and whose conclusions mirror their expectations and strong aprioristic beliefs.

7.6 Text Clustering and Categorization

Analysis of data can reveal interesting and often important structures or trends in the data that reflect a natural phenomenon. Discovering regularities in data can be used to gain insight, interpret certain phenomena, and ultimately make appropriate decisions in various situations. Finding inherent but invisible regularities in data is the main subject of research in data mining, machine learning, and pattern recognition.

Data clustering is a data mining technique that enables the abstraction of large amounts of data by forming meaningful groups or categories of objects, formally known as clusters, such that objects in the same cluster are similar to each other, and those in different clusters are dissimilar. A cluster of objects indicates a level of similarity between objects; when we can consider them to be in the same category, our reasoning about them is considerably simplified.

These objects can be simple values, as shown earlier. There are also instances when the values (condition indicators) are fused with expert knowledge, events, manufacturer data, and so on, in order to define a situation not just with a numerical value but also with a set of circumstances, thereby allowing the user to identify what is happening in a much more accurate way. Simply stated, there are many variables providing more information.

For this purpose, disparate data sources, including text sources in the form of work orders, reports, and OEM data must be considered. Text categorization is defined as the process of assigning predefined class labels to new text documents based on what the classifier learns from the training set of documents.

Text clustering approaches can be classified according to various independent dimensions. For instance, different starting points, methodologies, algorithmic points of view, clustering criteria, and output representations usually lead to different taxonomies of clustering algorithms. Properties of clustering algorithms can be described as follows:

1. Nonhierarchical methods:
   a. *k*-means and extensions (spherical *k*-means, kernel *k*-means, and bisecting *k*-means)
   b. Buckshot
   c. Leader–follower algorithm
   d. Self-organizing map (SOM)
2. Hierarchical methods:
   a. Agglomerative
   b. Divisive
3. Generative algorithms:
   a. Gaussian model
   b. Expectation maximization
   c. Von Mises–Fisher
   d. Model-based *k*-means
4. Spectral clustering:
   a. Divide and merge
   b. Fuzzy coclustering
5. Density-based clustering:
   a. A cluster is composed of well-connected dense regions. Density-based spatial clustering of applications with noise (DBSCAN) is a typical density-based clustering algorithm; it works by expanding clusters to their dense neighborhood (Ester et al., 1996).

6. Phase-based models:
   a. Suffix tree clustering
   b. Document index graph

Major categorization techniques are K nearest-neighbor (KNN), SVM, naïve Bayesian (NB), naïve Bayesian multinomial (MNB), decision tree, hidden Markov model (HMM), maximum entropy, and NN. The following modules are generally used in any clustering technique:

1. Preprocessing and feature extraction
2. Dimensionality reduction (e.g., PCA, nonnegative matrix factorization, soft spectral coclustering, and lingo)
3. Similarity measures
4. Clustering algorithm
5. Evaluation using internal and external validity measures

To perform text categorization, a set of modules must be applied. Table 7.1 shows the best-known text categorization algorithms alongside the required modules.

7.6.1 Applications and Performance

Clustering is used in a wide range of applications, such as marketing, biology, psychology, astronomy, image processing, and text mining.

Document clustering techniques are widely used in

1. Information retrieval
   a. Improve precision and recall
   b. Organize results
   c. Online search engines: clusty.com and iboogie.com

TABLE 7.1

Text Categorization Components

Modules	SVM	KNN	MNB
Feature extraction (stemming and stop word removal)	Yes	Yes	Yes
Feature selection	Yes	Yes	Yes
Document representation	Yes	Yes	Yes
Learning	Yes	Yes	Yes

2. Organizing documents
   a. Find nearest documents to a specific document
   b. Automatically generate hierarchical clusters of documents
   c. Browse a collection of documents, corpus exploration
3. Indexing and linking

Text categorization techniques are used in many applications, including:

1. Filtering email
2. Mail routing
3. Spam filtering
4. News monitoring
5. Sorting through digitized paper archives
6. Automated indexing of scientific articles
7. Classifying news stories
8. Searching for interesting information on the internet
9. Classifying business names by industry
10. Classifying movie reviews as favorable, unfavorable, and neutral
11. Classifying websites of companies by Standard Industrial Classification (SIC) code
12. Providing packages for clustering and categorization

Dependency between technologies

1. Clustering and feature selection (e.g., use of frequent item sets and closed frequent item sets)
2. Clustering and social network analysis
3. Clustering and image processing
4. Clustering and outlier detection
5. Clustering and gene expression analysis
6. Clustering and semantic-based techniques

Table 7.2 compares clustering techniques.

7.7 Contextual Clustering

Words are a basic form of data in social science research, largely because of their use as the most common medium of social exchange. For many

TABLE 7.2

Comparison of Clustering Techniques

Technique	Dis(similarity) Matrix	Number of Clustering	Sensitive to Outliers	Shape of Clusters
Nonhierarchical methods	No	Known	Yes	Spherical-shaped clusters
Hierarchical methods	Yes	Known	Not as nonhierarchical approaches	Elongated-shaped clusters
Generative algorithms	No	Known	Yes	Spherical-shaped clusters
Spectral clustering	No	Known	No	Arbitrary-shaped clusters
Density-based clustering	Yes	Unknown	No	Arbitrary-shaped clusters
Phase-based models	Yes	Known	No	Arbitrary-shaped clusters

purposes, insight into meanings can be obtained by examining profiles of ideas and contextual information contained in text. The social science research problem of systematically coding textual data (i.e., making quantitative distinctions between texts varying in the pattern of emphasis on sets of ideas and in the context or social perspective from which these ideas are studied) can be addressed by using scores to describe comparative patterns of meaning in textual data, to generate traditional statistical analyses with other nontextual variables, and to organize and focus further qualitative analysis (McTavish and Pirro, 1990).

By "text," we are referring to a transcript of naturally occurring verbal material. These are widely diverse and include conversations, written documents (i.e., diaries, organization reports, books), written or taped responses to open-ended questions, media recordings, and verbal descriptions of observations. Ultimately, a transcript will comprise a computerized file of conventional words and sentences for one or more cases.

Given the inherent diversity and the many possibilities, we need methodologies to directly, systematically, and efficiently handle textual data. Trained coders have traditionally been used, but serious validity, reliability, and practical problems are often the result. Although computer approaches have available since the 1960s and permit more systematic and reliable coding of themes and meanings, they still are not widely used in social science research. We agree with Markoff and Weitman (1975), when they say content analysis must be integrated with traditional methodology. Brifely stated, our approach extends computer content analysis, so that it becomes a useful complement to traditional social science methodology.

7.7.1 Overview of Contextual Content Analysis

Contextual content analysis, implemented in the Minnesota Contextual Content Analysis (MCCA) computer program, builds on computer content analysis methodology in a number of ways.[3]

First, in the social sciences, *all* words in one or more texts can be divided into many idea categories (including the category "not elsewhere classified"), based on a conceptual "dictionary." A dictionary groups words (or word meanings) into categories considered to express (singly or in patterns) ideas of interest to an investigator. Several conceptual dictionaries have been used in computer-based content analysis; the organization of each follows a different theoretical perspective.[4]

The MCAA contextual–conceptual dictionary is oriented toward frequently used words whose meanings can be organized in a number of mutually exclusive categories of general social science interest (Pierce and Knudsen, 1986). Words with multiple meanings are disambiguated in a text, and relative emphasis on each category is normed with respect to a standard (i.e., the expected emphasis, accounting for expected variability in use over a number of social contexts), a process described later. The resulting vector of normed scores ("emphasis" scores or E-scores) allows investigators to examine over and underemphasis relative to the norm of expected usage. Broader concepts and themes can be identified from scores for sets of related categories. Finally, investigators can make quantitative distinctions between texts by looking at the overall profile of emphasis on idea categories.

Second, MCCA picks up the last point and incorporates the hypothesis that social contexts (groups, institutions, organizational cultures, or other socially defined situations) can be identified by the overall profile of relative emphasis on idea categories used in their respective communications. The idea emphasis profile contains valuable information that helps researchers distinguish and characterize social contexts.

Four general "marker" contexts can be used to aid in interpreting contextual information in these profiles: "traditional," "practical," "emotional," "analytic." Each is an experimental, empirically derived profile of relative emphasis on each idea category. In addition, each characterizes the perspective typical of a general social or institutional context. As a set, the four markers act as dimensions to define a social context space. MCCA computes these contextual scores (or C-scores). Texts can be scored and differentiated on these four dimensions. For example, a more "traditional" concern for breach of norms and appropriate sanctions in a religious discussion can be distinguished from a more "practical" concern for failure to successfully achieve goals and consequences in a business discussion. Even similar ideas may be discussed in quite different ways in different social contexts; these scores may be important parameters of social contexts.

Third, MCCA links the strengths of qualitative and quantitative social science research. An investigator can examine transcribed conversational interviews on a topic for a large representative sample of cases, for example, while quantitative scores can guide comparative, qualitative analysis of social meanings in textual data, thereby adding qualitative depth to quantitative analyses.

Fourth, because computerized content analysis eliminates problems of coder reliability, measurement and validity issues are much less salient.

In short, the approach has much to offer systematic analyses. For example, norming provides a basis for examining topical emphasis (including distinctive omissions) in a text, formerly a problematic task in coding open-ended response data. More specifically, through the use of normed, idea-emphasis scores (E-scores) and scores emphasizing the four marker contexts (C-scores), any naturally occurring textual material can be "coded" to reflect meanings of interest to an investigator. In addition, the set of C-scores and E-scores for each text can be combined for traditional quantitative, statistical analysis with independent and dependent variables measured in other ways. This type of "contextual content analysis" differs from more traditional hand and computer content analytic approaches. Most notably, it provides social science research with a broad framework for characterizing social meanings in text and a practical, systematic means for scoring textual data.

The following sections elaborate on the MCCA and provide illustrations of its use (McTavish and Pirro, 1990).

7.7.2 Meanings in Text

As in any research, the meaning attributed to a text depends on the researcher's theory. There is no general answer to the question of what a textual passage "really" means. Nor is there usually a research interest in capturing "all" meanings in a text. Rather, the specific research problem and the investigator's personal use of theory will specify the relevant meanings in the appropriate text.

Markoff and Weitman (1975) distinguish between a situation where subjects want to share meanings and one where subjects intend to manipulate the investigator's understanding. From our point of view, manipulative intentions by subjects do not invalidate an analysis of what is said. It does suggest, however, that explanatory theories should also include the possibility of intentional manipulation. Similarly, subcultural and individualistic uses of words should be entertained in explanatory research uses of text (McTavish and Pirro, 1990).

7.7.3 Measuring Context

By "context," we are referring to a shared meaning or social definition of a situation of interaction. The context will explain the underlying orientation of any subsequent actions.[5] Context operates on many levels. Broad social contexts may be all encompassing, such as the meaning of being human or a member of a culture, a subculture, or a nation. We find shared contexts on certain aspects of life, such as work, family, or leisure.[6] Briefly stated, the meaning of social context plays an important role in a number of explanatory perspectives in social science.[7]

Other types of analysis can rest on context. Typically, in content analysis, context information is assumed or determined intuitively (e.g., "we are in a work context, so we will look at meanings of job satisfaction, not religious satisfaction") or uses information outside the actual communication (e.g., the speaker's status or the conditions under which the text was prepared) (Krippendorff, 1980). Unfortunately, the description or characterization of a communication can be confused with the explanatory problem of determining its causes and consequences. To avoid this, we could focus on measurement, for example, coding descriptive information about ideas and context expressed in a text and using some of these codes in explanatory analysis, in addition to independently measured variables.

Words typically indicate the context, but other signs and symbols may do the same thing. For example, someone may say, "Tell me about your work." By framing the request in this fashion, the speaker has established the limits and direction of the conversation through his/her choice of one context (an economic or work context) and relative exclusion of others (e.g., religion, family, and leisure).

Context is indicated by the range of vocabulary used in a social encounter or in discussions of a topic (i.e., the number of unique words and the total number of words used). More frequently used words carry important information that allows us to distinguish between general social contexts. In other words, out of all possible words and constructions, a specific subset is chosen because of its ability to encode that particular communication.[8]

Middle-range words carry interesting contextual information. These words are generally known and used, and they appear in different social contexts. Yet their relative use varies widely across social contexts. On the one hand, they include the general classes of nouns, verbs, adjectives, and adverbs that allow description and evaluation across settings. On the other hand, they also include the pronouns, adverbs, and adjectives that specify and structure a situation. *Note*: MCCA pays particular attention to middle range and more widely used words.

Contextual information is also found in the preference for certain words or word groups over others, evidenced, for example, in probability distribution patterns across idea/word categories. Individuals use ideas/words in distinctive, patterned ways, reflective of their role and location in a social system, along with their individual socialization and other individual factors. Subcultures display overall patterns in the use of conceptual categories, as do specific social settings or contexts. Individuals learn certain patterns, and their speech shows changes in patterns in different social settings (e.g., as they move from church to job to recreation). Such usage patterns typify and distinguish institutionalized social settings (Cleveland and Pirro, 1974;

Namenwirth, 1968). Finally, we can find contextual information in the connectedness or co-occurrence of ideas.

The social meaning of a situation is important, as it provides the starting point for individual social interaction. In other words, if we know the social context, we can anticipate general kinds of activity and the appropriate behavior for that activity. Importantly, social context can be measured empirically with the content analysis framework, providing the basis for a more precise evaluation of the meaning of social contexts and a better comparison of communication across contexts. To this end, MCCA attempts to systematically code contextual information from textual data.

7.7.4 Issues of Validity and Reliability

Concerns about the validity and reliability of contextual content analysis are best addressed in specific research situations, like other measurement techniques (e.g., Likert scaling) whose concepts and their measures require examination. A number of authors have looked at questions of reliability and validity in content analysis (Andren, 1981; Holsti, 1969; Krippendorff, 1980; Weber, 1983).

Computers handle some important aspects of reliability in content analysis. Computer content analysis procedures process a given text file reliably, in accordance with instructions in a specific program. This allows investigators to realistically consider the inclusion of verbatim text in systematic research on larger or more representative samples and contrasting varieties of substantive topics.

A larger question is the reliability involved in the production of text (e.g., will two conversations with the same person on the same topic yield the same patterns, controlling for pretesting and other change factors?). Another interesting area of inquiry involves sampling pattern variability, given a certain number of respondents, documents, words, topics, and so on, which may require some reconceptualization before applying traditional sampling theory. In some instances, for example, we should consider the overtime branching pattern developed in a conversation on a topic or acknowledge that in some conversations, different respondents independently articulate the same, widely shared, aspect of culture. At any rate, stability in patterns is often reached with relatively few words (500–1000 on a topic) or modest respondent samples.

It is not usually appropriate to generalize validity findings to a whole measurement approach, but in this case, we can identify certain typical strengths and weaknesses. Few social science theories provide a deductive link with word patterns and constructs, so investigators using content analysis turn to the assumptions about shared meanings central to other measurement approaches (e.g., survey questionnaires). The assumption of wide areas of consensus on word meanings appears justified (illustrated by the usefulness of a standard dictionary) in many research situations. There is also widespread reliance on face (content) validity assessment of measures, emerging from informed judgment, *a priori* knowledge, intuitive plausibility tests, and so on.

Can we make a general conclusion as to the validity of content analytic measures, including those computed by MCCA? Our experience suggests direct links between content analysis measures and certain theoretical concepts. For one thing, content analysis allows us to operate directly on the expressed meanings and emphases of subjects; we do not need subjects to translate their experiences into structured statements closer to our research needs, thus preserving respondents' emphases and nuances. In addition, the data are more likely to be gathered as a part of the normal process of human communication rather than as "encoded" or "prestructured" conversation.

Various research projects have offered myriad opportunities for predictive validity checks. For example, a study comparing the open-ended conversations of husbands and wives about their relationship resulted in accurate classifications of couples in behavioral terms (seeking divorce, seeking outside help, coping). These could be compared with the independent judgments of clinicians having access to the couples (McDonald and Weidetke, 1979).

Admittedly, it is helpful to include traditional measures of key concepts in content analytic studies and separately measured predictors. Adding "criterion text" profiles to an analysis also aids in validity assessment. If the criterion text represents a relatively pure instance of a theoretical construct or position on a theoretical continuum, an analysis of its distance from other texts can assist analysis and validity assessment. Additionally, when alternative information can identify relevant criteria to predict, predictive and posterior studies are useful.

Finally, we can frequently identify expected relationships to examine in content analysis. During the analytic process, a series of relatively low level expectations can be generated and tested along the line of "if this is an accurate interpretation, then we would also expect that to be true..." The result is a series of small construct validity tests or "triangulations," which shed significant light on questions of validity (McTavish and Pirro, 1990).

Computer-based, contextual/conceptual content analysis augments traditional measurement approaches and provides a way to integrate the relative strengths of qualitative and quantitative research. It suggests realistic possibilities for the reliable and systematic analysis of a broad range of social science data, including historic documents, cross-cultural materials, interview transcripts, verbal processes, open-ended responses, and so on.

We hasten to add, however, that further theoretical and quantitative work is required on linkages between conceptual definitions of key social science variables and patterns of word usage, including comparative word patterns across cultures, societies, institutions, organizations, and historic time. That said, work to date suggests an ability to more directly deal with expressions of social meanings in a rigorous analytic framework is indeed possible and will prove fruitful in future social science investigations.

References

Al-Akwaa, F. M., 2012. *Analysis of Gene Expression Data Using Biclustering Algorithms, Functional Genomics.* In: Meroni, G. (Ed.), *InTech.* [Online] Available from: http://www.intechopen.com/books/functional-genomics/analysis-of-gene-expression-data-using-biclustering-algorithms.

Aldenderfer, M. and Blashfield, R., 1984. *Cluster Analysis.* Newbury Park, CA: Sage Publications.

Anderberg, M. R., 1973. *Cluster Analysis for Applications,* 1st edition. New York: Academic Press.

Andren, G., 1981. Reliability and content analysis. In: Rosengren, K. E. (Ed.), *Advances in Content Analysis.* Beverly Hills, CA: Sage, Chapter 2, pp. 43–67.

Anon., 2001. Fuzzy modeling of client preference from large data sets: An application to target selection in direct marketing. *IEEE Transaction on Fuzzy Systems,* 9(1), 153–163.

Anon., 2008. *Clustering and Classification Methods for Biologists.* [Online] Available at: http://www.alanfielding.co.uk/multivar/ca_alg.htm.

Bacher, J., 1996. *Clusteranalyse [Cluster Analysis].* Opladen [only available in German].

Backer, E. and Jain, A., 1981. A clustering performance measure based on fuzzy set decomposition. *IEEE Transactions on Pattern Analysis and Machine Intelligence,* 3(1), 66–75.

Baraldi, A. and Alpaydin, E., 2002. Constructive feedforward ART clustering networks—Part I and II. *IEEE Transactions on Neural Networks,* 13(3), 645–677.

Beale, C., 1969. Natural decrease of population: The current and prospective status of an emergent American phenomenon. *Demography,* 6, 91–99.

Berson, A., Smith, S., and Thearling, K., 2008. *An Overview of Data Mining Techniques.* [Online] Available at: http://www.thearling.com/text/dmtechniques/dmtechniques.htm.

Bezdsek, J., 1991. *Pattern Recognition with Fuzzy Objective Functions Algorithms.* New York: Plenum Press.

Bishop, C., 1995. *Neural Networks for Pattern Recognition.* New York: Oxford.

Blashfield, R. K., 1976. Mixture model tests of cluster analysis: Accuracy of four agglomerative hierarchical methods. *Psychological Bulletin,* 83(3), 377.

Bonner, R., 1964. On some clustering techniques. *IBM Journal of Research and Development,* 8, 22.

Bradley, P. and Fayyad, U. M., 1998. Refining initial points for *k*-means clustering. *15th International Conference on Machine Learning.* San Francisco: Morgan Kaufmann, pp. 91–99.

R.W. Buddemeier, C.J. Crossland, B.A. Maxwell, S.V. Smith, D.P. Swaney, J.D. Bartley, G. Misgna, C. Smith, V.C. Dupra, and J.I. Marshall, 2002. *Land-Ocean Interactions in the Coastal Zone Core Project of the IGBP.* Texel, The Netherlands: Summary report and compendium. LOICZ Reports & Studies.

Calinski, T. and Harabasz, J., 1974. Adendrite method for cluster analysis. *Communication Statistics,* 3(1), 1–27.

Carl, G. L., 2008. *Artificial Neural Network Structures.* Nevada: University of Nevada.

Center for the New Engineer, 1995. *Types of Clustering Algorithms.* [Online] Available at: http://cs.gmu.edu/cne/modules/dau/stat/clustgalgs/clust3_frm.html.

Cherkassky, V. and Mulier, F., 1998. *Learning from Data: Concepts, Theory, and Methods.* New York: John Wiley & Sons.

Claudia Isazaa, M.-V. L. L. J. A.-M., 2008. Diagnosis of chemical processes by fuzzy clustering methods: New optimization method of partitions. *18th European Symposium on Computer Aided Process Engineering,* Lyon, France.

Cleveland, C. E. M. D. and Pirro, E. B., 1974. *Contextual Content Analysis.* ISSC/CISS Workshop on Content Analysis in the Social Sciences, Pisa, Italy.

Cormack, R. M., 1971. A review of classification. *Journal of the Royal Statistical Society (Series A),* 14, 279–298.

Cunningham, K. M. and Ogilvie, J. C., 1972. Evaluation of hierarchical grouping techniques—A preliminary study. *The Computer Journal,* 15(3), 209–13.

Dalton, T., 1999. *Fuzzy Logic in Fault Diagnosis.* Duisburg, Germany: University of Duisburg.

Denton, A. M., 2003. *Fast Kernel-Density-Based Classification and Clustering Using P-Trees.* North Dakota: North Dakota State University.

Dubes, R. and Jain, A. K., 1979. Validity studies in clustering methodologies. *Pattern Recognition,* 11(4), 235–254.

Duda, R., Hart, P., and Stork, D., 2001. *Pattern Classification,* 2nd edition. New York: John Wiley & Sons.

El-Hamdouchi, A. and Willett, P. C., 1989. Comparison of hierarchic agglomerative clustering methods for document retrieval. *The Computer Journal,* 32(3), 220–227.

Ester, M., Kriegel, H. P., Sander, J., and Xu, X. 1996, August. A density-based algorithm for discovering clusters in large spatial databases with noise. *KDD,* 96, 226–231.

Everitt BJ, D. A. R. T., 2001. The neuropsychological basis of addictive behaviour. *Brain Research Reviews,* 36(2), 129–38.

Everitt, B. S., 1993. *Cluster Analysis.* London: Edward Arnold.

Fritzke, B., 1997. *Some Competitive Learning Methods.* [Online] Available at: http://www.neuroinformatik.ruhr—uni—bochum.de/ini/VDM/research/gsn/JavaPaper.

Gordon, A., 1981. *Classification: Methods for the Exploratory Analysis of Multivariate Data.* New York: Chapman & Hall.

Griffiths, A., Luckhurst, H. C., and Willett, P., 1986. Using inter-document similarity information in document retrieval systems. *JASIS,* 37, 3–11.

Han, J. and Kamber, M., 2001. *Data Mining: Concepts and Techniques.* San Diego, CA: Academic Press.

Hebb, D. O. (Ed.), 1949. The first stage of perception: Growth of the assembly. In: *The Organization of Behaviour.* New York: Wiley, pp. 60–78.

Holsti, O., 1969. *Content Analysis for the Social Sciences and Humanities.* Reading, MA: Addison-Wesley.

Jain, A. and Dubes, R., 1988. *Algorithms for Clustering Data.* Englewood Cliffs, NJ: Prentice Hall.

Jain, A., Murty, M., and Flynn, P., 1999. Data clustering: A review. *ACM Computer Surveys,* 31(3), 264–323.

Jain, A. K. and Dubes, R. C., 1988. *Algorithms for Clustering Data.* Michigan State University: Prentice Hall.

Jardine, N. and Sibson, R., 1971. *Mathematical Taxonomy*. London and New York: Wiley.

Keller, F. and Crocker, M., 2003. *Connectionist and Statistical Language Processing.* [Online] Available at: http://www.coli.uni-saarland.de/~crocker/Teaching/Connectionist/Connectionist.html.

Kirriemuir, E. and Willett, P., 1975. Identification of duplicate and near-duplicate full text records. *Program*, 29(3), 241–256.

Klerfors, D., 1998. *Artificial Neural Networks*. [Online] Available at: http://osp.mans.edu.eg/rehan/ann/Artificial%20Neural%20Networks.htm.

Kohavi, R., 1995. A study of cross-validation and bootstrap for accuracy estimation and model selection. *14th International Joint Conference on Artificial Intelligence*, San Francisco, CA, USA, pp. 338–345.

Krippendorff, K., 1980. *Content Analysis: An Introduction to Its Methodology*. Beverly Hills, CA: Sage.

Krzanowski, W. J., 1988. *Principles of Multivariate Analysis: A User's Perspective.* Clarendon: Oxford.

Krzanowski, W. J., Jonathan, P., McCarthy, W. V., and Thomas, M. R., 1995. Discriminant analysis with singular covariance matrices: Methods and applications to spectroscopic data. *Journal of the Royal Statistical Society. Series C (Applied Statistics)*, 44(1), 101–115.

Kuiper, F. K. and Fisher, L., 1975. 391: A Monte Carlo comparison of six clustering procedures. *Biometrics*, 31(3), 777–783.

Kural, Y. S., 1999. *Clustering Information Retrieval Search Outputs*. London: City University.

Lance, G. N. and Williams, W. T., 1967. A general theory of classificatory sorting strategies: II. Clustering systems. *The Computer Journal*, 10(3), 271–277.

Lerman, A., 1971. Time to chemical steady states in lakes and oceans. In: Hem, J. D. (Ed.), *Nonequilibrium Systems in Natural Water Chemistry*. Advances in Chemistry Series #106. Washington, DC: American Chemical Society, pp. 30–76.

Lin, C.- T. and George Lee, C. S., 1996. *Neural Fuzzy Systems*. New Jersey, NJ: Prentice-Hall.

Linnaeus, C., 1737. *Genera Plantarum*.

Markoff, J. S. G. and Weitman, S. R., 1975. Toward the integration of content analysis and general methodology. In: *Sociological Methodology*. San Francisco, CA: Jossey-Bass, pp. 1–58.

Marriott, F. H. C., 1971. Practical problems in a method of cluster analysis. *Biometrics*, 27(3), 501–514.

Marsili-Libelli, S., 1998. Adaptive fuzzy monitoring and fault detection. *International Journal of COMADEM*, 1(3), 31–37.

Matteucci, M., 2003. *A Tutorial on Clustering Algorithms*. [Online] Available at: http://home.deib.polimi.it/matteucc/Clustering/tutorial_html/index.html [Accessed 14 April 2014].

McDonald, C. and Weidetke, B., 1979. *Testing Marriage Climate*. Iowa City, IA: Iowa State University.

McTavish, D. G. and Pirro, E. B., 1990. Contextual content analysis. *Quality and Quantity*, 24(3), 245–265.

Milligan, G., 1980. An examination of the effect of six types of error perturbation on fifteen clustering algorithms. *Psychometrika*, 42, 325–342.

Murtagh, F., 1983. A survey of recent advances in hierarchical clustering algorithms. *The Computer Journal*, 26(4), 354–359.

Namenwirth, J. Z., 1968. *Contextual Content Analysis.* Mimeo: Yale University.

Nguyen, H. T., 1994. Fuzzy neutral systems research in Asia. In: *Materials Science Research in Japan.* Darby, PA: DIANE Publishing Company, pp. 61–64.

Pallavi, 2003. *Clustering.* [Online] Available at: ftp://www.ece.lsu.edu/pub/aravena/ ee7000FDI/Presentations/Clustering-Pallavi/CLUSTERING_report.doc.

Pierce, J. M. D. and Knudsen, K. R., 1986. The measurement of job design: A content analytic look at scale validity. *The Journal of Occupational Behaviour*, 7, 299–313.

Qasem, M., 2009. *Vector Quantization.* [Online] Available at: http://www.mqasem.net/ vectorquantization/vq.html.

Rahman, N. B. A., Saman, F. I. B., and Zianuddin, N. B., 2011. *Academic Achievement Prediction Model Using Artificial Neural Network: A Case Study.* Malaysia: Universiti Teknologi Mara.

Rennols, K., 2002. Classification. In: *Data Mining.* [Online] Available at: http://cms1. gre.ac.uk/research/cassm/algorithmic_data_mining.htm.

Rohlf, F. J., 1973. Hierarchical clustering using the minimum spanning tree. *The Computer Journal*, 16, 93–95.

Sibson, R., 1973. SLINK: An optimally efficient algorithm for the single-link cluster method. *The Computer Journal*, 16(1), 30–34.

Sneath, P. H. A. and Sokal, R. R., 1973. *The Principles and Practice of Numerical Classification.* San Francisco: W.H. Freeman and Company.

Sylviane, G., 2007. *Supervision des Procédés Complexes.* Lavoisier.

Voorhees, P., 1985. The theory of Ostwald ripening. *Journal of Statistical Physics*, 38(1–2), 231–252.

Ward, J., 1963. Hierarchical grouping to optimise an objective function. *Journal of the American Statistical Association*, 58, 236–244.

Weber, R. P., 1983. Measurement models for content analysis. *Quality and Quantity*, 17, 127–149.

Willet, P., 1988. Recent trends in hierarchic document clustering: A critical review. *IP&M*, 24(5), 577–597.

Wang, X. and Syrmos, V. L. 2005. Optimal cluster selection based on Fisher class separability measure. In: *American Control Conference, 2005. Proceedings of the 2005.* IEEE, pp. 1929–1934.

Xu, R. and Wunsch, D.C., 2009. *Clustering*, 1st edition. New Jersey: John Wiley & Sons.

8

Statistical Techniques

8.1 Use of Stochastic Distributions to Detect Outliers

A stochastic process is a mathematical concept used to characterize a sequence of random variables (stochastic) that evolve according to another variable, usually time. Each random variable has its own probability distribution function and the variables can either be correlated or not. Each variable or set of variables subject to influences or random effects is a stochastic process. Any time a process (deterministic or essentially probabilistic) is analyzable in terms of probability, it deserves to be called a stochastic process.

The stochastic process is currently used to detect outliers (Bakar et al., 2006) or atypical observations. Some outliers are the result of recording or measurement errors and should be corrected (if possible) or deleted from the data.

An outlier may be defined as a data point that is very different from the rest of the data based on some measure. That point, on occasion, contains useful information about the abnormal behavior of the system described by the data (Aggarwal and Yu, 2005). From the point of view of a clustering algorithm, outliers are objects not in the data set groups and are often called noise (Breunig et al, 2000). At the same time, they may carry important information. Detected outliers are candidates for aberrant data that may otherwise lead to model misspecification, biased parameter estimation, and incorrect results. It is, therefore, important to identify them before modeling and analysis (Davies and Gather, 1993).

At present, some studies have been done on the detection of outliers for large data sets (Aggarwal and Yu, 2005). Many data-mining algorithms have as their function the minimization of the influence of outliers or their total elimination, but this could result in the loss of important hidden information; for example, what is noise to one person could be a signal to another (Knorr et al., 2000). In other words, the outliers themselves may be of particular interest, such as fraud detection, where outliers may indicate fraudulent activity (Han and Kamber, 2001).

8.1.1 Taxonomy of Outlier Detection Methods

Outlier detection methods include parametric (statistical) and nonparametric methods that are model free. Parametric methods include the statistical approach, control chart technique (CCT), and deviation-based approach. Nonparametric methods include the linear regression technique (LRT), data-mining methods, also called distance-based methods, and the Manhattan distance technique (MDT).

The statistical approach is appropriate for one-dimensional samples. The analysis is applied to the control chart and LRTs.

A distance-based approach, such as the MDT, can counter the main limitations of the statistical approach (Williams 2002).

Outlier detection methods include univariate and multivariate methods.

8.1.2 Univariate Statistical Methods

Many methods of detecting outlier univariates are based on the assumption of an underlying known distribution of the data, namely that the data are independent and identically distributed (iid). Equally, many discordance tests for detecting univariate outliers assume certain distribution parameters and certain outliers are expected. Needless to say, in real-world data-mining applications, these assumptions are often violated.

A central assumption in statistical-based methods for outlier detection is a generating model that allows a small number of observations to be randomly sampled from distributions $G_1,...,G_k$, differing from the target distribution F, often taken to be a normal distribution $N(\mu, \sigma^2)$. The outlier identification problem is then translated into the problem of identifying those observations lying in the so-called outlier region. This leads to the following definition.

For any confidence coefficient α, $0 < \alpha < 1$, the α-outlier region of the $N(\mu, \sigma^2)$ distribution is defined by

$$\text{out}\left(\alpha,\mu,\sigma^2\right) = \left\{x : |x - \mu| > z^\sigma_{1-\alpha/2}\right\} \tag{8.1}$$

where z_q is the q quintile of the $N(0,1)$. A number x is an α-outlier with respect to F if $x \in \text{out}(\alpha, \mu, \sigma^2)$. However, the normal distribution has been used as the target distribution. Therefore, this definition can be easily extended to any unimodal symmetric distribution with positive density function, including the multivariate case.

Note: The outlier definition cannot identify which observations are contaminated, that is, those resulting from distributions $G_1,...,G_k$; rather, it indicates those observations that lie in the outlier region.

8.1.2.1 Single Step versus Sequential Procedures

Davies and Gather (1993) make an important distinction between single step and sequential procedures for outlier detection. Single-step procedures identify all outliers at once as opposed to successive elimination or addition of data. In the sequential procedures, at each step, one observation is tested as an outlier. Following Equation 8.1, a common rule for finding the outlier region in a single-step identifier is given by

$$\text{out}\left(\alpha_n, \hat{\mu}_n, \hat{\sigma}_n^2\right) = \left\{x : |x - \hat{\mu}_n| > g\left(n, \alpha_n\right)\hat{\sigma}_n\right\} \tag{8.2}$$

where n is the size of the sample; $\hat{\mu}_n$ and $\hat{\sigma}_n$ are the estimated mean and standard deviation of the target distribution based on the sample, respectively; α_n denotes the confidence coefficient following the correction for multiple comparison tests; and $g(n, \alpha_n)$ defines the limits (critical number of standard deviations) of the outlier regions.

Traditionally, $\hat{\mu}_n, \hat{\sigma}_n$ are estimated, respectively, by the sample mean, $\bar{x}_n$, and the sample standard deviation, S_n. Since these estimates are highly affected by the presence of outliers, they are often replaced by more robust procedures. The multiple-comparison correction is used when several statistical tests are performed simultaneously. That is, while an α-value may be appropriate to decide whether a single observation is located in the region of outliers (i.e., a single comparison), this is not the case for a set of several multiple comparisons. To avoid false positives, the α-value needs to be lowered to account for the number of performed comparisons. Bonferroni's correction, a simple and conservative approach, sets the α-value for the whole set of n comparisons equal to α, by taking the α-value for each comparison equal to α/n? Another correction uses $\alpha_n = 1 - (1 - \alpha)^{1/n}$. *Note*: The traditional Bonferroni's method is "quasi-optimal" when the observations are independent, and this is unrealistic in most cases. The critical value $g(n, \alpha n)$ is often specified by numerical procedures, such as Monte Carlo simulations, for different sample sizes (Maimon and Rokach, 2010).

8.1.2.2 Inward and Outward Procedures

Sequential identifiers can be further classified into inward and outward procedures. In inward testing, or forward selection methods, at each step of the procedure, the "most extreme observation," that is, the one with the largest measure of outlyingness, is tested for being an outlier. If it is declared an outlier, it is deleted from the data set and the procedure is repeated. If it is declared a nonoutlying observation, the procedure terminates.

In outward-testing procedures, the sample of observations is first reduced to a smaller sample (e.g., by a factor of 2), with the removed observations placed in a reservoir. The statistics are calculated based on the reduced sample and the observations in the reservoir are tested in reverse order to determine whether they are outliers. If an observation is declared an atypical case, it is removed from the reservoir. If an observation is declared a nonoutlying observation, it is removed from the reservoir and added to the reduced sample; the statistics are recalculated and the process is repeated with a new observation. The outward-testing procedure is terminated when no more observations are left in the reservoir.

The classification into inward and outward procedures also applies to multivariate outlier detection methods (Ben-Gal, 2005).

8.1.2.3 Univariate Robust Measures

Traditionally, the sample mean and the sample variance give a good estimation of data location and data shape if they are not contaminated by outliers. When the database is contaminated, these parameters may deviate and significantly affect the outlier detection performance.

Hampel (1971, 1974) introduced the concept of the breakdown point as a measure for the robustness of an estimator against outliers. The breakdown point is defined as the smallest percentage of outliers that can cause an estimator to take arbitrary large values. Thus, the larger the breakdown point of an estimator, the more robust it is. For example, the sample mean has a breakdown point of $1/n$ since a single large observation can make the sample mean and variance cross any bound. Accordingly, Hampel suggested the median and the median absolute deviation (MAD) are robust estimates of the location and the spread. The Hampel identifier is often very effective in practical cases.

Other early work addressing the problem of robust estimators was by Tukey (1977) who introduced the Boxplot as a graphical display on which outliers can be indicated. The Boxplot has been used in many cases, and is based on the distribution quadrants. The first and third quadrants, Q_1 and Q_3, are used to obtain the robust measures for the mean, $\hat{\mu}_n = (Q_1 + Q_3)/2$, and the standard deviation, $\hat{\sigma}_n = Q_3 - Q_1$.

Another solution to obtain robust measures is to replace the mean with the median and compute the standard deviation based on $(1 - \alpha)$ percent of the data points where, typically, $\alpha = 5\%$.

Liu proposed an outlier-resistant data filter–cleaner (Liu et al., 2004), based on the earlier work of Martin and Thomson (1982). The proposed data filter–cleaner includes an online outlier-resistant estimate of the process model and combines it with a modified Kalman filter to detect and "clean" outliers. The method does not require *a priori* knowledge of the process model. It detects and replaces outliers online while preserving all other information in the data. The authors demonstrated that the proposed filter–cleaner is efficient in outlier detection and data cleaning for autocorrelated and even nonstationary process data.

8.1.2.4 Statistical Process Control

The field of statistical process control (SPC) draws on univariate outlier detection methods. It considers the case where the univariable stream of measures represents a stochastic process, and the detection of the outlier is required online. SPC methods have been applied for over 50 years and extensively studied in the statistics literature.

Ben-Gal categorizes SPC methods by two major criteria: (i) methods for independent data versus methods for dependent data; and (ii) methods that are model specific versus methods that are model generic (Ben-Gal et al., 2003). Model-specific methods require *a priori* assumptions on the characteristics of the process, usually defined by an underlying analytical distribution or a closed-form expression. Model-generic methods attempt to estimate the underlying model with minimum *a priori* assumptions.

Traditional SPC methods, such as Shewhart, cumulative sum (CUSUM), and exponential weighted moving average (EWMA), are model specific for independent data. These methods are applied widely in industry, although the independence assumptions are frequently violated in practice.

The majority of model-specific methods for dependent data are based on time series and often follow this format: find a time-series model that best captures the autocorrelation process, use this model to filter the data, and apply traditional SPC schemes to the stream of residuals. In particular, the ARIMA (autoregressive integrated moving average) family of models is widely implemented to estimate and filter process autocorrelation. Under certain assumptions, the residuals of the ARIMA model are independent and approximately normally distributed; traditional SPC can, therefore, be applied. Furthermore, ARIMA models, especially the simple ones such as autoregressive (AR), can effectively describe a wide variety of industry processes.

8.1.3 Multivariate Outlier Detection

In many cases, multivariable observations cannot be identified as outliers, as each variable is considered independently. Outlier detection is only possible when multivariate analysis is performed, and the interactions among different variables are compared within the class of data. A simple example appears in Figure 8.1. The figure shows data points with two measures on a

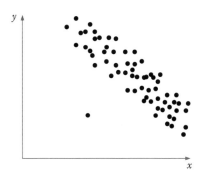

FIGURE 8.1
Two-dimensional space with one outlying observation (lower left corner).

two-dimensional space. The bottom left observation is clearly a multivariate outlier and not a univariate one. When considering each measure separately with respect to the spread of values along the *x* and *y* axes, we see they fall close to the center of the univariate distributions. Thus, the test for outliers must take into account the relationships between the two variables which, in this case, appear abnormal.

Data sets with multiple outliers or clusters of outliers are subject to masking and swamping effects. Although not mathematically rigorous, the following definitions give an intuitive understanding of these effects:

Masking effect: One outlier masks a second outlier if the second outlier can be considered an outlier only by itself, not in the presence of the first outlier. Therefore, after suppressing the first outlier, the second instance arises as an outlier. Masking occurs when a cluster of peripheral observations skews the mean and the covariance estimates toward it, and the resulting distance of the outlying point from the mean is small.

Swamping effect: One outlier swamps a second observation when the latter can be considered an outlier only in the presence of the first one. In other words, after the deletion of the first outlier, the second observation becomes a nonoutlying observation. Swamping occurs when a group of outlying instances skews the mean and the covariance estimates toward it and away from other nonoutlying instances, and the resulting distance from these instances to the mean is large, making them look like outliers.

8.1.3.1 Statistical Methods for Multivariate Outlier Detection

Multivariate outlier detection procedures can be divided into two broad categories: statistical methods that are based on estimated distribution parameters and data-mining-related methods that are typically parameter free.

Statistical methods for detecting multivariate outliers often indicate these observations are located relatively far from the center of the data distribution. Several distance measures can be applied to perform this task.

The shape and size of multivariate data are quantified by the covariance matrix. A well-known distance measure that takes into account the covariance matrix is the Mahalanobis distance, a familiar criterion that depends on estimated parameters of the multivariate distribution. Given *n* observations from a *p*-dimensional dataset (often *n–p*), let us denote the sample mean vector by $\bar{x}_n$ and the sample covariance matrix by V_n, where

$$V_n = \frac{1}{n-1} \sum_{i=1}^{n} (x_i - \bar{x}_n) * (x_i - \bar{x}_n)^T \qquad (8.3)$$

The Mahalanobis distance for each multivariate data point i, $i = 1, ..., n$, is denoted by M_i and given by

$$M_i = \left(\sum_{i=1}^{n} (x_i - \bar{x}_n)^T * V_n^{-1} (x_i - \bar{x}_n)^T \right)^{1/2} \tag{8.4}$$

Those observations with a large Mahalanobis distance are indicated as outliers. Masking and swamping effects play an important role in the adequacy of the Mahalanobis distance as a criterion for outlier detection. On the one hand, masking effects might decrease the Mahalanobis distance of an outlier. This might happen, for example, when a small cluster of outliers attracts $\bar{x}_n$ and inflates V_n in its direction. On the other hand, swamping effects might increase the Mahalanobis distance of nonoutlying observations, for example, when a small cluster of outliers attracts $\bar{x}_n$ and inflates V_n away from the pattern of the majority of the observations (Ben-Gal, 2005).

8.1.3.2 Multivariate Robust Measures

As in one-dimensional procedures, the distribution mean (measuring the location) and the variance–covariance (measuring the shape) are the two most commonly used statistics for data analysis in the presence of outliers. The use of robust estimates of the multidimensional distribution parameters can often improve the performance of the detection procedures in the presence of outliers. Hadi (1992) addressed this problem and proposed replacing the mean vector by a vector of variable medians and computing the covariance matrix for the subset of those observations with the smallest Mahalanobis distance. Caussinus and Roiz (1990) proposed a robust estimate for the covariance matrix, based on weighted observations according to their distance from the center. The authors also suggested a method for a low-dimensional projection of the dataset using generalized principle component analysis (GPCA) to reveal those dimensions displaying outliers. Other robust estimators of the location (centroid) and the shape (covariance matrix) include the minimum covariance determinant (MCD) and the minimum volume ellipsoid (MVE).

8.1.3.3 Preprocessing Procedures

Various paradigms have been suggested to improve the efficiency of different data-analysis tasks including outlier detection. One possibility is reducing the size of the data set by assigning the variables to several representative groups. Another option is eliminating some variables from the analyses using methods of data reduction such as principal components and factor analysis. Another way to improve the accuracy and the computational tractability of multiple outlier detection methods is the use of biased sampling (Kollios et al., 2003).

8.2 Issues Related to Data Set Size

In the past, small data fit the conceptual structure of classical statistics. Small always referred to the sample size, not the number of variables, even though these were kept to a handful. Depending on the method employed, small was seldom fewer than 5, sometimes between 5 and 20, frequently between 30 and 50, and 50 and 100, and rarely between 100 and 200 individuals. In contrast to today's big data, small data are a tabular display of rows (observations or individuals) and columns (variables or features) that fits on a few sheets of paper.

In addition to the compact area they occupy, small data are neat and tidy. They are "clean," in that they contain no unexpected values, except for those due to primal data entry error. They do not include the statistical outliers and influential points, or the exploratory data analysis (EDA) far-flung and outside points.

They are in the "ready-to-run" condition required by classical statistical methods.

There are two sides to big data. First, classical statistics considers big as simply not being small. Theoretically, big is the sample size after which asymptotic properties of the method "kick in" for valid results. Second, contemporary statistics considers big in terms of lifting observations and learning from the variables. Although it depends on who is analyzing the data, a sample size >50,000 individuals can be considered "big." Thus, calculating the average income from a database of 2 million individuals requires heavy-duty lifting (number crunching). In terms of learning or uncovering the structure among the variables, big can be considered 50 variables or more. Regardless of which side the data analyst is on, EDA scales up for both rows and columns of the data table.

The data size discussion raises the following question: how large should a sample be? Simply stated, sample size can be anywhere from folds of 10,000 up to 100,000.

It has been observed that the less experienced and less well-trained statistician/data analyst uses sample sizes that are unnecessarily large. Analyses have been performed on models built from samples that are too large by factors ranging from 20 through 50. Although the PC can perform the heavy calculations, the extra time and cost of getting the larger data out of the data warehouse and then processing them and thinking about them are almost never justified. Of course, the only way a data analyst learns that extra big data are a waste of resources is by performing small versus big data comparisons (Ratner, 2004).

8.2.1 Data Size Characteristics

There are three distinguishable characteristics of data size: condition, location, and population.

Condition refers to the state of readiness of the data for analysis. Data that require minimal time and cost to clean before reliable analysis can be performed are well conditioned; data that involve a substantial amount of time and cost are ill conditioned. Small data are typically clean and thus well-conditioned. Big data are an outgrowth of today's digital environment, with data flowing continuously from all directions at an unprecedented speed and volume. These data almost always require cleaning; they are considered "dirty" mainly because of the merging of multiple sources. The merging process is inherently a time-intensive process, as multiple passes of the sources must be made to get a sense of how the combined sources fit together. Because of the iterative nature of the process, the logic of matching individual records across sources is at first "fuzzy," then fine-tuned to soundness; at that point, unexplainable, seemingly random, nonsensical values result. Thus, big data are ill conditioned.

Location refers to where the data reside. Unlike the rectangular sheet for small data, big data reside in relational databases consisting of a set of data tables. The link among the data tables can be hierarchical (rank or level dependent) and/or sequential (time or event dependent). The merging of multiple data sources, each consisting of many rows and columns, produces data of even greater number of rows and columns, clearly suggesting bigness.

Population refers to a group of individuals with qualities or characteristics in common. Small data ideally represent a random sample of a known population; we do not expect to encounter changes in the composition of such data in the foreseeable future. The data are collected to answer a specific problem, permitting straightforward answers from a given problem-specific method. In contrast, big data often represent multiple, nonrandom samples of unknown populations, shifting in composition in the short term. Big data are "secondary" in nature; that is, they are not collected for an intended purpose. They are available from a plethora of marketing information for use on any post hoc query, and may not yield a straightforward solution.

It is interesting to note that Tukey never talked specifically about big data. However, he did predict that the cost of computing, both in time and dollars, would be cheap, which arguably suggests he knew big data were coming. The cost of today's PC proves him right (Ratner, 2004).

8.2.2 Small and Big Data

A valuable characteristic of "big" data is that they contain more patterns and interesting anomalies than "small" data. Thus, organizations can gain greater value by mining large data volumes than small ones. But big data involve more complex data, have greater processing and data storage requirements, and call for intensified filtering and analysis. Thus, while users can detect the patterns in small data sets using simple statistical methods, ad hoc query and analysis tools, or by simply eyeballing the data, they need sophisticated techniques to mine big data (Eckerson, 2011). Researchers are increasingly preoccupied with the question of how to analyze large amounts of data that cannot be analyzed using traditional tools.

The goal of small data is to organize information to make it understandable and actionable. The analysis and processing of these simpler data directly affect a business; for one thing, business opportunities are identified more efficiently as the data are in daily use. By focusing only on small data, a company obtains actionable information but loses the possibilities inherent in research on a large scale.

Simply stated:

- Small data focus on converting more actionable data into knowledge; big data permit large-scale research.
- Big data analyze and predict behavior patterns; small data are more qualitative.

8.2.3 Big Data

The scale, distribution, diversity, and/or timeliness of big data require the use of new technical architectures and analytics to enable insights into new sources of business value (EMC, 2012; Eckerson, 2011). Big data require the following:

- New data architectures, analytic sandboxes.
- New tools.
- New analytical methods.
- Integration of multiple skills into the new role of the data scientist.

There are multiple characteristics of big data, but four stand out as defining characteristics:

- Huge *volume* of data (for instance, tools that can manage billions of rows and millions of columns).
- Speed or *velocity* of new data creation.
- *Variety* of data created (structured, semistructured, and unstructured).
- *Complexity* of data types and structures, with an increasing volume of unstructured data (80%–90% of the data in existence are unstructured); part of the Digital Shadow or "Data Exhaust."

Because of the above, big data, due to their size or level of structure, cannot be efficiently analyzed using only traditional databases or methods. Think of these characteristics as "V3 + C."

There are many examples of the emerging opportunities in big data. Here are a few:

- Netflix suggesting your next movie rental.

- Dynamic monitoring of embedded sensors in bridges to detect real-time stresses and longer-term erosion.
- Retailers analyzing digital video streams to optimize product and display layouts and promotional spaces on a store-by-store basis.

Of course, such opportunities require new tools/technologies to store and manage the data in order to realize a business benefit. Big data necessitate new architectures supported by new tools, processes, and procedures that enable organizations to create, manipulate, and manage very large data sets and the storage environments that house them (Figure 8.2).

Big data can come in multiple forms, from highly structured financial data, to text files, to multimedia files, and genetic mappings. The high volume of the data is a consistent characteristic of big data. As a corollary to this, because of the complexity of the data themselves, the preferred approach for processing big data is in parallel computing environments and massively parallel processing (MPP), which enable simultaneous, parallel ingestion and data loading and analysis. As shown in Figure 8.3, most big data are unstructured or semistructured, thus requiring different techniques and tools for processing and analysis.

Big data is a relative term. For some organizations, terabytes of data may be unmanageable; other organizations may find that petabytes of data are overwhelming. If you cannot process your data with your existing capabilities, you have a big data problem.

The most prominent feature of big data is their structure. The graphic in Figure 8.3 portrays different types of data structures, with 80%–90% of the future data growth coming from nonstructured data types (semi, quasi, and unstructured).

Although the image shows four separate types of data, in reality, these can be mixed together at times. For instance, we may have a classic relational

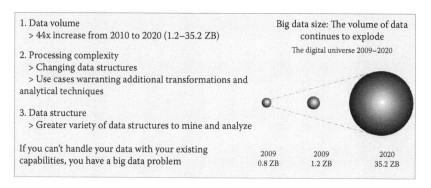

FIGURE 8.2
Key characteristics of big data. (From EMC Corporation, 2012. Big Data Overview. http://uk.emc.com/collateral/campaign/global/forum2012/ireland/big-data-overview-big-data-transforms-business-cvr.pdf)

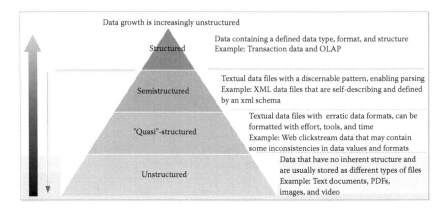

FIGURE 8.3
Unstructured data: fueling big data analytics. (From EMC Corporation, 2012. *Big Data Overview*.)

database management system (RDBMS) storing call logs for a software support call center. In this case, we may have typical structured data, such as date/time stamps, machine types, problem type, and operating system, probably entered by the support desk person from a pull-down menu graphical user interface (GUI).

Additionally, we will likely have unstructured or semistructured data, such as free form call log information, taken from an e-mail ticket or an actual phone call description of a technical problem and the solution. The most salient information is often hidden.

Another possibility is voice logs or audio transcripts of the actual call associated with the structured data. Until recently, most analysts would analyze the most common and highly structured data in the call log history RDBMS, since the mining of textual information is labor intensive and not easily automated (Figure 8.4).

People tend to both love and hate spreadsheets. With their introduction, business users were able to create simple logic on data structured in rows and columns and create their own analyses of business problems. Users do not need intensive training as a database administrator to create spread sheets; hence, business users can set these up quickly and independently of IT groups. Two main benefits of spreadsheets are

- They are easy to share
- End users have control over the logic involved

However, the proliferation of spreadsheets (data islands) caused organizations to struggle with "many versions of the truth," it was impossible to determine if they had the right version of a spreadsheet with the most current data and logic in it. Moreover, if a user lost a laptop or it became corrupted, this was the end of the data and their logic. Many organizations still

Data repositories: An analyst's perspective

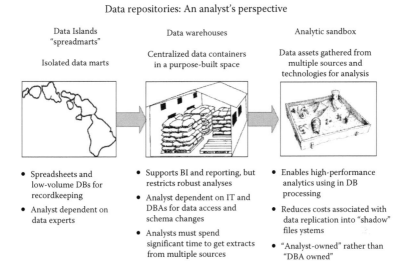

Data Islands "spreadmarts"	Data warehouses	Analytic sandbox
Isolated data marts	Centralized data containers in a purpose-built space	Data assets gathered from multiple sources and technologies for analysis

- Spreadsheets and low-volume DBs for recordkeeping
- Analyst dependent on data experts

- Supports BI and reporting, but restricts robust analyses
- Analyst dependent on IT and DBAs for data access and schema changes
- Analysts must spend significant time to get extracts from multiple sources

- Enables high-performance analytics using in DB processing
- Reduces costs associated with data replication into "shadow" files ystems
- "Analyst-owned" rather than "DBA owned"

FIGURE 8.4
Data repositories: an analyst's perspective. (From EMC Corporation, 2012. *Big Data Overview.*)

face this challenge (Excel is used on millions of PCs worldwide), creating the need to centralize data (EMC Corporation, 2012).

8.3 Parametric Techniques

Statistical parametric methods either assume a known underlying distribution of the observations or, at least, are based on statistical estimates of unknown distribution parameters. These methods flag as outliers those observations deviating from the model's assumptions. They are often unsuitable for high-dimensional data sets or for arbitrary data sets without prior knowledge of the underlying data distribution (Maimon and Rokach, 2010).

8.3.1 Statistical Approach

The statistical approach to outlier detection assumes a distribution or probability model for the given data set and identifies outliers from the model using a discordancy test (Han and Kamber, 2001). A statistical approach has five phases:

1. *Data collection:* Data are assumed to be part of a working hypothesis.
2. *Compute average value/compute linear regression equation:* The average value is computed to determine the centerline for the CCT. Otherwise, linear regression equation is calculated to determine the linear regression line.

3. *Compute upper and lower control limits (LCLs)/compute upper and lower bound value:* The upper control limit (UCL) and LCL for the control graph technique are based on a particular formula (see Equations 8.6 through 8.9 in Section 8.3.2); the upper and lower bound for the LRT are based on 95% of the linear regression equation (line).

4. *Data testing:* Actual data, the centerline, UCL, and LCL are plotted on the control graph, and actual data, the linear regression line, and the upper and lower bound are plotted on a linear regression graph. Outlier data can be identified from these graphs. Data plotted outside the UCLs and LCLs/bounds are detected as outlier data.

5. *Analysis and comparison of output:* The output from data testing is used to compare and analyze these techniques. The purpose is to determine the best technique for detecting outlier data based on the statistical approach. Each data object in the data set is compared to the working hypothesis and is either accepted in the working hypothesis or rejected as discordant and placed into an alternative hypothesis (outliers) (Abu et al., 2006).

8.3.2 Control Chart Technique

CCT is generally used to determine whether a process is statistically controlled. The main goal of a control chart is to detect any unwanted changes in the process. Such changes will be signaled by abnormal (outlier) points on the graph. The control chart has three basic components:

1. A centerline, usually the mathematical average of all the samples plotted.

2. UCLs and LCLs that define the constraints of common cause variations.

3. Performance data plotted over time.

To calculate the mean of the data points in order to determinate the centerline of a control chart, the formula is

$$\bar{X} = \frac{\sum_{i=1}^{n} X_i}{n} \tag{8.5}$$

where
$\bar{X}$ = mean/average value
X_i = every data value
$(X_i...X_n)$, n = total number of data

Now, calculate the UCL and LCL by

$$UCL(calculated) = \bar{X} + Z\sigma_x \tag{8.6}$$

$$LCL(calculated) = \bar{X} - Z\sigma_x \tag{8.7}$$

$$\sigma_x = \frac{\sigma}{\sqrt{n}} \tag{8.8}$$

$$\sigma = \text{standard derivation} = \left[\frac{\sum(X_i - \bar{X})^2}{n-1}\right]^{1/2} \tag{8.9}$$

In a three-sigma system, Z is equal to 3. Three-sigma control limits balance the risk of error because for normally distributed data, data points will fall inside three-sigma limits 99.7% of the time when a process is in control. This makes witch hunts infrequent, while still making it likely that unusual causes of variation will be detected.

Finally, plot the data on the chart; those data not within UCL and LCL are detected as outlier data. Figure 8.5 shows a control chart with one data point outside the UCL. This is an example of outlier data (Abu et al., 2006) (Figure 8.6).

8.3.3 Deviation-Based Approach

Given a set of data points (local group or global set), outliers are points that do not fit the general characteristics of that set; that is, the variance of the set is minimized when the outliers are removed. Outliers are the outermost points of the data set. In other words, an element disturbing a series of similar data is considered an exception.

In the deviation-based approach, the model is given a smoothing factor SF(I) that computes for each $I \subseteq$ DB how much the variance of DB is decreased when I is removed from DB. With equal decreases in variance, a smaller exception set is better. Outliers are the elements of the exception set $E \subseteq$ DB for which the following holds:

$$SF(E) \geq SF(I) \quad \text{for all } I \subseteq DB$$

The deviation-based approach resembles classical statistical approaches ($k = 1$ distributions) but is independent of the chosen kind of distribution.

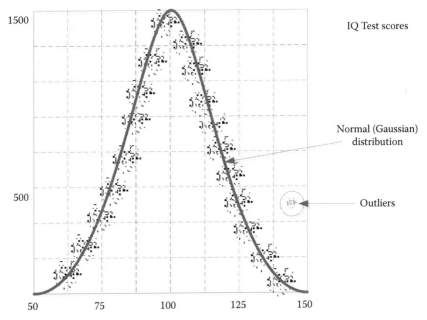

Working hypothesis: H: $o_i \in F$, where $i = 1,2,...,n$

Discordancy test: is o_i in F within standard deviation = 15

Alternative hypothesis:
—Inherent distribution: Ĥ: $o_i \in G$ where $i = 1,2,....n$
—Mixture distribution: Ĥ: $o_i \in (1 - \lambda)F + \lambda G$ where $i = 1,2,....n$
—Slippage distribution: Ĥ: $o_i \in (1 - \lambda)F + \lambda F'$ where $i = 1,2,....n$

FIGURE 8.5
Statistical approach to outlier detection.

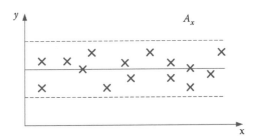

FIGURE 8.6
Control chart with outlier data. (From Abu, B. Z. et al. 2006. *A Comparative Study for Outlier Detection Techniques in Data Mining.* Member IEEE.)

In addition, heuristics such as random sampling or the best first search can be applied. Originally designed as a global method, it is applicable to any data type (depends on the definition of SF) and its output is labeling.

8.4 Nonparametric Techniques

Data-mining methods, also called distance-based methods, are nonparametric outlier detection methods. These methods are usually based on local distance measures and are capable of handling large databases. Another class of outlier detection methods is based on clustering techniques, where a cluster of small groups are considered clustered outliers. A related class of methods uses spatial detection techniques. These search for extreme examples or local instabilities with respect to neighboring values; however, these observations may not be significantly different from the entire population.

8.4.1 Linear Regression Technique

Statistical concepts used in data mining include point estimation, Bayes theorem, and regression. We use LRT in our analysis because it is appropriate to evaluate the strength of a relationship between two variables. In general, regression is the problem of estimating a conditional expected value, and "linear" refers to the assumption of a linear relationship between y (response variable) and x (predictor variable). Thus, in statistics, linear regression is a method of estimating the linear relationship between the input and output data (Montgomery, 2012). The common formula for a linear relationship used in this model is (Han and Kamber, 2001)

$$Y = \alpha + \beta x \tag{8.10}$$

where the variance of Y is assumed to be constant, and α and β are regression coefficients specifying the Y-intercept and slope of the line, respectively.

Given s samples or data points of the form $(x_1, y_1), (x_2, y_2)..., (x_s, y_s)$, then α and β can be estimated with the following equations:

$$\beta = \frac{\sum_{i=1}^{s} (x_i - \bar{x})(y_i - \bar{y})}{\sum_{i=1}^{s} (x_i - \bar{x})^2} \tag{8.11}$$

$$\alpha = \bar{y} + \beta\bar{x} \tag{8.12}$$

where x is the average of $x_1, x_2, ..., x_s$, and y is the average of $y_1, y_2,..., y_s$. The coefficients α and β often provide good approximations of otherwise complicated regression equations (Abu et al., 2006).

8.4.2 Manhattan Distance Technique

Distances can be based on a single dimension or multiple dimensions. It is up to the researcher to select the right method for his/her specific application. For our outlier detection analysis, we use MDT because the data are in a single dimension. The general formula for MDT is

$$d(t_i, t_j) = \sum_{h=1}^{k} \left| (t_{ih} - t_{jh}) \right| \tag{8.13}$$

where $t_i = <t_{i1}, ..., t_{ik}>$ and $t_j = <t_{j1}, ..., t_{jk}>$ are tuples in a database (Abu et al., 2006).

8.4.3 Data-Mining Methods for Outlier Detection

Data mining involves exploring and analyzing large amounts of data to find patterns in those data. The techniques come from the fields of statistics and artificial intelligence (AI), with a bit of database management thrown into the mix. Generally, the goal of data mining is either classification or prediction. In classification, the idea is to sort data into groups. For example, a marketer might be interested in the characteristics of those who responded versus those who did not respond to a promotion. These are two classes. In prediction, the idea is to predict the value of a continuous (i.e., nondiscrete) variable. For example, a marketer might be interested in predicting those who will respond to a promotion.

Typical algorithms used in data mining include the following (Hurwitz et al., 2013):

- *Classification trees:* A popular data-mining technique used to classify a dependent categorical variable based on measurements of one or more predictor variables. The result is a tree with nodes and links between the nodes that can be read to form if-then rules.
- *Logistic regression:* A statistical technique that is a variant of standard regression but extends the concept to deal with classification. It produces a formula predicting the probability of the occurrence as a function of the independent variables.
- *Neural networks:* A software algorithm modeled after the parallel architecture of animal brains. The network consists of input nodes, hidden layers, and output nodes. Each unit is assigned a weight. Data are given to the input node, and by a system of trial and error, the algorithm adjusts the weights until it meets a certain stopping criterion. Some people have likened this to a black-box (you do not necessarily know what is going on inside) approach.
- *Clustering techniques such as k-nearest neighbors:* A technique that identifies groups with similar records. The k-nearest-neighbor technique

calculates the distances between the record and the points in the historical (training) data. It then assigns this record to the class of its nearest neighbor in a data set.

For an example of a classification tree, consider the situation where a telephone company wants to determine which residential customers are likely to disconnect their service. The company has the following information: how long customers have had the service, how much they spend on the service, whether they have had problems with the service, whether they have the best calling plan for their needs, where they live, how old they are, whether they have other services bundled together with their calling plan, competitive information about other carriers' plans, and most importantly, whether they still have the service or have disconnected it. Of course, the company can find many more variables, if it wishes, but the last is the outcome variable; the software will use it to classify the customers into one of two groups—perhaps calling them stayers and flight risks.

The data set is broken into training data and a test data set. The training data consist of observations (called attributes) and an outcome variable (binary in the case of a classification model), in this case, the stayers or the flight risks. The algorithm is run over the training data and comes up with a tree shape that can be read like a series of rules. For example, if the customers have been with the company for more than 10 years and they are over 55 years old, they are likely to remain loyal customers.

These rules are then run over the test data set to determine how good this model is on "new data." Accuracy measures are provided for the model. For example, a popular technique is the confusion matrix, a table that provides information on how many cases were correctly or incorrectly classified. If the model looks good, it can be deployed on other data, as they are available (i.e., using it to predict new cases of flight risk). On the basis of the model, the company might decide, for example, to send out special offers to those customers whom it thinks are flight risks (Hurwitz et al., 2013).

8.4.3.1 Methods to Manage Large Databases from High-Dimensional Spaces

Data-mining-related methods are often nonparametric, not assuming an underlying generating model for the data. These methods are designed to manage large databases from high-dimensional spaces. The category includes distance-based methods, clustering methods, and spatial methods.

8.4.3.1.1 Distance-Based Methods

An inconvenience of the statistical approach is that it requires knowledge about the parameters of the data set, such as the data distribution. In many cases, this may not be known (Han and Kamber, 2001), but a distance-based approach solves the problem. The criterion for obtaining outliers using this method is having two parameters, parameter (p) and distance (d), which can be derived in advance using knowledge about the data or changed during the iterations to choose the most representative outliers.

A distance-based approach has nine phases:

1. *Collect data:* As discussed in Section 8.3.1.

2. *Compute distances between data* (d_1): The distance between data is computed for every single data point.

3. *Identify maximum distance value of data* (d_2): The maximum distance value is identified to find a range for threshold distance value (d_3).

4. *Determine threshold distance value* (d_3): This value is based on the maximum distance value (d_2), but threshold distance value (d_3) should be smaller than (d_2). Otherwise, comparisons cannot be made.

5. *Compare d_3 and d_1* (p): In this phase, parameter value (p) is determined by comparing d_3 and d_1 where p is equal to $d_1 \geq d_3$.

6. *Determine threshold value* (t): Threshold value (t) is assigned to indicate the research space.

7. *Compare t and p:* Threshold value is compared to the result of phase 5.

8. *Data testing:* Outlier data are identified.

9. *Output:* The output from data testing is used for comparison and analysis.

An observation is defined as an outlier based on distance, as long as a fraction β of the observations in the data set are farther than r from it. This definition is based on a single, global criterion determined by the parameters r and β. It frequently has problems, however, such as the determination of r or the lack of a classification for outliers. The time complexity of the algorithm is $O(pn^2)$, where p is the number of features and n is the sample size. Therefore, it is not a recommended for use with large databases. There can also be problems when the data set has both dense and sparse regions.

Alternatively, the following definition is suggested: given two integers υ and $l(\upsilon, l)$, outliers are determined to be the top l-sorted observations with the greatest distance to their υth nearest neighbor. One shortcoming of this definition is that it only considers the distance to the υth neighbor and ignores information on closer observations. Another alternative is to define outliers as those observations with a large average distance to the υth nearest neighbors. The drawback in this case is that it takes longer to be calculated (Abu et al., 2006).

8.4.3.1.2 Clustering Methods

Clustering-based methods consider a small cluster, including a single observation, as clustered outliers. Examples include partitioning around medoids (PAM), a fractal dimension-based method, and clustering large applications

(CLARAs). A modified version of the latter for spatial outliers is called clustering large applications based on randomized search (CLARANS). Since their main objective is clustering, these methods are not always optimal for outlier detection. In most cases, the outlier detection criteria are implicit and cannot easily be inferred from the clustering procedures (Raymond and Han, 1994).

8.4.3.1.2.1 Clustering Algorithms Based on Partitioning

Partioning around Medoids: In the past 30 years, cluster analysis has been used in many areas, including medicine (classification of diseases), chemistry (group of compounds), social sciences (classification of statistical results), and so on. The main objective in all cases is to identify structures or groups present in the data. Because there is no general definition of a cluster, we have developed algorithms for finding various types of clusters: spherical, linear, drawn out, and so on.

Out of the myriad possibilities, we have selected the k-medoid method for our algorithm because it is very robust to the existence of outliers (i.e., data points that are far from the other data points). In addition, the clusters found by k-medoid methods do not depend on the order in which the objects are examined. They are also invariant with respect to translations and orthogonal transformations of data points. Last but not least, experiments have shown that the k-medoid methods described below can handle large amounts of data efficiently.

PAM was developed by Kaufman and Housseeuw (1990). To find k clusters, PAM determines a representative object for each cluster. This representative object, called a medoid, is meant to be the most centrally located object within the cluster. Once the medoids have been selected, each nonselected object is grouped with the medoid to which it is the most similar. More precisely, if O_j is a nonselected object, and O_i is a (selected) medoid, O_j belongs to the cluster represented by O_i, if $d(O_j, O_e) = \min O_e\, d(O_j, O_e)$, where the notation $\min O_e$ denotes the minimum over all medoids O_e, and the notation $d(O_a, O_b)$ denotes the dissimilarity or distance between objects O_a and O_b. All dissimilarity values are given as inputs to PAM. Finally, the quality of a clustering (i.e., the combined quality of the chosen medoids) is measured by the average dissimilarity between an object and the medoid of its cluster.

To find the k-medoids, PAM begins with an arbitrary selection of k objects. Then in each step, a swap between a selected object O_i and a nonselected object O_h is made, as long as such a swap will result in an improvement of the quality of the clustering. To calculate the effect of such a swap between O_i and O_h, PAM computes costs C_{jih} for all nonselected objects O_j. Depending on which of the following cases O_j is in, C_{jih} is defined by one of the equations below.

First case: Suppose O_j currently belongs to the cluster represented by O_i. Furthermore, let O_j be more similar to $O_{j,2}$ than O_h; that is, $d(O_j, O_h) \geq d(O_j, O_{j,2})$, where $O_{j,2}$ is the second most similar medoid to O_j. Thus, if O_i is replaced by O_h as a medoid, O_j would belong to the cluster represented by $O_{j,2}$. Hence, the cost of the swap as far as O_j is concerned is

$$C_{jih} = d(O_j, O_{j,2}) - d(O_j, O_i) \tag{8.14}$$

This equation always gives a nonnegative C_{jih}, indicating a nonnegative cost incurred in replacing O_i with O_h.

Second case: O_j currently belongs to the cluster represented by O_i. But this time, O_j is less similar to $O_{j,2}$ than O_h; that is $d(O_j, O_h) < d(O_j, O_{j,2})$. So if O_i is replaced by O_h, O_j will belong to the cluster represented by O_h. Therefore, the cost for O_j is given by

$$C_{jih} = d(O_j, O_h) - d(O_j, O_i) \tag{8.15}$$

Unlike Equation 8.14, here C_{jih} can be positive or negative, depending on whether O_j, is more similar to O_i or to O_h.

Third case: Assume O_j actually belongs to a cluster not represented by O_i. Let $O_{j,2}$ be the representative object of that group. Additionally, let O_j be more similar to $O_{j,2}$ than O_h. Then even if O_i is replaced by O_h, O_j will stay in the cluster represented by $O_{j,2}$, making the cost

$$C_{jih} = 0 \tag{8.16}$$

Fourth case: O_j currently belongs to the cluster represented by $O_{j,2}$. But O_j is less similar to $O_{j,2}$ than O_h. In this case, replacing O_i with O_h will cause O_j to skip to the group of O_h from that of $O_{j,2}$. Therefore, the cost is

$$C_{jih} = d(O_j, O_h) - d(O_j, O_{j,2}) \tag{8.17}$$

and is always negative.

The combination of the four previous cases, the total cost of replacing O_i with O_h is given by

$$TC_{ih} = \sum_j C_{jih} \tag{8.18}$$

The PAM algorithm is formulated as follows:

1. Select k-representative objects arbitrarily.
2. Compute TC_{ih} for all pairs of objects O_i, O_h where O_i is currently selected, and O_h is not.
3. Select the pair O_i, O_h that corresponds to $\min O_i$, O_h TC_{ih}. If the minimum TC_{ih} is negative, replace O_i with O_h, and go back to Step (2).
4. Otherwise, for each nonselected object, find the most similar representative object.
5. Halt.

Experimental results show PAM works satisfactorily for small data sets (e.g., 100 objects in five clusters), but it is not efficient in dealing with medium and large data sets. This is not surprising. In Steps (2) and (3), there are altogether $k(n - k)$ pairs of O_i, O_h. For each pair, computing TC_{ih} requires the examination of $(n - k)$ nonselected objects. Thus, Steps (2) and (3) combined produce $O(k(n - k)^2)$. And this is the complexity of only one iteration. It is obvious that PAM is too costly for large values of n and k. This motivated the development of CLARA (Ratner, 2004).

Clustering Large Applications: Designed by Kaufman and Rousseeuw (1990) to handle large data sets, CLARA relies on sampling. CLARA takes a sample data set, but instead of finding representative objects for the entire data set, it applies PAM to the sample and finds its medoids. If the sample is taken in a sufficiently random manner, the medoids of the sample will approximate medoids across the data set. For better approximations, CLARA elicits multiple samples and offers the best clustering as the output. For accuracy, we base the quality of a clustering on the average dissimilarity of all objects in the entire data set, not merely the objects in the samples. Experiments reported in Kaufman and Rousseeuw (1990) indicate that five samples of size $40 + 2k$ give satisfactory results.

The CLARA algorithm is formulated as follows:

1. For $i = 1 - 5$, repeat the following steps.
2. Remove a sample of $40 + 2k$ random objects from the entire data set and use the PAM algorithm to find k-medoids of the sample.
3. For each object O_j in the whole data set, determine which of the k-medoids is the most similar to O_j.
4. Calculate the average dissimilarity of the clustering obtained in the previous step. If this value is less than the current minimum, use it as the current minimum and retain the k-medoids found in Step (2) as the best set of medoids obtained so far.
5. Return to Step (1) to start the next iteration.

Supplementing PAM, CLARA works well for large data sets (e.g., 1000 objects in 10 clusters). As mentioned in Section 8.4.3.1.2.1, each iteration of PAM is $O(k(n - k)^2)$. However, for CLARA, by applying PAM only to the samples, each iteration is $O(k(40 + k)^2 + k(n - k))$. This explains why CLARA is more efficient than PAM for large values of n (Raymond and Han, 1994).

8.4.3.1.3 Spatial Methods

Spatial methods are closely related to clustering methods. A spatial outlier is a spatially referenced object whose nonspatial attribute values are significantly different from the values of its neighborhood. The methods of spatial statistics can be generally classified into two subcategories: quantitative and graphic approaches. Quantitative methods distinguish spatial outliers from the remainder of data; two representative approaches are the scatterplot and the Moran scatterplot. Graphic methods are based on visualization of spatial data to highlight spatial outliers; examples are variogram clouds and pocket plots.

Some suggest using a multidimensional scaling (MDS) that represents the similarities between objects spatially, as on a map. MDS seeks to find the best configuration of the observations in a low-dimensional space. As indicated above, CLARANS is a clustering method for spatial data mining based on a randomized search. Two spatial data-mining algorithms that use CLARANS have been suggested. First, Shekhar and Lu introduced a method for detecting spatial outliers in a graph data set (Shekhar et al., 2002). The method is based on the distribution property of the difference between an attribute value and the average attribute value of its neighbors. Then, Lu proposed a set of spatial outlier detection algorithms to minimize the false detection of spatial outliers when the neighborhood contains true spatial outliers (Abu et al., 2006).

8.4.3.1.3.1 Spatial Data   The distinction between spatial and nonspatial data can easily become the subject of extensive discussion. In general, observations for which absolute location and/or relative positioning (spatial arrangement) are taken into account can be considered spatial data. These data can be subdivided into two major categories representing discrete and continuous phenomena. On the basis of the former classification, also called the entity view, spatial phenomena are described using zero-dimensional objects such as points, one-dimensional objects such as lines, or two-dimensional objects such as areas. If space is described using continuous phenomena, as in the case of temperature or topography, this is called the field view. In practice, the latter is usually measured based on sampling discrete entities such as locations in space.

The entity view allows spatial objects to have attributes. Spatial analysis is typically aimed at the spatial arrangement of the observed units, but can also take into account attribute information. An analysis conducted only on the basis of the attributes of the observed units ignoring the spatial relationships is not considered spatial data analysis (Pfeiffer, 1996).

Spatial Data Analysis: The methods used in spatial data analysis can be broadly categorized as those concerned with visualizing data, with exploratory data analysis, and with developing statistical models. During most analyses, a combination of techniques is used, with the data first being displayed visually, followed by exploration of possible patterns and possibly modeling.

Data Visualization: One of the first steps in any data analysis should be an inspection of the data. Visual displays of information using plots or maps provide the basis for generating hypotheses and, if required, help assess the fit or predictive ability of models. Recently, interactive computer packages have been developed to allow dynamic displays of the data. Geographic information systems can be used to produce maps; these allow the exploration of spatial patterns in an interactive manner.

Exploratory Data Analysis: Data exploration is aimed at developing hypotheses and makes extensive use of graphical views of the data on maps or scatter plots. Exploratory data analysis makes few assumptions about the data and should be robust to extreme data values. Simple analytical models can also be used in this analysis phase.

8.4.3.1.3.2 A Clustering Algorithm Based on Randomized Search This section presents the clustering algorithm CLARANS, describes the details of the algorithm, and explains its use in spatial data mining (Pfeiffer, 1996).

*Motivation of CLARANS: A Graph Abstraction*Given n objects, the process described above of finding k-medoids can be viewed abstractly as searching through a graph. On this graph, denoted by $G_{n,k}$, a node is represented by a set of k objects $(O_{m1},...,O_{mk})$, intuitively indicating that $O_{m1},...,O_{mk}$ are the selected medoids. The set of nodes on the graph is the set $\{(O_{m1},...,O_{mk})$ | where $O_{m1},...,O_{mk}$ are objects in the data set$\}$.

Two nodes are neighbors (i.e., connected by an arc) if their sets differ by only one object. More formally, two nodes $S_1 = (O_{m1},...,O_{mk})$ and $S_2 = (O_{w1},...,O_{wk})$ are neighbors if and only if the cardinality of the intersection of S_1 and S_2 is $k-1$; that is, $|S_1 \cap S_2| = k-1$.

We can clearly see that each node has $k(n-k)$ neighbors. Since a node represents a collection of k-medoids, each node belongs to a cluster. Therefore, each node can be assigned a cost which is defined as the total dissimilarity between each object and the medoid of its cluster. If objects O_i, and O_h are the differences between neighbors S_1 and S_2 (i.e., O_i, $O_h \notin S_1 \cap S_2$, but $O_i \in S_1$ and $O_h \in S_2$), respectively, the cost differential between the two neighbors is given by TC_{ih} as defined in Equation 8.18.

By now, it is obvious that PAM can be viewed as a search for a minimum on the graph $G_{n,k}$. At each step, all neighbors of the current node are examined. The current node is replaced by the neighbor with the greatest decrease in costs, and the search continues until a minimum is obtained. For large values of n and k (such as $n = 1000$ and $k = 10$), examining all $k(n-k)$ neighbors of a node is time consuming. This accounts for the inefficiency of PAM for large data sets.

CLARA tries to examine fewer neighbors and restricts the search to subgraphs that are much smaller than the original graph $G_{n,k}$. However, the subgraphs examined are entirely defined by the objects in the samples. Let S_a be the set of objects in a sample. The subgraph $G_{Sa,k}$ consists of all nodes that are subsets (of cardinalities k) of S_a. Even though CLARA thoroughly examines $G_{Sa,k}$ via PAM, the search is fully confined within $G_{Sa,k}$. If M is the minimum node in the original graph $G_{n,k}$, and if M is not included in $G_{Sa,k}$, M will never be found in the search of $G_{Sa,k}$, regardless of how thorough the search is. To atone for this deficiency, many, many samples must be collected and processed.

Like CLARA, the CLARANS algorithm does not check every neighbor of a node. But unlike CLARA, it does not restrict its search to a particular subgraph. In fact, it searches the original graph $G_{n,k}$. One key difference between CLARANS and PAM is that the former only checks a sample of the neighbors of a node. But unlike CLARA, each sample is extracted dynamically in the sense that no nodes corresponding to particular objects are eliminated outright. In other words, while CLARA takes a sample of nodes at the beginning of a search, CLARANS takes a sample of neighbors in each step of a search. This has the benefit of not confining a search to a localized area. As will be seen below, a CLARANS search gives higher-quality clustering than CLARA and fewer searches are required.

Clustering Large Applications Based on Randomized Search: The CLARANS algorithm (Raymond and Han, 1994) is formulated as follows:

1. Input parameters numlocal and maxneighbor. Initialize i to 1, and mincost to a large number.
2. Set the current node to an arbitrary node in $G_{n,k}$.
3. Set j to 1.
4. Consider an aleatory neighbor S of the current node, and based on Equation 8.18, calculate the cost differential of the two nodes.
5. If S has a lower cost, set the current node to S and proceed to Step (3).
6. Otherwise, increment j by 1. If $j \leq$ maxneighbor, go to Step (4).
7. Otherwise, when $j >$ maxneighbor, compare the cost of the current node with mincost. If the former is less, set mincost to the cost of the current node, and set bestnode to the current node.
8. Increment i by 1. If $i >$ numlocal, output bestnode and halt.
9. Otherwise, go to Step (2).

Steps (3) through (6) search for nodes with progressively lower costs. If the current node has already been compared with the maximum number of the neighbors of the node (specified by maxneighbor) and continues to be lowest cost, it is declared a "local" minimum. Then in Step (7), the cost of this local minimum is compared to the lowest cost obtained to this point. The lower of the costs is stored in mincost. The process is repeated to find other local minima until numlocal has been found.

As indicated above, CLARANS has two parameters: the maximum number of neighbors examined (maxneighbor), and the number of local minima obtained (numlocal). This means that the higher the value of maxneighbor, the closer CLARANS is to PAM, and the longer each search for a local minima becomes. But the quality of such a local minima is higher and fewer minima need to be obtained (Raymond and Han, 1994).

8.4.3.1.3.3 Spatial Data Mining Based on Clustering Algorithms
Spatial Dominant Approach: SD(CLARANS) There are many approaches to spatial data mining. The kind considered here assumes that a spatial

database consists of both spatial and nonspatial attributes, with the latter stored in relations. Our general approach uses clustering algorithms to deal with the spatial attributes and employs other learning tools to take care of the nonspatial counterparts.

DBLEARN is the tool for mining nonspatial attributes. As inputs, it takes relational data, generalization hierarchies for attributes, and a learning query specifying the focus of the mining task to be carried out. From a learning request, DBLEARN first extracts a set of relevant tuples via structured query language (SQL) queries. Then based on the generalization hierarchies of the attributes, it iteratively generalizes the tuples.

For example, suppose the tuples corresponding to a given learning query have the major attribute ethnic group. Also assume the generalization hierarchy for ethnic group has Indians and Chinese generalized to Asians. In this case, a generalization operation on the attribute ethnic group causes all tuples of the form (m, Indian) and $(m, \text{Chinese})$ to be merged into the tuple (m, Asians). This fusion has the effect of reducing the number of remaining (generalized) tuples. Each tuple has a system-defined count that keeps track of the number of original tuples (stored in the relational database) as represented by the current (generalized) tuple. This attribute enables DBLEARN to output such statistical statements as 8% of all students specializing in sociology are Asians. A generalization hierarchy may have multiple levels (e.g., Asians further generalized to non-Canadians), and a learning query may require more than one generalization operation before the final number of generalized tuples drops below a certain threshold. Finally, such declarations as 90% of all arts students are Canadians may be returned as results of the learning query.

Having explained what DBLEARN does, the next issue to address is how to extend DBLEARN to deal with spatial attributes. In what follows, we present two ways to combine clustering algorithms with DBLEARN. The algorithm shown below, called SD(CLARANS), combines CLARANS and DBLEARN in a spatially dominant manner. That is, spatial clustering is performed first, followed by nonspatial generalization of every cluster.

The SD algorithm (CLARANS) is formulated as follows:

1. Given a learning request, find the initial set of relevant tuples using the appropriate SQL queries.
2. Apply CLARANS to the spatial attributes and find the most natural number k_{nat} of clusters.
3. For each of the k_{nat} clusters obtained above,
   a. Collect the nonspatial components of the tuples included in the current cluster
   b. Apply DBLEARN to this collection of nonspatial components.

Note that algorithms SD(PAM) and SD(CLARA) can also be obtained, but CLARANS is more efficient than either.

Determining k_{nat} for CLARANS: Step (2) of algorithm SD(CLARANS) tries to find k_{nat} clusters, where k_{nat} is the most natural number of clusters for the given data set. However, recall that CLARANS and all partitioning algorithms require the number of k clusters to be given as input. Thus, an immediate question is whether SD(CLARANS) knows beforehand what k_{nat} is and can simply pass the value of k_{nat} to CLARANS. Unfortunately, the answer is no. In fact, determining k_{nat} is one of the most difficult problems in cluster analysis, for which no unique solution exists.

SD(CLARANS) adopts the heuristics of computing the silhouette coefficients, first developed by Kaufman and Rousseeuw (1990). For space considerations, we do not include the formulas for computing silhouettes but concentrate on how silhouettes are used in the algorithms. Intuitively, the silhouette of an object O_j, a dimensionless quantity between –1 and 1, indicates how much O_j really belongs to the cluster in which O_j is classified. The closer the value is to 1, the greater the degree of membership. The silhouette width of a cluster is the average silhouette of all objects in the cluster. After extensive experimentation, Kaufman and Rousseeuw (1990) proposed the following interpretation of the silhouette width of a cluster (Table 8.1).

For a given number $k \geq 2$ of clusters, the silhouette coefficient for k is the average silhouette width of the k clusters. Note that the silhouette coefficient does not necessarily decrease monotonically as k increases. If the value k is too small, some distinct clusters are incorrectly grouped together, leading to a small silhouette width. When k is very large, some natural clusters may be artificially split, again leading to a small silhouette width. Thus, the most natural k is the one whose silhouette coefficient is the highest. However, our experiments on spatial data mining indicate that simply using the highest silhouette coefficient does not necessarily yield intuitive results. For example, some clusters may not have reasonable structures, that is, widths ≤0.5. Thus, we suggest using the following heuristics to determine the value k_{nat} for SD(CLARANS):

1. Find the value k with the highest silhouette coefficient.

2. If all k clusters have silhouette widths ≥0.51, $k_{nat} = k$, and halt.

3. Otherwise, remove the objects in those clusters whose silhouette widths are below 0.5, provided the total number of objects removed

TABLE 8.1

Silhouette Width versus Interpretation

Silhouette/Width	Interpretation
0.71–1	Strong cluster
0.51–0.7	Reasonable cluster
0.26–0.5	Weak or artificial cluster
≤0.25	No cluster found

so far is less than a threshold (e.g., 25% of the total number of objects). The objects removed are considered outliers or noise. Go back to Step (1) for the new data set without the outliers.

4. If in Step (3), the number of outliers to be removed exceeds the threshold, simply set $k_{nat} = 1$, indicating, in effect, that no clustering is reasonable (Raymond and Han, 1994).

8.4.3.2 Data-Mining Tasks

In general, data-mining tasks can be classified as *descriptive data mining* or *predictive data mining*. The first describes the data set in a concise manner and provides some interesting general properties; the second constructs one or a set of models, making inferences about the available set of data, and attempting to predict the behavior of new data sets.

A data-mining system may accomplish one or more of the following datamining tasks:

1. *Class description*: Class description provides a concise and succinct summary of a collection of data and distinguishes them from other data. The summarization of a collection of data is called "class characterization," the comparison of two or more collections of data is called "class comparison" or "discrimination." Class description should cover properties of data dispersion, such as variance, quartiles, and so on. For example, class description can be used to compare European and Asian sales of a company, identifying important factors that discriminate the two classes and presenting a summarized overview.

2. *Association*: Association can be defined as the discovery of relationships or correlations among a set of items. This is frequently expressed in the rule that shows the attribute-value conditions occurring frequently in a given set of data. An association rule in the form of $X \rightarrow Y$ can be interpreted as "database tuples that satisfy X are likely to satisfy Y." Association analysis is widely used in transaction data analysis for direct marketing, catalog design, and other business decision-making processes. Substantial research has been performed recently on association analysis and efficient algorithms have been proposed, including the level wise *a priori* search, mining multiple-level, multidimensional associations, mining associations for numerical, categorical, and interval data, metapattern directed or constraint-based mining, and mining correlations.

   Classification: Classification studies a group of training data (i.e., a group of objects whose class label is known) and constructs a model for each class, based on the features in the data. A decision tree, or a set of classification rules, is generated that can be used to understand each class better in the database and to classify future data.

For example, diseases can be classified based on the symptoms of patients. Classification methods have been developed in the fields of machine learning, statistics, databases, neural networks, and rough sets. They have been applied to customer segmentation, business modeling, and credit analysis.

3. *Prediction*: This mining function predicts the possible values of certain missing data, or the value distribution of certain attributes in a set of objects. It involves finding the set of attributes relevant to the attribute of interest (e.g., by statistical analysis) and predicting the value distribution based on the set of data similar to the selected objects. For example, an employee's potential salary can be predicted based on the salary distribution of similar employees in the company. To date, regression analysis, generalized linear model, correlation analysis, and decision trees have been useful tools in quality prediction. Genetic algorithms and neural network models have also enjoyed popular use.

   Clustering: Clustering analysis identifies groups embedded in the data, where a cluster can be defined as a group of data objects that are "similar" to each other. Similarity can be expressed by distance functions, as specified by users or experts. A good clustering method produces high-quality clusters to ensure that the inter-cluster similarity is low and the intracluster similarity is high. For example, we could cluster houses according to house category, floor area, and geographical location. To date, data-mining research has concentrated on high quality and scalable clustering methods for large databases and multidimensional data warehouses.

4. *Time-series analysis*: Time-series analysis considers large sets of time-series data to determine their regularity and interesting characteristics. This includes searching for similar sequences and mining sequential patterns, periodicities, trends, and deviations. For example, we might predict the trend of the stock values for a company based on its stock history, business situation, competitors' performance, and the current market.

Other data-mining tasks include outlier analysis, and so on (Ratner, 2004).

References

Abu, B. Z., Mohemad, R., Ahmad, A., and Mat, D. M., 2006. *A Comparative Study for Outlier Detection Techniques in Data Mining*. Member IEEE.

Aggarwal, C. C. and Yu, S. P., 2005. An effective and efficient algorithm for high-dimensional outlier detection. *The VLDB Journal*, 14, 211–21.

Bakar, Z., Mohemad, R., Ahmad, A., and Deris, M., 2006. *A Comparative Study for Outlier Detection Techniques in Data Mining*. Bangkok, pp. 1–6.

Ben-Gal, I., 2005. Outlier detection. In: Maimon, O. and Rockach, L. (Eds.), *Data Mining and Knowledge Discovery Handbook*. Heidelberg, Berlin: Springer, pp. 131–146.

Ben-Gal, I., Morag, G., and Shmilovici, A., 2003. CSPC: A monitoring procedure for state dependent processes. *Technometrics*, 45(4), 293–311.

Breunig, M. M., Kriegel, H. P., Ng, R. T., and Sander, J. 2000. LOF: Identifying density-based local outliers. In *ACM Sigmod Record* 29(2), 93–104.

Caussinus, H. and Roiz, A., 1990. Interesting projections of multidimensional data by means of generalized component analysis. In: Momirović, K. and Mildner, V. (Eds.), *Compstat 90*. Heidelberg: Physica, pp. 121–126.

Davies, L. and Gather, U., 1993. The identification of multiple outliers. *Journal of the American Statistical Association*, 88(423), 782–92.

Eckerson, W., 2011. *Big data analytics: Profiling the use of analytic platforms in user organizations*. [Online] Available at: http://pt.slideshare.net/weckerson/big-data-analytics-webinar.

EMC, 2012. *Data Science and Big Data Analytics*. [Online] Available at: http://education.emc.com/guest/certification/framework/stf/data_science.aspx.

EMC Corporation, 2012. *Big Data Overview*. [Online] Available at: http://uk.emc.com/collateral/campaign/global/forum2012/ireland/big-data-overview-big-data-transforms-business-cvr.pdf.

Hadi, A., 1992. Identifying multiple outliers in multivariate data. *Journal of the Royal Statistical Society. Series B*, 54, 761–71.

Hampel, F., 1974. The influence curve and its role in robust estimation. *Journal of the American Statistical Association*, 69, 382–93.

Hampel, F. R., 1971. A general qualitative definition of robustness. *Annals of Mathematics Statistics*, 42, 1887–96.

Han, J. and Kamber, M., 2001. *Data Mining Concepts and Techniques*. USA: Morgan Kaufmann.

Hurwitz, J., Nugent, A., Halper, F., and Kaufman, M., 2013. *Big Data for Dummies*. New Jersey: John Wiley & Sons.

Kaufman, L. and Rousseeuw, P., 1990. *Finding Groups in Data: An Introduction to Cluster Analysis*. New York: Wiley.

Knorr, E. M., Ng, R. T., and Tucakov, V., 2000. Distance-based outliers: Algorithms and applications. *The VLDB Journal—The International Journal on Very Large Data Bases*, 8(3–4), 237–253.

Kollios, G., Gunopulos, D., Koudas, N., and Berchtold, S., 2003. Efficient biased sampling for approximate clustering and outlier detection in large data sets. *IEEE Transactions on Knowledge and Data Engineering*, 15(5), 1170–87.

Liu, H., Shah, S., and Jiang, W., 2004. On-line outlier detection and data cleaning. *Computers and Chemical Engineering*, 28, 1635–47.

Maimon, O. and Rokach, L., 2010. *Data Mining and Knowledge Discovery Handbook*, 2nd edition. New York: Springer.

Martin, R. D. and Thomson, D. J., 1982. *Robust-Resistant Spectrum Estimation*. pp. 1097–115.

Montgomery, D. C., Peck, E. A., and Vining, G. G. 2012. *Introduction to Linear Regression Analysis*. Vol. 821. John Wiley & Sons.

Pfeiffer, D., 1996. *Issues Related to Handling of Spatial Data*. Palmerston North, New Zealand: Australian Veterinary Association, pp. 83–105.

Ratner, B., 2004. *Statistical Modeling and Analysis for Database Marketing: Effective Techniques for Mining Big Data*. Florida: CRC Press.

Raymond, T. N. and Han, J., 1994. *Efficient and Effective Clustering Methods for Spatial Data Mining*. Santiago, Chile.

Shekhar, S., Lu, C. T., and Zhang, P., 2002. Detecting graph-based spatial outlier. *Intelligent Data Analysis: An International Journal*, 6(5), 451–68.

Tukey, J., 1977. *Exploratory Data Analysis*. New Jersey: Addison-Wesley.

Williams, G., 2002. *A Comparative Study of RNN for Outlier Detection in Data Mining*. Maebashi City, Japan, pp. 709–12.

9

Information Theory–Based Techniques

9.1 Introduction

The word "information" comes from the Latin "informare" meaning "give form to." In this Aristotelian concept of substance, form informs matter, while matter materializes form, thereby becoming a substance (Dodig-Crnkovic, 2006). The various concepts of information comprise a complex body of knowledge able to accommodate many different views, as Maijuan puts it, "Inconsistencies and paradoxes in the conceptualization of information can be found through numerous fields of natural, social and computer science" (Marijuan, 2003, p. 214).

The corresponding question "What is information?" is the topic of much discussion, so much so that a special issue of the *Journal of Logic, Language and Information* (Volume 12, No 4, 2003) was dedicated to precisely this topic. There is also a handbook available for consultation: see *A Handbook on the Philosophy of Information* (Van and Adriaans, 2006).

For their part, Capurro and Hjørland (2003) say information is a constructive tool whose theory dependence is a typical interdisciplinary concept. To substantiate their view, they note contributions to the theory of information from physicists, biologists, systems theorists, philosophers, and documentalists (library and information science) during the past 25 years.

Capurro et al. (1999) posit the possibility of a unified theory of information (UTI), suggesting that UTI is an expression of a metaphysical quest for a unifying principle, a quest also found in energy and matter.

For those who take a reductionist unification approach, reality is an information-processing phenomenon: "We would then say: whatever exists can be digitalized. Being is computation" (Capurro et al., 1999, p. 214). A networked structure of various information concepts that retain their specific fields of application is a possible alternative to UTI.

9.1.1 Informational Universe—Pan-Informationalism

The present informatization of society is the result of an increasingly ubiquitous use of computers as information and communication technology. In fact, as Baeyer (2003) notes, information is rapidly replacing matter/energy as the primary constitutive principle of the universe. So much so that it will become the unifying framework for describing and predicting reality in the twenty-first century.

At a fundamental level, information characterizes the world, as we gain all our knowledge through information. Yet we are only beginning to understand what information means (Benthem, 2005). This chapter defines some basic concepts of information used in computing (Dodig-Crnkovic, 2005).

9.1.2 Information as a Structure: Data–Information–Knowledge–Wisdom–Weaving

In Stonier's definition (1997), raw data (also source or atomic data) have not been processed for a given use. We might also call these data "unprocessed" in an operational sense, in that no effort has been made to interpret or understand the data before use. They are recorded as "facts of the world," either given/chosen, the result of an observation or measurement process, or the output of a previous data generating process (especially in the case of computer data). Although we have already been widely referring to the word data, it is worth noting at this point that "data" is the plural of Latin "datum," "something given," or "atomic facts." Information is the end product of data processing, and knowledge is the end product of information processing. Just as raw data are used as input and processed to get information, information too itself becomes the input for a process resulting in knowledge.

Data are a series of disconnected facts and observations that may be converted to information by organizing them in some manner: analyzing, cross-referring, selecting, sorting, summarizing, and so on. Patterns of information, in turn, can be converted into a coherent body of knowledge. Knowledge comprises an organized body of information, such information patterns, and this forms the basis of the kinds of insights and judgments we call wisdom.

Stonier provides the following useful analogy: "Consider spinning fleece into yarn, and then weaving yarn into cloth. The fleece can be considered analogous to data, the yarn to information and the cloth to knowledge. Cutting and sewing the cloth into a useful garment is analogous to creating insight and judgment (wisdom). This analogy emphasizes two important points: (1) going from fleece to garment involves, at each step, an input of work, and (2) at each step, this input of work leads to an increase in organization, thereby producing a hierarchy of organization" (Stonier, 1997).

Put another way, the work added at each subsequent organizational level represents the input of new information at lower levels (Dodig-Crnkovic, 2006).

9.1.3 Different Schools of Information

Stonier takes a structuralist point of view, but there are many other schools of information. Thus, information is variously defined by different theorists

(Holgate, 2002). The following is a brief list of the many possible schools of thought, along with their definitions of information:

- The communication school (or quotidian school, documentalism) defines information as communicated knowledge, or any "notifying matter" (Machlup, 1983), "telling something or to the something that is being told."
- Documentalists (library and information science) define information as evidentiary documentation that must be managed (Michael Buckland's information-as-thing) or as a searching behavior whereby an individual navigates a textual universe using information storage and retrieval tools.
- The Batesonian school sees information as the pattern or "formation" (formative interaction) taken by data in the "difference that makes a difference" (Bateson, 1972).
- Information dialectics sees information as linked to patterned organization and the reduction of uncertainty (an organizing principle in nature, Collier's "symmetry breaking" [1996]). The informatory dialectic pits presence against absence, potential, and expression (Javorszky, 2003).
- The logic school posits information can be inferred from data, knowledge, and so on (Leyton's process grammar of inferred shapes, Floridi's information logic, Popper's logical positivism), with the data/information/knowledge pyramid as the underlying model. It says how meaningful, contextualized data (information) become knowledge or wisdom is unresolved.
- The Hermeneutic school has various influences: Rafael Capurro's diachronic form of information (molding); Descartes' "forms of thought which inform the spirit"; the quantum school (Weizsacker, Lyre); information defined as a "double image" (Wittgenstein's duck/rabbit), that is, simultaneously form and relationship. To this, Capurro (2002) adds the following: "Information as a second order category (not as quality of things but a quality ascribed to relationships between things) in the sense of 'selection' that takes place when systems interact and choose from what is being offered."
- The Heraclitian school sees information as a process: a "continuous present" (Matsuno), an "information flow" (Dretske), situational semantics (Barwise, Perry, Israel), a process philosophy (A.N. Whitehead), or as a "dynamic process of formation" (Hofkirchner/Fleissner).

Within this school, Pedro Marijuan regards information as a self-regulating entity moving "in formation." In this definition, he is adopting Michael Conrad's sense of a "vertical information flow" circulating through molecules–cells–organisms–niches and linking the computational with the biological such that: "The living entity is capable of continuously keeping itself in balance, always in formation, by appropriately communicating with its environment and adequately combining the creation of new structures with the trimming down of obsolete, unwanted parts" (Marijuan, 2003, p. 216).

- The semiotic school adherents say information represents data/sign/ structure in an environmental context for an interpreting system; such theorists come from cybersemiotics (Søren Brier), physico-chemical semiosis (Edwina Taborsky), infological systems (Mark Burgin).

As an example of the semiotic way of thinking, Mahler (1996, p. 72) comments: "Information can only be defined within the scenario; it is not just out there." To this, Norbert Frenzl adds: "Signs are differences of input and they need to be 'interpreted' by the receiver to be information FOR the receiving system. If the organization pattern, the logic of its structural organization, enables the open system to react to the incoming signs (to actualize its own inner structural information), we can say that the system processes the signs to information."

- The stimulus school sees information as stimulus/trigger/ignition (Karpatschoff), for example, in a neural net activation in cognitive neurology.

Karpatschof (2000) argues: "It is a relational concept that includes the source, the signal, the release mechanism and the reaction as its relatants."

- Adherents of the mechanicists school believe computation and AI can fill the void left by the postmodern deconstruction of human reason (Katherine Hayles posthumanism, AI, robotic cognition).

For example, Hayles comments: "Located within the dialectic of pattern/ randomness and grounded in embodied actuality rather than disembodied information, the posthuman offers resources for rethinking the articulation of humans with intelligent machines".

- Skeptic school thinkers include Rifkin, Bogdan, Miller, Spang-Hanssen, and Maturana.

Capurro and Hjørland (2003, p. 18) give the following definition: "These concepts of information are defined in various theories such as physics, thermodynamics, communication theory, cybernetics, statistical information theory, psychology, inductive logic and so on. There seems to be no unique idea of information upon which these various concepts converge and hence no proprietary theory of information." Humberto Maturana (the Vienna school) reveals a similar skepticism when he says "information" lies outside the closed system that is autopoiesis. However, cybersemiotics (Søren Brier adopting Charles Pierce's sign theory) attempts to rescue "information" by expressing it as a possibility of "openness."

- The phenomenological school considers information to lie in situated action/interaction/experience (Niklas Luhmann [Husserl]). As part of this school, we can include Merleau-Ponty's "lived experience" and the "horizon of numerous perspectival views" (von Bertalanffy's perspectivism, Brenda Dervin's structural multiperspectivity) (see also Dodig-Crnkovic, 2005).

9.1.4 Theories of Information

A brief review of several characteristic theories of information will likely be helpful at this point (following Collier).

9.1.4.1 Syntactic Theories of Information

In syntactic approaches, information content is determined by the structure of language; it has nothing to do with the meaning of messages.

9.1.4.1.1 Statistical Shannon's Communications Theory

Shannon's theory of communications suggests the probability of transmitting messages with a specified accuracy in the presence of noise, including transmission failure, distortion, and so on. A statistical interpretation of information assumes an ensemble of possible states, each having a definite probability. Information can be expressed as the sum of the base 2 log of the inverse of the probability of each weighted by the probability of the state

$$H = \sum \text{prob}(si)\log(1/\text{prob}(si))$$

an expression similar to that for entropy in Boltzmann's statistical thermodynamics.

9.1.4.1.2 Wiener's Cybernetics Information

The cybernetics theory of information was formulated by Norbert Wiener, based on the amount of information, entropy, feedback, and background noise as essential in the characterization of the human brain. Wiener (1948, p. 18) says:

> The notion of the amount of information attaches itself very naturally to a classical notion in statistical mechanics: that of entropy. Just as the amount of information in a system is a measure of its degree of organization, so the entropy of a system is a measure of its degree of disorganization.

Wiener (1948, p. 76) goes on to define information as an integral, or an area of probability measurements: "The quantity that we here define as amount of information is the negative of the quantity usually defined as entropy in similar situations."

He also considers information to contain a structure that has a meaning: "It will be seen that the processes which lose information are, as we should expect, closely analogous to the processes which gain entropy" Wiener (1948, p. 78).

For Wiener, then, information is closely related to communication and control. For system theorists who build on his concept, information is something used by a mechanism or organism to steer a system toward a predefined goal. The actual performance is compared to the goal; signals are sent back to the sender if the performance deviates from it (feedback). The idea of feedback has become a powerful control mechanism.

9.1.4.1.3 *Complementarity of Wiener and Shannon Definitions*

Clearly, we can make an important difference between Shannon and Wiener. While Wiener sees information as negative entropy, that is, a "structured piece of the world," Shannon's information is the opposite, positive entropy.

The difference might be explained by the fact that Shannon's information describes the phenomenon of information transfer, or information communication, whereas Wiener's information is a structure, pattern, or order in a medium (biological organism, human brain), literally Marshall McLuhan's "The Medium is the Message." Focusing on a structure, negative entropy measures the degree of order.

During the process of communication via message transmission, the background settings represent the originally structured state, while message transmitted through the channel causes "disorder" in the background structure. To explain, we can use the analogy of a figure-background question: either the figure may be defined by black dots on the white background, or white dots on the background while the rest of the points are black.

9.1.4.2 *Algorithmic Information Theory (Kolmogorov, Chaitin)*

The algorithmic information theory was developed by Kolmogorov, Solomonoff, and Chaitin. There are now several formulations of Kolmogorov complexity or algorithmic information. Algorithmic information theory combines the ideas of program-size complexity with recursive function theory. The complexity of an object is measured by the size in bits of the smallest program for computing it.

Kolmogorov suggested program-size complexity explicates the concept of information content of a string of symbols, an interpretation later adoped by Chaitin. The intuitive idea behind this theory is that the more difficult an object is to specify or describe, the more complex it is. We define the complexity of a binary string s as the size of the minimal program that, when given to a Turing machine T, prints s and halts. To formalize Kolmogorov–Chaitin complexity, we have to specify exactly the types of programs. Fortunately, it does not really matter: we could take a particular notation for Turing machines, or LISP programs, or Pascal programs, and so on.

9.1.4.3 Fisher Information

Statistician R.A. Fisher defined a measure of information in a sample as the value of a parameter in the population, provided the first moment exists.

Put otherwise, Fisher information is defined as the amount of information an observable random variable X has about an unobservable parameter θ upon which the probability distribution of X depends. Since the score's expectation is zero, the variance also represents the second moment of the score. Thus, the Fisher information can be written as

$$\mathcal{I}(\theta) = E\left[\left[\frac{\partial}{\partial \theta} \ln f(X;\theta)\right]^2\right],$$

where f is the probability density function of random variable X, and, thus, $0 \le \mathcal{I}(\theta) < \infty$. In other words, Fisher information is the expectation of the square of the score. If a random variable carries high Fisher information, the absolute value of the score is frequently high.

Frieden (2014) states a certain amount of Fisher information, the physical information, is always lost during the observation of a physical effect. Accordingly, Frieden expands the physical information (*note*: usually this information is minimized) by varying the system probability amplitudes, that is, the principle of extreme physical information (EPI), thereby deriving differential equations and probability density functions describing the physics of the source effect. *Note*: Frieden's coauthors use Fisher's information theory to derive a number of contemporary physical theories, laws of biology, chemistry, economics, and so on.

9.1.5 Semantic Theories of Information

Shannon (1948, p. 3) is clear "Semantic aspects of communication are irrelevant to the engineering problem." His approach, while often seen as a mathematical theory of information, is widely considered to describe the semantic information content of a message. As Bar-Hillel (1980, p. 97) says, however, "It is psychologically almost impossible not to make the shift from the one sense of information, that is, information = signal sequence, to the other sense, information = what is expressed by the signal sequence."

The semantic theory of information theorizes the information content of messages, or in other words, what they express. Initiated by Carnap and Bar-Hillel, the theory has been developed and generalized by Hintikka. Proponents of the semantic approach see information as the content of a representation.

Fifty years ago, Carnap and Bar-Hillel (1964) used inductive logic to define the information content of a statement in a given language in terms of those states it rules out. The more possibilities (possible states of affairs) a sentence rules out, the more informative it is; in other words, information represents the elimination of uncertainty. The information content of a statement is, thus, language related, with evidence taking the form of observation statements (Carnap's "state descriptions," Hintikka's "constituents") containing information through the class of state descriptions ruled out by the evidence. Underlying this definition is the assumption that observation statements can be unambiguously related to experience.

Carnap and Bar-Hillel (1964) also suggested two possible measures of information. The first measure of the information contained in a statement S is the content measure, cont(S), defined as the complement of the *a priori* probability of the state of affairs expressed by S, expressed as

$$\text{cont}(S) = 1 - \text{prob}(S)$$

Content measure is not additive, and violates certain natural intuitions of conditional information.

The second measure, the information measure, inf(S) in bits, is given by

$$\text{inf}(S) = \log 2(1/(1 - \text{cont}(S))) = -\log 2 \text{prob}(S)$$

Again, prob(S) is the probability of the state of affairs expressed by S, not the probability of S in some communication channel. Bar-Hillel suggests cont(S) measures the substantive information content of sentence S, while inf(S) measures the surprise value, or the unexpectedness, of sentence H. Although inf(S) may satisfy additivity and conditionalization, it has the following property. If some evidence E is negatively relevant to a statement S, it holds that the information measure of S conditional on E will be greater than the absolute information measure of S. However, this violates common intuition whereby the information of S, given E, must be less than or equal to the absolute information of S, leading Floridi (2004) to label it the Bar-Hillel semantic paradox.

A more serious problem is the linguistic relativity of information and the logical empiricist program supporting it, such as the theory-ladenness of observation (Collier, 1990).

For some recent semantic theories (i.e., Barwise and Perry, 1980; Devlin, 1991), refer to Collier, http://www.nu.ac.za/undphil/collier/information/information.html.

9.1.5.1 Dretske's Information

In *Knowledge and the Flow of Information*, Dretske (1981) develops a theory of epistemology and philosophy of mind based on Shannon's mathematical theory of communication. Information is an objective commodity built on the dependency relations between distinct events, knowledge is information-caused belief, and perception is the delivery of this belief in analog form (experience) for conceptual use by cognitive mechanisms. In Dretske's theory, meaning (i.e., belief content) plays an information-carrying role.

9.1.5.2 Situated Information

In their theory of situation semantics, Barwise and Perry (1980) say situated information, or information specific to a particular situation, is analogous to situated knowledge. Some methods of generating information, including trial and error, or learning from experience, create highly situational information. Barwise and Parry define a situation as a projection of the external environment onto the agent's senses via some sense medium, making a situation an agent-centered notion.

The authors comment: "Reality consists of situations—individuals having properties and standing in relations at various spatiotemporal locations" (Barwise and Perry, 1980, p. 17). In this view, individuals, properties, relations, and locations constitute uniformities across various situations. In turn, living organisms adjust to various uniformities, depending on their biological needs. Meanings are seen as a kind of uniformity: for example, in the meaning of a simple declarative sentence there is a certain uniformity in the relationship between the utterance situation in which the sentence is produced and the situation it describes. For Barwise and Perry (1980, p. 670), this is "the relation theory of meaning."

Situation semantics relocates meaning in the world (i.e., the environment) instead of in the human brain: "We believe linguistic meaning should be seen within [a] general picture of a world teeming with meaning, a world full of information for organisms appropriately attuned to that meaning" (Barwise and Perry, 1980, p. 16). Such thinkers go beyond the dichotomy between natural and nonnatural meaning, simply seeing linguistic meaning as an especially complex set of regularities of information flow. Instead of being located in an abstract world of sense, meaning is found in the flow of information between situations; at the same time, however, these situations are abstractions.

9.1.5.3 Leyton's Information

Michael Leyton defines information as identical to shape, applied concept especially applicable to natural information. Leyton (1992, p. 19) says: "We should note that there is a possibility that a third term information is equivalent to those other two. Certainly, in statistical information theory, the term information is defined as variety, and that makes the term similar to the term asymmetry which we are defining as distinguishability."

Variety in a set of data is also measured by algorithmic information theory. Thus, in Leyton's view, that particular ingredient in the present from which we extract the past is information in the abstract sense of statistical or algorithmic information theory. "Therefore," Leyton, concludes, "we might be able to regard the terms shape and information as identical terms. That is, we might be able to regard the mathematical study of shape as a general, and more thorough, information theory than has been attempted in the current approaches to the study of information measurement" (Leyton, 1992, p. 28).

9.1.6 What Is the Difference That Makes a Difference? Syntactic versus Semantic Information

Haugeland (1989, p. 115) famously said "If you take care of the syntax, the semantics will take care of itself." Another commonly cited definition of information in the same category is "the difference that makes the difference" (Bateson, 1973, p. 318). When it comes to making a difference, the most fundamental decisions are whether a sensory input is identical to its background or different, whether it is one object or several objects, whether a set of objects is a collection of separated individual things (here, we recognize the differences between objects in the collection) or groups that share properties (here, we recognize the similarities between objects in the collection). In other words, the two elementary processes are differentiation and integration.

On the one hand, a system might be described by its "state;" on the other hand, we might describe that state and call the description "data." In any system of states, then, "information" represents the difference between any two states, and a collection of such differences permits us to consider "patterns of information." In these definitions, "state," "information," or "patterns" do not require complex interpretation; a mechanical interpretation is sufficient. "Recognition" (as opposed to "comparison") involves complex transformations in organisms.

Within the syntactic–semantic distinction, theories of information can be grouped as:

- Syntactic information theories (Chaitin–Kolmogorov, Shannon–Weaver, Wiener, Fisher) are quantitative, mathematical, and "objective." The semantics is tacit, and syntax is explicated.
- Semantic information theories (Bar-Hillel, 1980; Barwise and Perry, 1980; Dretske, 1981; Devlin, 1991) consider someone's interpretation of information; the syntax is tacit, and the semantics is explicated.

The dichotomy between semantic and syntax corresponds to the dichotomy of form/content. Semantic information on the individual level is subjective.

If semantic information is to become "objective," it must be intersubjectively negotiated through communication. Different communities of people exchanging the same information may have different uses (views). The same phenomenon may have different meanings for individuals or for groups. "Information" is a typical example.

An interesting feature of the concept of information is that it describes an entity common to many domains of human inquiry, making it a bridge over the gaps between various fields. Although scientific and scholarly fields have all formed views of information, a core of common intuitions relates them, as in Wittgenstein's family resemblance.

9.1.7 No Information without Representation! Correspondence Models versus Interactive Representation

Landauer (1996, p. 188) tells us: "Information is not a disembodied abstract entity; it is always tied to a physical representation. It is represented by engraving on a stone tablet, a spin, a charge, a hole in a punched card, a mark on paper, or some other equivalent. This ties the handling of information to all the possibilities and restrictions of our real physical world, its laws of physics, and its storehouse of available parts."

Throughout the tradition of Western thought, information has been understood in conjunction with representation. In correspondence theory, the mind is identical with the consciousness that is carrying out passive input processing.

Figure 9.1 shows an informational correspondence model, with a symbolic process of information transmission via several steps of physical transformations. Step three illustrates information in the brain. As this "correspondence scheme" does not need to imply any special kind of transformation, or any type of encoded information, step three may stand for an emergent result of

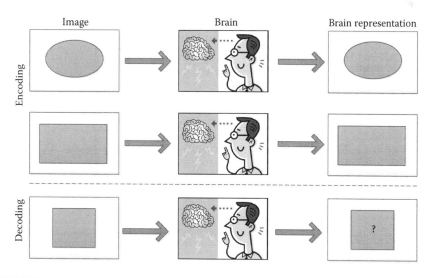

FIGURE 9.1
Informational correspondence model: A symbolic process of information transmission via several steps of physical transformations. Step 3 symbolizes information in the brain.

a dynamic process in the brain. The transformations are usually supposed to be causally related (Dodig-Crnkovic, 2005).

There are several versions of the correspondence (encoding–decoding) models of representation:

- Models of isomorphic correspondence relationships, that is, the physical symbol system hypothesis (Newell and Simon, 1972)
- Models of trained correspondences with activation vectors, that is, connectionist models (Rumelhart, McClelland)
- Models of transduced or causal or nomological relationships, that is, physical or logical (Fodor)
- Models of function and representation as function (Godfrey-Smith, Millikan).

A traditional point of view says information is caused by some external past event. However, explaining what exactly produced the representation in the animal or machine is problematic.

Bickhard and Terveen (1995, pp. 129–130) provide an explanation: "Some state or event in a brain or machine that is in informational correspondence with something in the world must in addition have content about what that correspondence is with in order to function as a representation for that system—in order to be a representation for that system. Any such correspondence, for example, with this desk, will also be in correspondence (informational, and causal) with the activities of the retina, with the light processes, with the quantum processes in the surface of the desk, with the desk last week, with the manufacture of the desk, with the pumping of the oil out of which the desk was manufactured, with the growth and decay of the plants that yielded the oil, with the fusion processes in the sun that stimulated that growth, and so on all the way to the beginning of time, not to mention all the unbounded branches of such informational correspondences. Which one of these relationships is supposed to be the representational one? There are attempts to answer this question too, but, again, none that work."

Intentionality is important to the formation of representations. The informational content of the world is infinite, with each object becoming a part of an all-encompassing network of causation and physical interaction. The agent extracts (registers) some specific information from the world because it acts in the world, pursuing a number of goals, the most basic being survival; in this way, an agent actively chooses which information is of interest.

Over the past century, pragmatic theory has developed as an alternative to the correspondence model of representation (Joas, Rosenthal, Bickhard). Pragmatism sees interaction as the most appropriate framework for understanding mind, including representation.

The interactive model of representation and standard correspondence approaches differ in several important ways. Interactive explanation is future oriented, as the agent is concerned with anticipated future possibilities of interaction. Actions are oriented internally to the system, thus optimizing their internal outcome. In the interactive case, the environmentprimarily represents resources for the agent. That being said, correspondence with the environment is basic to interactive systems as well. Ultimately, although they are attractive, connectionist models are not sufficient to completely account for representation; see http://www.lehigh.edu/~interact/isi2001/isi2001.html.

To reiterate, representation emerges in the anticipatory interactive processes of both natural and artificial agents who are pursuing their goals while communicating with the environment.

Goertzel (1994) suggests every mind is a superposition of a structurally associative memory (heterarchical network) and a multilevel control hierarchy (perceptual-motor hierarchical network) of processes. In this dual-aspect framework, the former corresponds to information structure, while the latter corresponds to a computational process network.

Goertzel's hypothesis supports an interactivist view of representation. More specifically, "The 'complex function' involved in the definition of intelligence may be anything from finding a mate to getting something to eat to building a transistor or browsing through a library. When executing any of these tasks, a person has a certain goal, and wants to know what set of actions to take in order to achieve it. There are many different possible sets of actions—each one, call it X, has certain effectiveness at achieving the goal. This effectiveness depends on the environment E, thus yielding an 'effectiveness function' $f(X, E)$. Given an environment E, the person wants to find X that maximizes f—that is maximally effective at achieving the goal. But in reality, one is never given complete information about the environment E, either at present or in the future (or in the past, for that matter). So there are two interrelated problems: one must estimate E, and then find the optimal X based on this estimate" (Goertzel, 1994).

Interactive models are becoming prominent in contemporary artificial intelligence (AI), cognition, cognitive robotics, consciousness, language and interface design, a development paralleling the new interactive computing paradigms and approaches to logic (dialogic logic, game-theoretic approaches to logic).

9.2 Information Contained in Maintenance Data

Traditional condition monitoring (CM) is based on decisions on the evolution of certain features. These features are statistical parameters of a signal that show information embedded into the signal, but are not visible at a glance. This technique combines both time domain and frequency domain features to compose a featured space, specifically, a hyperspace suitable to monitor the degradation of the machine or identify faulty states.

However, these features, though effective and efficient, often lack knowledge of the surrounding environment of the sensor and the context of the events. This chapter discusses the feature selection and extraction process in terms of information transformation to a featured space and shows how the trend is evolving for more accurate diagnostics and prognostics.

9.2.1 Feature Selection

9.2.1.1 Need for Feature Reduction

Many factors affect the success of machine learning on a given task. The representation and quality of the example data is first and foremost. Today, the need to process large databases is becoming increasingly common. Full text database learners typically deal with tens of thousands of features; vision systems, spoken word, and character recognition problems all require hundreds of classes and may have thousands of input features. The majority of real-world classification problems require supervised learning, as the underlying class probabilities and class-conditional probabilities are frequently unknown, and each instance is associated with a class label. Because relevant features are often unknown *a priori* in real-world situations, candidate features may be introduced, as theoretically, having more features should result in more power to discrimiate (Tato et al., 2002). However, practical experience with machine learning algorithms has shown that this is not always the case; current machine learning toolkits are insufficiently equipped to deal with contemporary data sets, and many algorithms cannot deal with complexity. Furthermore, when faced with many noisy features, some algorithms take an inordinately long time to converge, or never converge at all. And even if they do converge, conventional algorithms will tend to construct poor classifiers (Kononenko, 1994).

Many features introduced during the training of a classifier are partially or completely irrelevant/redundant to the target concept; an irrelevant feature does not affect the target concept in any way, while a redundant feature adds nothing new. In many applications, a data set is so large that learning might not work very well unless these unwanted features are removed. Recent research shows common machine learning algorithms are adversely affected by irrelevant and redundant training information. The simple nearest-neighbor algorithm is sensitive to irrelevant attributes; its sample complexity (number of training examples needed to reach a given accuracy level) grows exponentially with the number of irrelevant attributes (Aha et al., 1991; Langley and Sage, 1994a,b, 1997). Sample complexity for decision tree algorithms can grow exponentially on some concepts (such as parity) as well. The naive Bayes classifier can be adversely affected by redundant attributes due to its assumption that attributes are independent given the class (Langley and Sage, 1994a,b, 1997). Decision tree algorithms such as C4.5 (Quinlan, 1986, 1993) can sometimes over-fit training data, resulting in large trees. In many cases, removing irrelevant and redundant information can result in C4.5 producing smaller trees (Kohavi and John, 1996). Neural networks are supposed to cope with irrelevant and redundant features when the amount of training data is enough to compensate for this drawback; otherwise, they are also affected by the amount of irrelevant information.

Reducing the number of irrelevant/redundant features drastically reduces the running time of a learning algorithm and yields a more general concept. This yields better insight into the underlying concept of a real-world classification problem. *Feature selection* methods try to pick up a subset of features relevant to the target concept.

9.2.1.2 Feature Selection Process

The problem introduced in the previous section can be alleviated by preprocessing the data set to remove noisy and low-information-bearing attributes. Kira and Rendell define feature selection as the following: "Feature selection is the problem of choosing a small subset of features that ideally is necessary and sufficient to describe the target concept." As the terms "necessary" and "sufficient" imply, feature selection attempts to select the minimally sized subset of features according to certain criteria.

Ideally, feature selection methods search through the subsets of features to find the best one among 2^N candidate subsets according to some evaluation function. This procedure is an exhaustive one, as it tries to find only the best one. It may even be too costly and/or prohibitive in a practical sense for a medium-sized feature set. Other methods use heuristic or random search methods to reduce computational complexity even though they may compromise performance (Tato et al., 2002).

Feature selection should accomplish the following:

1. Classification accuracy will not significantly decrease
2. Class distribution, given only values for the selected features, will be as close as possible to the original class distribution, given all features.

9.2.1.2.1 General Criteria for a Feature Selection Method

Feature selection needs a criterion for stopping to prevent an exhaustive search.

A typical feature selection method has four steps:

1. *Starting point*: Choosing a point in the feature subset space from which to start a search can affect the subsequent direction of the search. We can begin with no features and successively add attributes; this type of search is said to proceed forward through the search space. Or we can start with all features and successively remove them; this search proceeds backwards through the search space. Another

possibility is to start somewhere in the middle and move outwards from that point.

2. *Search organization*: An exhaustive search of the feature subspace is generally prohibitive; with N initial features, there are 2^N possible subsets. Heuristic search strategies are more feasible than exhaustive ones and yield good results, but do not guarantee finding the optimal subset.

3. *Evaluation strategy*: The method of evaluating feature subsets is the biggest differentiating factor among feature selection algorithms for machine learning. One model, the *filter* (Kohavi, 1995; Kohavi and John, 1996), operates independently of any learning algorithm; it filters undesirable features from the data before the start of learning. This type of algorithm uses heuristics based on general characteristics of the data to evaluate the value of feature subsets. Other theorists argue that the bias of a particular induction algorithm should be considered when selecting features. One example, the *wrapper* (Kohavi, 1995; Kohavi and John, 1996), uses an induction algorithm and a statistical re-sampling technique such as cross-validation to determine the accuracy of feature subsets.

4. *Stopping criterion*: A feature selector must decide when to stop searching through feature subsets. Depending on the evaluation strategy, a feature selector might stop adding or removing features when no alternative improves on a current feature subset. Or the algorithm might continue to revise the feature subset as long as the merit does not degrade. Another option is to continue generating feature subsets until the opposite end of the search space is reached and then select the best.

Many learning algorithms can be seen as making a (biased) estimate of the probability of the class label, given a certain set of features. This is a complex, high-dimensional distribution. Unfortunately, induction often involves limited data, making the estimation of numerous probabilistic parameters difficult. Many algorithms use Occam's Razor (Gamberger and Lavrac, 1997) bias to build a simple model that achieves an acceptable level of performance on the training data but does not over-fit them. This bias may lead an algorithm to prefer a small number of predictive attributes over a large number of features that are fully predictive of the class label if used in the proper combination. With too much irrelevant and redundant information or with noisy and unreliable data, learning during the training phase is more difficult.

Feature subset selection involves identifying and removing as much irrelevant and redundant information as possible to reduce the dimensionality of the data and allow learning algorithms to operate faster and more effectively. At times, future classification can be made more accurate; at other times, we get a more compact, easily interpreted representation of the target concept.

9.2.1.3 *Feature Selection Methods Overview*

Feature subset selection has long been a research area in statistics and pattern recognition (Devijver and Kittler, 1982; Miller, 1990). It is not surprising that feature selection is as much of an issue for machine learning as it is for pattern recognition, as both fields share the common task of classification. In pattern recognition, feature selection can have an impact on the economics of data acquisition and on the accuracy and complexity of the classifier (Devijver and Kittler, 1982). This is also true of machine learning, which has the added concern of distilling useful knowledge from data. Fortunately, feature selection has been shown to improve the comprehensibility of extracted knowledge (Kohavi and John, 1996).

There is an enormous number of feature selection methods. A study by Dash and Liu (1997) presents 32 different methods grouped according to the types of generation and evaluation function used. If the original feature set contains N number of features, the total number of competing candidate subsets to be generated is 2N. This is a huge number even for medium-sized N. *Generation procedures* are approaches to solving this problem; these include: *complete*, in which all the subsets are evaluated; *heuristic*, where subsets are generated by adding/removing attributes (incremental/decremental); and *random*, in which a certain number of randomly generated subsets are evaluated.

The aim of an *evaluation function* is to measure the discriminating ability of a feature or a subset to distinguish the class labels. There are two common approaches: a *wrapper* uses the intended learning algorithm itself to evaluate the usefulness of features, while a *filter* evaluates features according to heuristics based on general characteristics of the data. The wrapper approach is generally considered to produce better feature subsets, but runs much more slowly than a filter (Hall and Smith, 1999). Dash and Liu (1997) divide evaluation functions into five categories: *distance*, which evaluates differences between class conditional probabilities; *information*, based on the information of a feature; *dependence*, based on correlation measurements; *consistency*, where an acceptable inconsistency rate is set by the user; and *classifier error rate*, which uses the classifier as an evaluation function. Only the last evaluation function, *classifier error rate*, can be counted as a wrapper.

Table 9.1 sums up the classification of methods in Dash and Liu (1997). The blank boxes in the table signify that no method exists for these combinations. Since in-depth analysis of each of the feature selection techniques is not our present purpose, references for further information are given in the table. Hall and Smith (1999) present an approach to feature selection, correlation-based feature selection (CFS), using a correlation-based heuristic

TABLE 9.1

Different Feature Selection Methods as Stated by M. Dash and H. Liu

Generation	Heuristic	Complete	Random
Evaluation			
Distance	Relief [Kir92], Relief-F [Kon94], Segen [Seg84]	Branch & Bound [Nar77], BFF [XuL88], Bobrowski [Bob88]	
Information	DTM [Car93], Koller & Sahami [Kol96]	MDLM [She90]	
Dependency	POE1ACC [Muc71], PRESET [Mod93]		
Consistency		Focus [Alm92], Schlimmer [Sch93], MIFES-1 [Oli92]	LVF [Liu96]
Classifier error rate	SBS, SFS [Dev82], SBS-SLASH [Car94], PQSS, BDS [Doa92], Schemata search [Moo94], RC [Dom96], Queiros & Gelsema [Que84]	Ichino & Sklansky [Ichi84] [Ichi84b]	LVW [Liu96b], GA [Vaf94], SA, RGSS [Doa92], RMHC-PF1 [Ska94]

Source: From Dash, M. and Liu, H., 1997. *Intelligent Data Analysis,* 1(3), 131–156.

to evaluate the worth of features. We have not used this method here but have extracted various ideas about feature selection, that is, using correlation measurements between features and between features and output classes. Consequently, a brief overview of this method of feature selection appears in the following section.

9.2.1.4 Correlation-Based Feature Selection

A CFS algorithm relies on a heuristic for evaluating the worth or merit of a subset of features. This heuristic takes into account the usefulness of individual features for predicting the class label, along with the level of inter-correlation among them. The hypothesis on which the heuristic is based can be stated as "Good feature subsets contain features highly correlated with (predictive of) the class, yet uncorrelated with (not predictive of) each other" (Hall and Smith, 1999, p. 236). Similarly, Gennari et al. (1989, p. 51) state: "Features are relevant if their values vary systematically with category membership." Briefly stated, then, a feature is only useful if it is correlated with or predictive of the class; otherwise it is irrelevant.

Empirical evidence provided in the literature on feature selection suggests that along with irrelevant features, redundant information should also be eliminated (Kohavi, 1995; Kohavi and John, 1996; Langley and Sage, 1994a,b, 1997). A feature is considered redundant if one or more other feature is highly correlated with it. The above definitions for relevance and redundancy suggest the best features for a given classification are highly correlated with one of the classes and have an insignificant correlation with the rest of the features in the set.[*]

If the correlation between each component and the outside variable is known, and the intercorrelation between each pair of components is given, the correlation between a composite consisting of the summed components and the outside variable can be predicted as per Ghiselli (1964), Hogarth (1977), and Zajonic (1962) as

$$r_{zc} = \frac{k\overline{r_{zi}}}{\sqrt{k + k - (k-1)\overline{\overline{r_{ii}}}}} \tag{9.1}$$

where
r_{zc} = correlation between the summed components and the outside variable
k = number of components (features)
$\overline{r_{zi}}$ = average of the correlations between the components and the outside variable
$\overline{\overline{r_{ii}}}$ = average intercorrelation between components.

Equation 9.1 represents Pearson's correlation coefficient, where all variables have been standardized. The numerator can be thought of as indicating how predictive of the class a group of features is; the denominator shows the extent of redundancy among them. Thus, Equation 9.1 indicates the correlation between a composite and an outside variable is a function of the number of component variables in the composite and the magnitude of the intercorrelations among them, together with the magnitude of the correlations between the components and the outside variable. The following conclusions can be extracted from Equation 9.1:

- When there is a higher correlation between the components and the outside variable, the correlation between the composite and the outside variable is correspondingly higher.
- As the number of components in the composite increases, the correlation between the composite and the outside variable also increases.
- The lower the intercorrelation among the components, the higher the correlation between the composite and the outside variable.

Theoretically, when the number of components in the composite increases, the correlation between the composite and the outside variable also increases. However, it is unlikely that a group of components highly correlated with

---

[*] Subset of features selected for evaluation.

the outside variable will, at the same time, have low correlations with each other (Ghiselli, 1964). Furthermore, Hogarth (1977) notes that when the inclusion of an additional component is considered, low intercorrelation with the already selected components may well predominate over high correlation with the outside variable.

9.2.2 Feature Extraction

9.2.2.1 Features from Time Domain

The measurement used for failure diagnosis is called a feature. This type of measurement is crucial to failure diagnosis. The existing failure diagnosis of rolling element machines focuses on the frequency domain, for example, using Fourier transform, or time–frequency domain such as wavelet transform. In the early stage of failure development, damage is not significant; therefore, the defect signal is not significant, but is mixed into the noise signal. The periodicity of the signal is not significant either. Therefore, spectral analysis may not be effective. Yet the periodicity is significant; thus, using the feature from the time domain is necessary, as the normal and defect signals differ in their statistical characteristics in the time domain. Using time domain features accompanied by other domain features can improve diagnosis accuracy.

Some features of state-of-the-art uses of the time domain are listed in Table 9.2. Among these, kurtosis is an important feature used in rolling element machines. Kurtosis defines the peakedness of the amplitude in the

TABLE 9.2

State-of-the-Art Time Domain Features

	Feature	Definition		Feature	Definition
1	Peak value	$Pv = (1/2)$ $[\max(x_i) - \min(x_i)]$	6	Clearance factor	$Clf = \dfrac{Pv}{\left((1/n)\sum_{i=1}^{n}\sqrt{|x_i|}^{2}\right)}$
2	RMS	$RMS = \sqrt{\dfrac{1}{n}\sum_{i=1}^{n}(x_i)^2}$	7	Impulse factor	$Imf = \dfrac{Pv}{(1/n)\sum_{i=1}^{n}|x_i|}$
3	Standard deviation	$Std = \sqrt{\dfrac{1}{n}\sum_{i=1}^{n}(x_i - \bar{x})^2}$	8	Shape factor	$Shf = \dfrac{RMS}{(1/n)\sum_{i=1}^{n}|x_i|}$
4	Kurtosis value	$Kv = \dfrac{(1/n)\sum_{i=1}^{n}(x_i - \bar{x})^4}{RMS^4}$	9	NNL value	$NNLV = -\ln L;$ $L = \prod_{i=1}^{N} f(x_i, \mu, \sigma)$
5	Crest factor	$Crf = Pv/RMS$	10	Beta parameter	The parameters in beta function

signal. When the amplitude follows a normal distribution, the kurtosis value is a fixed value, 3. Beta parameters are the shape and scale parameters in the beta distribution when assuming the amplitude of signal follows beta distribution (Fuqing et al., u.d.). Beta distribution is a flexible distribution and most signals can fit it. As the parameters in the beta distribution for normal signal and defect signal differ, these parameters have been used to diagnose failure. However, Heng and Nor (1998) argue that the beta offers no significant advantage over kurtosis and the crest factor for rolling element bearings.

The features kurtosis, crest factor, and impulse factor are all nondimensional features and are independent of the magnitude of the signal power.

However, RMS, peak value, standard deviation, and normal negative likelihood (NNL) value depend on the power of the signal. Some nuisance factors, such as poor quality or location, can also influence the power of a sensor's signal. The advantage of the nondimensional features is that they can be immune from such nuisance factors. Nevertheless, RMS is an important feature in signal processing; it measures the power of the signal and can be used to normalize the signal. Some features are derived from RMS. Other available features include beta-kurtosis (Wang et al., 2001), Weibull negative likelihood value (Abbasion et al., 2007; Sreejith et al., 2008), kurtosis ratio (Vass et al., 2008), and so on.

9.2.2.1.1 Normal Negative Likelihood

NNL is used by some researchers as a feature to diagnose failure (Sreejith et al., 2008). In NNL, the amplitudes of the signal are assumed to follow normal distribution. The parameters u and σ are estimated using the maximum likelihood estimator method. This research proves the NNL is equivalent to another computationally cheaper feature.

Let the amplitudes of the signal be denoted by a series $x_1, x_2, ..., x_n$ discretely. When parameters u and σ are known, the negative likelihood function of this series is

$$f(x_1, x_2, ..., x_n | \mu, \sigma^2) = \prod_{i=1}^{n} f(x_i | \mu, \sigma^2) = \left(\frac{1}{2\pi\sigma^2}\right)^{n/2} \exp\left(-\frac{\sum_{i=1}^{n}(x_i - \mu)^2}{2\sigma^2}\right)$$

(9.2)

The maximum likelihood estimator of u and σ is

$$\hat{u} = \bar{x} = \sum_{i=1}^{n} x_i/n \text{ and } \hat{\sigma}^2 = \frac{\sum_{i=1}^{n}(x_i - \hat{\mu})}{n}$$

(9.3)

Substituting Formula 9.3 into Formula 9.2 and simplifying it, we obtain the following equation:

$$f(x_1, x_2, \ldots, x_n \mid \mu, \sigma^2) = \left(\frac{1}{2\pi}\right)^{n/2} \exp(-n/2) \cdot \sigma^{-n} \tag{9.4}$$

Thus, the negative likelihood can be rewritten as

$$Ln = \ln\left(\left(\frac{1}{2\pi}\right)^{n/2} \exp(-n/2)\right) - n \ln \sigma \tag{9.5}$$

Evidently, the NNL is essentially equivalent to $\ln \sigma$.

Obviously, $\ln \sigma$ is not a nondimensional feature. By normalizing it with the energy of the signal, one new feature can be derived as

$$NNNL = \ln \frac{\sigma}{RMS} \tag{9.6}$$

The above feature is called the normalized normal negative likelihood value (NNNL) (Sreejith et al., 2008).

9.2.2.1.2 Mean Variance Ratio

The distribution of amplitude differs from the normal and defect signal. The density function of a normal signal without defect is dominated by some noise signals. The peakedness of the signal, thus, tends to be a peak. The distribution of a defect signal has wider amplitude, so the variance is bigger than normal. The left figure in Figure 9.2 indicates a normal signal of a bearing; the other figure shows a defect signal. It is evident that the defect signal

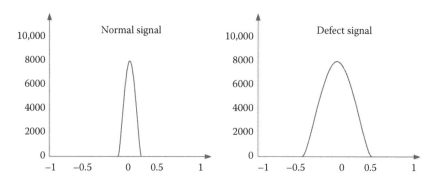

FIGURE 9.2
Normal and defect signal. (From Fuqing, Y., Kumar, U., and Galar, D., u.d. *Reliability Engineering and System Safety.*)

differs from the normal signal and has a wider variance. Therefore, the mean and variance ratio (MVR) could be a feature used to discriminate them. The definition of MVR is

$$\text{MVR} = \frac{(1/n)\sum_{i=1}^{n} x_i}{\sqrt{(1/n)\sum_{i=1}^{n} (x_i - \bar{x})^2}} \tag{9.7}$$

Obviously, MVR is also a nondimensional feature independent of signal power. MVR implies the degree of scatter for the distribution of signal amplitude.

9.2.2.1.3 Symbolized Sequence Shannon Entropy

All features described in Table 9.1 are statistical features (Fuqing et al., u.d.), as they consider statistical characteristics of the amplitude distribution. However, they all ignore the information on the spacious patterns of the amplitude. For example, when there is a defect in a rolling element machine, the amplitude tends to be periodic. This periodicity cannot be reflected in the statistical features. Figure 9.3 shows a simple example to verify this argument. This figure is composed of 100 samples. The amplitude of each sample is composed of {1, 2, 3, 4}, where each has equal possibility. The upper and lower figures in Figure 9.3 are plots of the same samples, but have different

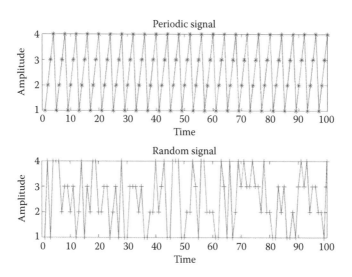

FIGURE 9.3
Periodic signal and random signal. (From Fuqing, Y., Kumar, U., and Galar, D., u.d. *Reliability Engineering and System Safety.*)

space distribution. In the upper periodic signal, the amplitude is distributed deterministically with a sequence of 1234, 1234,... iteratively. In the random signal, the amplitude is distributed randomly.

From Figure 9.3, it is evident that the two signals are different. But their statistical feature values listed in Table 9.1 are the same; that is, the statistical features are not able to discriminate them.

Shannon entropy can measure the uncertainty of a random process. A rolling element machine without failure tends to have a more random signal, while the machine with failure usually tends to have a more deterministic signal; that is, their Shannon entropy will be different. To extract the periodicity in the signal, we propose a feature called symbolized Shannon entropy (SSE). In this feature, we first symbolize the signal and then use the Shannon entropy. SSE has been used to detect weak signals (Finney et al., 1998; Tang et al., 1995). The procedure of calculating SSE is as follows.

Discretize the signal and predefine a threshold. The amplitude below the threshold is coded as 0 and the amplitude above the threshold is coded as 1.

Thus, the signal is discretized into a binary sequence, denoted by

$$b_1, b_2, b_3, \dots .b_i, \dots$$

Now segment the binary signal with equal length L. For example, segment the binary sequence 110010010 into 110, 010, 010 with length $L = 3$. Calculate the decimal value of each segment; "6," "2," "2" in this example.

Calculate the probability of each segment. The probability is considered the frequency of each segment. For "6" in this example, the probability is 1/3 and for "2" it is 2/3.

Next calculate the entropy using the following Shannon entropy formula:

$$H = -\frac{1}{\log N} \sum_i p_i \log p_i \tag{9.8}$$

where N is the total number of observed events and p_i is the probability of this event.

In a periodic signal, some sequences will occur frequently, and the Shannon entropy is, thus, lower. Shannon entropy values vary with the signal and can be used as a feature to measure the characteristics of a signal. For a random signal, the Shannon entropy value is 1. For a deterministic signal, the entropy is between 0 and 1 (Finney et al., 1998). Applying the above procedure to the example given in Figure 9.4, the SSE of the periodic signal is 0 and the random signal is 0.905. These two signals can be significantly discriminated.

9.2.2.2 Performance of Features by Simulation

This section discusses simulating signals to test features, similar to Heng and Nor (1998). The feature values are shown in Table 9.2 for various signals.

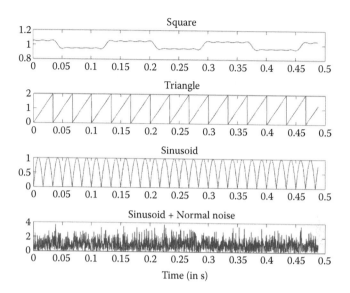

FIGURE 9.4
Simulated signals. (From Fuqing, Y., Kumar, U., and Galar, D., u.d. *Reliability Engineering and System Safety*. Submitted.)

As the table shows, the values differ for different signals. In other words, the features can discriminate the signals.

As Table 9.3 shows, for deterministic signals, such as square, sin, and triangle, the entropy is small. For random signals, such as Gaussian random and uniform random, the entropy values are above 0.9. For the singular signal, two white Gaussian noise $N(0,0.1)$ and $N(0,1)$ are added to the determinist and the random signals, respectively. With this addition, all the time domain features differ, implying they are sensitive to noise.

TABLE 9.3

State-of-the Art Time Domain Features

Signals	Kurtosis	Crest Factor	NNNL	MVR	Entropy	Shape	Clear
Square	1.0866	0.059	3.02	20.42	0.12	1.00	0.06
Triangle	1.8	0.87	0.69	1.73	0.19	1.15	1.12
N(0,1)	3.91	2.08	0.50	1.32	0.90	1.25	3.07
U(0,1)	1.79	0.86	0.70	1.74	0.91	1.15	1.12
Sin	1.93	0.71	0.83	2.07	0.26	1.11	0.86
Sin + N(0,0.1)	1.97	0.95	0.81	2.02	0.46	1.11	1.17
Sin + N(0,1)	3.33	1.78	0.52	1.35	0.90	1.24	2.60

As with time domain features, some researchers try to extract information from the frequency domain, dividing the bandwidth of the signal into several subbands and filtering the targeted signal. Once this signal is chopped out of different signals (i.e., time domain transformed into the inverse of the outputs of bank filters), new time domain features can be extracted. For this reason, the frequency domain can contribute numerous time domain features from the various subbands of a signal where some failures can be more visible than others.

However, as noted earlier, this information transformation tries to find hidden aspects of the signal, but does not merge the information contained in that signal with the surrounding environment to eventually provide new information and segregate the signals (*note*: with a different environment these may have totally different meanings) (Fuqing et al., u.d.).

9.2.3 Contextual Information

Today's advanced networks and the widespread use of mobile devices, such as PDAs and smartphones, make the computing paradigm more distributed and pervasive. In this environment, users expect to get useful services and information via their mobile devices. However, because the number of services and the amount of information have also been increasing rapidly, it is difficult and time consuming for users to search for proper services or information.

A context-aware system actively and autonomously adapts and provides the most appropriate services or information to users, using people's contextual information and requiring little interaction. As this is a key driver for solving the problems presented above, there has been much research on the topic since the early 1990s.

In this section of the chapter, we survey the existing research on context awareness, especially focusing on context-aware systems and frameworks (Lee et al., 2011).

9.2.3.1 General Process in Context-Aware Systems

Context-aware systems are usually complicated systems, responsible for many jobs, such as representation, management, reasoning, and analysis of context information. They provide their functionalities through the collaboration of many different components in a system.

There are various types of different context-aware systems; thus, it is hard to generalize a context-aware system process; however, generally, a context-aware system follows four steps.

The first step is acquiring context information from sensors. Sensors convert real-world context information into computable context data. By using physical and virtual sensors, the system can acquire various types of context-aware information. After acquiring context information, the system stores acquired context data in its repository. When storing context data, the kind of data model used to represent context information is very important. Context

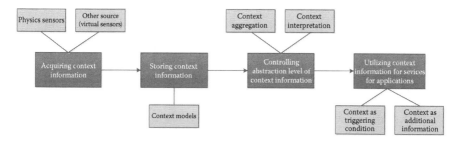

FIGURE 9.5

A general process in context-aware systems. (From Lee, S.-K., 2011. *A Survey of Context-Aware System—The Past, Present, and Future.* [Online] Available at: ids.snu.ac.kr/w/images/d/d7/Survey(draft).doc.)

models are diverse, and each has its own characteristics. To easily use the stored context data, the system controls the abstraction level of these data by interpreting or aggregating them. Finally, the system utilizes the abstracted context data for context-aware applications.

In the following subsections, we explain how a general context-aware system processes each step and discuss related issues. Figure 9.5 shows a general process in a context-aware system.

9.2.3.2 Acquiring Context Information

Because of the diversity of the types of context information, context information can be acquired in many ways. Physical sensors are hardware devices that convert physical analog properties into computable digital data, and can be used for acquiring context information. Depending on the type of context information, many different physical sensors can be used. For example, to attain location information, the global positioning system (GPS), active badge system, and IR sensors are appropriate. Microphones can be used to get audio context, and cameras can be used to acquire visual context. In addition, motion sensors or pressure sensors can be applied in context-aware systems for various purposes.

However, using physical sensors is not the only way to acquire context information. A context-aware application can recommend a music playlist based on a user's preference or give the local weather conditions in in the user's current location. In this situation, the user's preference can be acquired by analyzing his/her music play history, and the weather conditions of the current location can be attained by querying a web service provided by a forecasting site. Although these contexts can be acquired without using physical sensors, we need software modules that perform as virtual sensors. Virtual sensors acquire context information by analyzing various data or querying external sources. Explicitly provided context information by users can also be used in a context-aware system.

Context Model	Examples
Object-Oriented Model	Hydrogen
Logic-Based Model	McCarthy's Approach
Key-Value Model	Schilit's Approach, Context Toolkit
Mark-up Scheme Model	CC/PP, UAProf, CSCP, GPM
Ontology Model	SOCAM, CoBra, CASS, CoCA
Graphical Model	Context Extension of ORM, Vector Space Model

FIGURE 9.6

Context models. (From Lee, S.-K., 2011. *A Survey of Context-Aware System—The Past, Present, and Future.* [Online] Available at: ids.snu.ac.kr/w/images/d/d7/Survey(draft).doc.)

9.2.3.3 Storing Context Information

Most context systems store acquired context data in their repository; in fact, context models are closely related to context storing. Many factors, such as expressiveness, flexibility, generality, and the computational cost of processing context-aware data, depend on what kind of context model is used in the system (Lee et al., 2011). See Figure 9.6 for an example.

Context information can be represented in many ways from very simple data models such as key-value models to complex ontological models. Schilit et al. (1994) and Salber et al. (1999) used a key-value model to express context information. This model can be easily adopted for context-aware systems, but it is limited in its ability to represent complex structures of context information. McCarthy and Buvac (1997) introduced a logic-based context model that represents context information using facts and rules. This model has high mathematical formality and support for inferences. Structured context data models include the markup scheme context model, object-oriented context mode, or context extension of Unified Modeling Language (UML), object-relational mapping (ORM).

Recently, the ontological context model has been widely used. Many context-aware systems, such as SOCAM, CoBrA, CoCA, have adopted ontology as their context data model. The ontological context model has high expression power and formality at the same time, and it is a suitable data model for representing the relationships among context information and entities (Lee et al., 2011).

9.2.3.4 Controlling the Context Abstraction Level

A context-aware system is responsible for controlling the abstraction level of context information and performs context abstraction in two ways—context aggregation and context interpretation. Context aggregation means the system aggregates many low-level signals (raw data) into a manageable amount of high-level information. For example, a context-aware system converts thousands of temperature signals into several keywords (e.g., "hot," "cold," "moderate," "cool," and "warm") by context aggregation. Context interpretation is a method of interpreting context information and adding semantics. For example, a context-aware system can interpret a GPS signal into a street name.

It is hard for context-aware systems to directly use the raw data provided by sensors. Accordingly, context-aware systems translate sensed signals into meaningful data so they can understand and use context data more easily. Additionally, context-aware systems can reduce the number of context data and achieve better performance by controlling the level of context abstraction. If the context abstraction is separated from a context-aware application, the context-aware application does not have to know the details of sen-sors but still can use the sensed context data collected by the sensors.

9.2.3.5 Using Context Information

Utilizing acquired and abstracted context information as useful information for services or applications is the last step of the general context-aware system process. Context-aware systems use context information for two purposes—as a triggering condition or as additional information.

A context-aware system can use context information as an action triggering condition when it wants to trigger actions and the current context satisfies a specific situation.

To enhance the quality of the service or application, context information can be used as additional information. For example, assume a user sends queries to an information server; the user's current context information can be used as additional information to obtain better results in the queries. These two purposes of context information usage can be combined.

We can use context information for many types of context-aware applications. Several examples of context-aware application categories are presented below (Lee, 2011).

9.2.3.6 Design Considerations of Context-Aware Systems

When designing context-aware systems, we need to consider a number of aspects. These systems can be implemented many ways and can have different structures, depending on the development focus of the system. In this section, we discuss some design considerations, not mentioned in previous sections, including architecture style, handling dynamicity, privacy protection, and performance and scalability.

9.2.3.6.1 Architecture Style

A context-aware system's representative architecture can be categorized into three styles: stand-alone, distributed, and centralized architecture. Figure 9.7

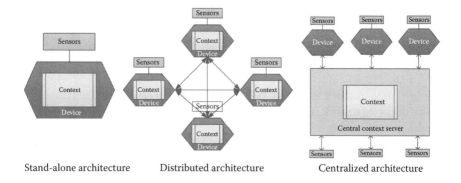

Stand-alone architecture   Distributed architecture   Centralized architecture

FIGURE 9.7
Context-aware systems architecture styles. (From Lee, S.-K., 2011. *A Survey of Context-Aware System—The Past, Present, and Future*. [Online] Available at: ids.snu.ac.kr/w/images/d/d7/ Survey(draft).doc.)

shows a simplified diagram of the architecture of each category. The characteristics, advantages, and disadvantages of each are explained below.

Stand-alone architecture: This basic architecture directly accesses sensors and does not consider context sharing of devices. It can be relatively easily implemented but has limitations in that it cannot process device collaboration. This architecture is appropriate for small, simple applications or domain-specific applications.

Distributed architecture: This architecture can store context information in many separate devices, with no additional central server. Each device is independent of other devices; thus, a context-aware system can ignore failure or bottleneck problems and still continue context-aware operations. Each device manages its own context information and shares context information with other devices by communicating with them; thus, an ad hoc communication protocol is required. However, it is hard for a device to know the overall situation of every device when using ad hoc communication protocols. Mobile devices usually lack resources and computation power, so a distributed architecture has limitations dealing with computationally intensive applications.

Centralized architecture (context server) sensors and devices: These are connected to a centralized context server with adequate resource and computational power; context information is stored in a centralized server. If a device needs to get another device's context information, the device queries the centralized server. In this architecture, all communication is performed by querying the context server, so the communication protocol can be relatively simpler than for distributed architecture. By using a computationally powerful device as a centralized server, many applications requiring high resources and cost can be performed. A disadvantage of this approach is that it is extremely problematic if the centralized server fails or a bottleneck problem occurs.

9.2.3.7 Handling Dynamicity

Handling dynamicity is an important consideration if a context-aware system is expected to process sophisticated context-aware applications. Entities (people, devices, sensors, etc.) varying from simple sensors and resource-poor mobile devices, to central servers with high performance requirements, may want to process context-aware applications. At the same time, connections and disconnections of the devices may occur dynamically. A context-aware system should be able to deal with heterogeneous entities and provide support for resource discovery (Lee, 2011).

9.2.3.8 Privacy Protection

Supporting privacy protection is another consideration for context-aware systems. Context-aware systems autonomously gather information from the users; some users may feel uncomfortable in that the system can use or open their information without notice. Thus, a context-aware system should allow users to express their privacy needs. Such systems should protect a user's context information from illegal access and guarantee anonymity.

9.2.3.9 Performance and Scalability

In most cases, operations for context-aware applications should be processed in real time, and a sophisticated context-aware system requires reasoning and inference functionalities, which necessitate high computational costs and resources. Performance and scalability of a context-aware system is important if the system is to instantly respond to the current context of users. The communication protocol must also perform adequately with acceptable scalability.

9.2.4 Context as Complex Information Content in Maintenance: An Example of Health Assessment for Machine Tools

In the context of maintenance activities, maintainers rely on machine information, including their past breakdowns, repair methods and guidelines, as well as new research in the area. They get access to information and knowledge by using information systems (nondestructive testing [NDT] or CM), local databases, e-resources, or traditional print media. Basically, it can be assumed that the amount of available information affects the quality of maintenance decision making.

Machine health information retrieval is the application of information retrieval concepts and techniques to the operation and maintenance domain. Contextual information retrieval is a subarea of information retrieval that incorporates context features in the search process to facilitate improvement. Both areas have been gaining interest in the research community, especially as they may be able to perform more accurate prognostics based on specific scenarios and real circumstances.

This chapter will show the effects of the interaction of context features on machine tools' health information. This interaction context and health assessment is bidirectional in the sense that health information-seeking behavior can also be used to predict context features that can be used without disturbing the operational environment or disrupting production (Johansson et al., 2014).

The extraction of multiple features from multiple sensors, already deployed in this type of machinery, may constitute snapshots of the current health of certain machine components. The mutation status of these snapshots has been proposed as a prognostic marker in machine tools' problems. But only the spindle fingerprint mutation has been validated independently as a prognostic for overall survival and survival after relapse; the prognostic value of these mutations can be investigated in various contexts defined by stratifications of the machine population. At this point, the prognostic value of the rest of the components' mutation is still being investigated.

CM activities include obtaining data from sensors coming from the machine; the analysis of these data helps to measure and understand the machine tool performances. This approach allows computing some indicators at the local level to monitor the local health of a machine (Figure 9.8).

Classical machine tool monitoring techniques are related to acoustic and vibrations techniques (Abele et al., 2010). However, the various CM methods, such as vibration or acoustic monitoring, usually require expensive sensors (Alzaga et al., 2014). One way to achieve a proactive CM approach, with nonintrusive monitoring techniques, affordable in terms of both cost and effectiveness, is to use the current analysis to assess the health status of the machine, with special focus on the critical components: the spindle and the linear axis. Recent research has been directed toward electrical monitoring of the motor with emphasis on inspecting the stator current of the motor (Alzaga et al., 2014; Benbouzid, 2000; Kliman and Stein, 1992). This should reveal the relationship between electrical signals and the wear of the spindle and the linear axis. Based on the signature analysis results, the health index of these components could be computed and associated with different degradation modes of the components (e.g., gears that are missing teeth).

Context can be created by assessing the evolution of operating conditions to understand both the usage of the machine and the effect of the environment. The characterization of operational conditions helps us to understand the relationship between the health status and the usage of the machine (machine working for a long period with high loads, or high speed, etc.).

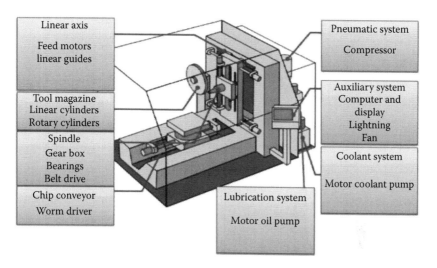

FIGURE 9.8

Machine tool (source Siemens). (From Johansson, C.-A., 2014. *The Importance of Operation Context for Proper Remaining Useful Life Estimation.*)

This, in turn, yields a better understanding of the impact of operating conditions on the evolution of the degradation modes (Medina-Oliva et al., 2013).

9.2.4.1 Fingerprint Data Use for Current Signature Analysis

Fingerprint data are collected in a standardized way every day/week/ month. This means the machine is running a special fingerprint program during the data collection where feed, speed, rpm, and so on are changed but no production is performed. There is also a possibility of using standard sequences in ordinary production programs such as tool changing for part of the fingerprint collection. The data are collected with a sampling frequency of many kilohertz.

Typical data collected and synchronized in time are

- Vibration data
- Motor power data for spindle and linear axis (current signal and motor current signature analysis)
- rpm and speed data for spindle and linear axis and axis position.

The data are analyzed in both time and frequency domains (Galar et al., 2012) and a number of features (Table 9.4) are calculated for each signal. In the frequency domain, the vibration level on known frequencies, such as gear mesh frequencies, bearing frequencies, rotational speed, and so on and their harmonics is followed.

TABLE 9.4

Time Domain Features

Feature	Definition
Peak value	$Pv = \dfrac{1}{2}[\max(x_i) - \min(x_i)]$
Root mean square	$RMS = \sqrt{\dfrac{1}{n}\sum_{i=1}^{n}(x_i)^2}$
Standard deviation	$Std = \sqrt{\dfrac{1}{n}\sum_{i=1}^{n}(x_i - \bar{x})^2}$
Kurtosis value	$Kv = \dfrac{\dfrac{1}{n}\sum_{i=1}^{n}(x_i - \bar{x})^4}{RMS^4}$
Crest factor	$Crf = {Pv}\big/{RMS}$
Clearance factor	$Clf = \dfrac{Pv}{\left(\dfrac{1}{n}\sum_{i=1}^{n}\sqrt{|x_i|}\right)^2}$
Impulse factor	$Imf = \dfrac{Pv}{\dfrac{1}{n}\sum_{i=1}^{n}|x_i|}$
Shape factor	$Shf = \dfrac{RMS}{\dfrac{1}{n}\sum_{i=1}^{n}|x_i|}$
NNL value	$NNL = -lnL; \; L = \prod_{i=1}^{N} f(x_i, \mu, \sigma)$

For faults/problems in the gear train, such as bearing and gear problems, the most sensitive features are chosen through testing in a test bench where different types of faults can be simulated and by testing faulty components sent in by the customer for repair.

A machine fingerprint is the electrical signature of that machine in a specific time domain. Fingerprint raw data are processed and the relevant signal features are extracted. These relevant features are used to compare the fingerprints through time to assess the machine health. To compute the health index of the machine, it is necessary to correlate any load and speed variation with current and voltage variations to reflect loads, stresses, and wear of components. But to do this, we must identify the healthy electrical signature of the machine, called the reference fingerprint, along with the degraded ones associated with the various failure modes (Figure 9.9).

In an industrial environment, it is necessary perform experiments on test benches to determine the potential correlation between electrical signals and wear, stresses, and load on the machine. Such test benches include

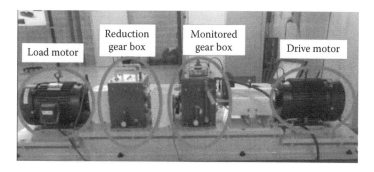

FIGURE 9.9
Test bench for gear train. (From Johansson, C.-A., 2014. *The Importance of Operation Context for Proper Remaining Useful Life Estimation.*)

the gearbox test bench, spindle test bench, and linear guides test bench, as explained in Alzaga et al. (2014).

A GPS test bench has been extensively used in experiments. This test bench is particularly convenient because of its flexibility. Different sets of sensors can be placed in different positions, and multiple combinations of speeds and loads can be established. Faults such as missing teeth, chipped teeth, eccentricity, and different degrees of surface degradation can be deliberately created in the gears. When test parameters are selected to emulate the working conditions of electromechanical actuators and machine tools, we can perform constant speed and transient tests. In the transient tests, fast speed changes are performed to produce acceleration, to investigate the concomitant changes produced in the signal. The analysis can be done in both the time domain and the frequency domain and, complementarily, using the wavelet decomposition. The results allow us to discern the various types of gear defects, thus allowing us to detect the fault conditions and assess the health state of the gearbox.

Once the fingerprint experimental phase is completed in the test benches, and the main signals, features, and the relationship with failure modes are established, the implementation can continue with an operational milling machine using external hardware.

With the data obtained from the fingerprint test, various features are computed, such as the mean, median, variance, and so on, for time domain and frequency analysis, as well as the health index. These results are then sent to the remote level. This way, different types of degradations can be detected early, including gear degradations (e.g., missing teeth and chipped teeth).

In a study by Chandran et al. (2012), the processed data were from the channel U of the drive motor. Two types of analysis were performed: one for the raw signal, in the time domain, and another for the time–frequency domain of the wavelet decomposition signal. In the case of the raw signal analysis, the researchers obtained 14 descriptors from the signal: rms, average, peak value, crest factor, skewness, kurtosis, median, minimum, maximum, deviation, variance, clearance factor, impulse factor, and shape factor

(Chandran et al., 2012). They were obtained from each repetition, and the median of all of the results was calculated.

Time–frequency domain analysis has also been performed (Peng and Chu, 2004; Cusido et al., 2008), with several levels being studied. At each level, the 14 descriptors noted above were achieved. Another descriptor has also been calculated; this descriptor represents the difference between one level and the next (Subasi, 2007).

9.2.4.1.1 Time Domain Analysis

After analyzing the several descriptors, we conclude that not all provide useful information. Out of the 14 descriptors, only half give results that are good enough to differentiate the good-condition gears from the gears with faults. The useful descriptors are average, deviation, maximum, median, peak value, root mean square, and variance.

9.2.4.1.2 Time–Frequency Domain Analysis

The results of the one-way analysis test reveals that the best levels for the decomposition are levels 1, 4, and 15. The most interesting variables for level 1 are crest factor, peak value, shape value, and variance. For level 4, they are average, skewness, and ratio. For level 15 decomposition, the best variables are the clearance factor, median, ratio, and variance.

The fingerprint represents the condition/status of the machine and the change in condition represents the degradation of the machine/component. The operational data represent the way the machine has been used between the fingerprints. By using data mining knowledge extraction techniques, we can find the correlation between operational data and changes in the fingerprint. This information may be used in prediction models for remaining useful life (RUL).

9.2.4.1.3 Operational Data for Inferring the Use of the Machine and for Contextualizing the Component/Machine Performances

Operational data are collected with a sampling frequency between 1 and 100 Hz via interfacing with the computer numerical control (CNC) controller of the machine tool. In the case of the Power-OM project, the GEM OA hardware from Artis are used for data collection (Figure 9.10).

Typical data collected are:

- Spindle power and rpm
- Motor power and position for linear axis
- Difference between commanded and actual position
- Temperatures
- Program number
- Tool number
- Alarms (sampled from the CNC or taken from the log file).

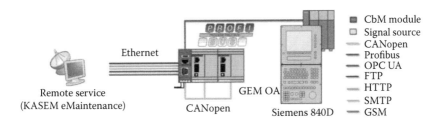

FIGURE 9.10
Local data collection unit. (From Johansson, C.-A., 2014. *The Importance of Operation Context for Proper Remaining Useful Life Estimation.*)

The operational data describe the way the machine has been used between fingerprints.

As a starting value, estimation can be based on the history of the machine tool such as:

- Age of the component/machine tool
- Designed life time for the component/machine tool
- Type of production/use (8 h/24 h/7 days, heavy, medium, low)
- Maintenance history
- Experience from similar machines in the fleet

Over time, the estimation is increasingly based on results from fingerprint and operational data.

The change of the value of different features between two fingerprints indicates the degradation of the component and depends on the previous condition and the way the machine has been used.

This means that the future condition, the feature value F_n, is a function of previous condition value F_{n-1} and the subsequent operational data.

$$F_n = f(F_{n-1}, \text{Operational data}) \tag{9.9}$$

Modern data mining algorithms can be used to extract knowledge from the fingerprint and operational data and find the correlation between the change in condition and the way the machine tool is used. The result is used to estimate the RUL for this machine/component and its uncertainty. An overview of different techniques for RUL estimation can be seen in Butler (2012).

The prediction of remaining useful life in engineering systems is affected by several sources of uncertainty, and it is important to correctly interpret this uncertainty to facilitate meaningful decision making. Thus, the uncertainty of the RUL depends on uncertainty in collected data and feature

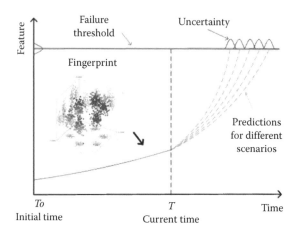

FIGURE 9.11
Uncertainty in RUL estimation. (From Johansson, C.-A., 2014. *The Importance of Operation Context for Proper Remaining Useful Life Estimation.*)

calculations, in prediction algorithms, in future operating conditions, and in threshold settings (Sankararaman and Goebel, 2013).

The probability distribution of the RUL can sometimes be extremely skewed; this can vary with the distance to "end of life."

By the classification of operational data into groups, for instance, based on power and speed, the user can make predictions for each group. In addition, if the program number is classified into the same groups, the user can estimate the RUL depending on what products/program number he/she plans to run (i.e., different scenarios) (Figure 9.11).

9.2.4.1.4 Need for e-Maintenance as a Context Manager Tool

In e-Maintenance, a remote server will receive data from different machines in different sites (the fleet), aggregate them, and make them semantically comparable, while considering their different contexts: technical differences (the machines are not exactly the same), operational conditions, historical failures, and so on. e-Maintenance solutions provide a mechanism that allows organizations to transfer data to make decisions from a system perspective. Such decisions are based on understanding data relationships and patterns. Materialized as a set of interoperable, independent, and loosely coupled information services, a framework with this type of inherent infrastructure (i.e., e-Maintenance Cloud—eMC) can provide fleet-wide, continuous, coordinated service support and service delivery functions for operation and maintenance. The e-Maintenance platform is collaborative, integrating engineering, proactive maintenance, decision making, and expertise tools. The foundation of the platform is Services-Oriented Architecture (SOA) and Enterprise Service Bus: a software and systems architectural principles, based

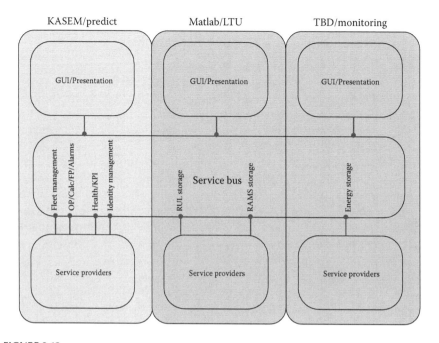

FIGURE 9.12
Infrastructure among user applications and service providers. (From Johansson, C.-A., 2014. *The Importance of Operation Context for Proper Remaining Useful Life Estimation.*)

on web services, to bring together a set of enterprise applications through an XML-based engine.

The e-Maintenance platform supports the infrastructure in a true SOA way, through services and data exchanges via web services between modules (Figure 9.12). This flexible infrastructure makes it possible for different partners to develop modules in different development environments.

An organization using e-Maintenance can:

1. Collect and store data from the local level, that is, the machines composing the fleet
2. Manage and store knowledge using a knowledge-based system
3. Share data from the units composing the fleet
4. Achieve large data analysis from stored, offline data from the fleet
5. Facilitate access, communication, user-friendly interpretation, and decision through remote services by means of web browser visualization to support fleet-wide services by providing the corresponding information. Information from different machines can be uploaded and analyzed in the platform.

9.3 Entropy and Relative Entropy Estimation

Entropy can be seen as the amount of disorder within a system. For example, a deck of cards just out of a box has low entropy. It is not zero because there is still potential energy in it, even when the cards are in order. As we shuffle the deck, we increase its entropy. Eventually, we reach a point where the deck is maximally disorganized; at this point, the deck has reached its maximum entropy (Leslie, 2011).

Entropy is a cornerstone of physics with consequences in many aspects of physics. It was first recognized as a phenomenon by German physicist Rudolf Clausius. His Clausius equation eventually became the second law of thermal dynamics. In physics, thermal dynamics is the study of how energy, work, heat, and so on interact with physical systems. In thermal dynamics, entropy is the potential amount of energy available to be converted into work that a system can do and how much of that energy will be dissipated by heat. This description only applies when describing a closed system that is in equilibrium; we will discuss other applications of entropy later.

The second law of thermal dynamics describes the propensity of heat to naturally flow from hot to cold items. In the Clausius equation, we see this law and entropy defined at their most basic level. The Clausius equation illustrates how the change in entropy can be represented by heat in Joules divided by the absolute temperature in Kelvin. This gives the change in entropy as a function of heat over temperature; however, for this to be true, heat must be transferred at a fixed temperature. If heat is added to the system, the entropy of the system increases, and if the heat is taken, the entropy decreases in equal measure (Walker, 2009) (Figure 9.13).

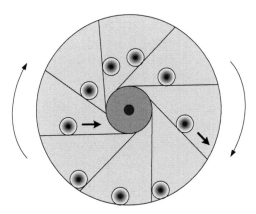

FIGURE 9.13
This diagram shows what a reversible system might look like, as it would operate the same no matter the direction in which it runs. (From Ducker, C., 2009. *Force, Work & Energy with the Ropes and Pulleys*. [Online] Available at: http://www.cdrucker.com/files/labsphys/forceworkenergy. html.)

Entropy in this situation is a state function: entropy as a variable is only dependent on the current status of the system, not on any other factor. The change in entropy can, thus, be determined even if the system is irreversible, as long as we know the starting and final states of the process. Calculating this change in entropy can be done by using a reversible process to connect the two states.

To satisfy the parameters of retaining entropy within a system, that system must be reversible, but no real system can be perfectly reversible. The process of moving heat from one part of a system to another will always bleed some portion of that heat; in other words, it will not retain its current entropy level. In a real engine, heat converted to work is less efficient at being converted when the heat is transferred from a hot reservoir than from a cold reservoir. We can determine, then, that the heat waste in an irreversible engine is greater than that of a reversible engine. Entropy change in an irreversible engine should always be positive. Consequently, the running of an irreversible process should increase entropy, as in the failure progression, deterioration, and degradation.

There are a couple of important things to note here, if we follow these points to their conclusion. In the universe as a whole, entropy should always be increasing as irreversible processes occur, in other words, all the time. The total entropy in the universe does not change from reversible processes, but all real processes are irreversible; otherwise, we would be able to have perpetual motion machines. This means the universe is always creating entropy, giving us one direction of motion, or what is often called the arrow of time.

This is the only theory we have which gives us a sense of the importance of the one directionality of time in the universe. All other thermal dynamic processes are, in theory, reversible; it does not matter which direction we run time—the equations and the theory work out the same. This is strange and baffling because in common day experience, it is obvious that processes do not run in reverse.

For example, without entropy, we could drop an egg on the floor; after it smashes, we could throw it back into the air and expect all of the broken pieces to come back together again—resulting in a whole egg. Of course, we cannot do this because of the one directionality, or arrow, of time. But what is it about the egg smashing into the ground that makes it impossible to put the egg back together again? Simply stated, it is the entropy created in the process of the energy of the falling egg being converted into work as it hits the floor. This creates complexity and disorder greater than what the egg had before it fell. In short, the entropy cannot be easily undone (Smoli, 2007).

It is important to disambiguate entropy from energy. It is appealing to think of entropy as energy, but it is not. Entropy in the universe is always increasing, while energy is constant and eternal. Energy cannot be created or undone, and this is a fundamental principle of the universe. We may want to represent entropy like energy by considering the original and final value of a system to be equal. This is usually not the case though; it is only true for

FIGURE 9.14
Cards being shuffled.

entropy when a system is reversible, and this cannot actually happen. It is only useful to think of a reversible system as a tool.

Nevertheless, it is possible for the entropy of a system to decrease as long as entropy is increased by a larger margin in another part of the system. This may make it seem as if some processes can decrease entropy. For example, in the analogy of a deck of cards, it is possible by random chance that part of the deck will become ordered again. If we shuffle the deck enough times, this is actually inevitable. So does that not suggest the entropy in the deck of cards has decreased? Well, actually it has, but the deck of cards is only a part of a larger system. The entropy in the hands of the person shuffling the deck would have increased from the process of doing the work of shuffling. This means the overall net entropy of the system has increased, even though it has been lowered in the deck of cards (Vedral, 2012) (Figure 9.14).

Another way to think of this concept is to compare it to putting a tray of water in a freezer. The water in the tray will eventually freeze, creating ice. This makes it seem as if the entropy has decreased because heat has been removed from the water to make ice, and we established in the Clausius equation that entropy decreases when heat is removed from an object. To freeze the ice, however, the freezer does work to draw out the heat in the water. If we measure the amount of heat the freezer releases into its surrounding environment, we will find it is greater than the amount of heat taken from the water. In reality, then, the overall entropy of the system has increased (Figure 9.15).

There are many ways to describe entropy. Entropy can be thought of as the amount of disorder in the universe, for example. If we think about the Clausius equation in terms of disorder, we see that a high-temperature reservoir separated from a low-temperature reservoir is a very orderly state. All of the low-energy molecules are grouped together, as are all of the high-energy molecules. Entropy, then, is the disordering of this system by dispersing the heat between the two extremes (Figure 9.16).

FIGURE 9.15
Ice cubes.

This leads to a new conclusion: if increasing entropy in a system is the same as randomizing the state of order in that system, entropy must be equal to the decrease of order in the system. Another way of looking at the second law of thermal dynamics now is that with time, the universe is continually decreasing in order.

In this chapter, entropy generally refers to information theory. These kinds of entropy are sometimes given other names, such as Shannon entropy, named for the theorist who developed information theory. This is not information in the traditional sense, though; it is information in the mathematical understanding of the word.

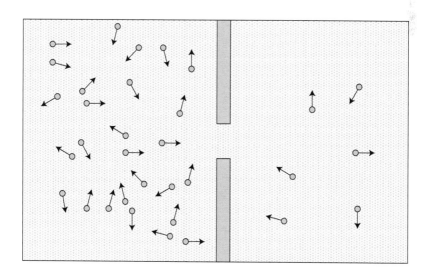

FIGURE 9.16
This diagram illustrates how two objects placed next to each other with heat variation are organized and why entropy will tend to want to bring them to equilibrium through random motions. (From Heylighen, F., 2003. *Entropy Increase: The Box With Two Compartments.* [Online] Available at: http://pespmc1.vub.ac.be/ENTROBOX.html.)

FIGURE 9.17
Flipping a coin.

Information theory is a branch of applied mathematics originally formulated to help understand the limits of processing data through compression or communicating it over distances. It has since been found to relate to many fields, including physics. In physics, entropy is represented by the amount of uncertainty we have about a random variable. For example, if we have a coin and flip it, we will have an equal probability of heads or tails, so the entropy will equal one. But if we flip the coin a second time, the resulting toss has a higher probability because we now know the result of the first toss, thereby changing the entropy to be lower. Clearly, entropy has many very different applications (Figure 9.17).

Over time, the universe increases its amount of entropy, but with this, it extinguishes its own ability to do work because in a reversible process, entropy is in equilibrium. A reversible system is maximally efficient and can always do the same amount of work. In an irreversible process, excess entropy is always created and exhausted, decreasing the amount of work that can be sustained within the system using the same amount of energy. This means if energy is eternal and cannot be created, eventually entropy will bleed out all available energy in the universe until there is no available energy left to do work. This is an alarming thought.

This energy is lost forever because to lower the entropy in one part of the universe, we must increase it somewhere else in greater measure, using even more energy. If there is no more available energy, then there is no energy that can be converted to try to retrieve some of the already lost energy. Because of this relationship, entropy is sometimes thought as a quality of energy. When energy is lost because of entropy, we call it the degradation of that energy. Energy degradation is an ongoing process in the cosmos.

Today, we can look up at the night sky and see a universe full of diversity and complexity. The universe is so inconceivably old that it is hard to think of it as anything but eternal. But if we have an arrow of time and a directionality defined by the amount of entropy, we must be able to rewind the cosmic clock back to a time with no entropy in the universe—a beginning.

What would this time and place look like? Physics calls this the Big Bang; a time when the universe was maximally organized into a single point containing all the energy of the universe, the same amount of energy that exists today.

9.3.1 Entropy Estimation

Entropy measures provide important tools to indicate variety in distributions at particular moments in time (e.g., presence of failures) or to analyze evolutionary processes over time (e.g., technical change). Importantly, entropy statistics are suitable to decomposition analysis, which renders the measure preferable to such alternatives as the Herfindahl index. There are several applications of entropy in industrial organization and innovation studies. The chapter contains two sections, one on statistics and one on applications. In the first section, we discuss, in this order (Frenken, 2013)

- Introduction to the entropy concept and information theory
- Entropy decomposition theorem
- Prior and posterior probabilities
- Multidimensional extensions

The second section discusses the following applications of entropy statistics:

- Concentration in industry
- Corporate diversity
- Regional industrial diversity
- Income inequality
- Organization theory

9.3.2 Entropy Statistics

The concept of entropy originated with Boltzmann (1877) and was later given a probabilistic interpretation in information theory by Shannon (1948). In the 1960s, Theil (1967) applied information theory to economics, and published the results in *Economics and Information Theory* (1967) and *Statistical Decomposition Analysis* (Theil, 1972).

9.3.2.1 Entropy Formula

The formula for entropy expresses the expected information content or uncertainty of a probability distribution. Let E_i stand for an event, for example, the adoption of a technology (i), and p_i be the probability of event E_i to occur. Assume n events $E_1,..., E_n$ with probabilities $p_1,..., p_n$ adding up to 1. Since the occurrence of events with smaller probability yields more information (because they are less expected), a measure of information h should be a decreasing function of p_i. Shannon's (1948) proposed logarithmic function to express information $h(p_i)$ is written as

$$h(p_i) = \log_2\left(\frac{1}{p_i}\right)$$
(9.10)

which decreases from infinity to 0 for pi ranging from 0 to 1. This means the lower the probability of an event to occur, the higher the amount of information of a message stating that the event has occurred. Information is expressed in bits, using 2 as a base of the logarithm; others express information in "nits" using a natural logarithm.

From n number of information values $h(p_i)$, the expected information content of a probability distribution, or entropy, can be derived by weighing the information values $h(p_i)$ by their respective probabilities, expressed as

$$H = \sum_{i=1}^{n} p_i \log_2\left(\frac{1}{p_i}\right)$$
(9.11)

where H stands for entropy in bits.

We customarily say (Theil, 1972, p. 5)

$$p_i \log_2\left(\frac{1}{p_i}\right) = 0 \quad \text{if } p_i = 0$$
(9.12)

and this accords with the limit value of the left-hand term for p_i approaching zero (Theil, 1972, p. 5).

When the entropy value H is nonnegative, and the minimum possible entropy value is zero, an event has the unit probability:

$$H_{min} = 1 \cdot \log_2\left(\frac{1}{1}\right) = 0$$
(9.13)

If all states are equally probable, that is, $(p_i = (1/n))$, the entropy value is maximum:

$$H_{max} = \sum_{i=1}^{n} \frac{1}{n} \log_2(n) = n\frac{1}{n}\log_2(n) = \log_2(n)$$
(9.14)

Thus, the maximum entropy increases with n, but decreasingly so.[*] *Note:* for the proof, see Theil (1972, pp. 8–10).

Entropy may be understood as a measure of uncertainty. The greater the uncertainty before a message arrives that an event occurred, the larger the amount of information conveyed by the arriving message, on average. Theil (1972, p. 7) says, in this aspect, the entropy concept is similar to the variance of a random variable whose values are real numbers, except that entropy applies to qualitative, not quantitative values and, as such, depends exclusively on the probabilities of possible events.

When a message is received that prior probabilities p_i are transformed in posterior probabilities q_i, we get the following (Theil, 1972, p. 59):

$$I(q \mid p) = \sum_{i=1}^{n} q_i \log_2\left(\frac{q_i}{p_i}\right) \tag{9.15}$$

which equals zero when the posterior probabilities equal the prior probabilities (i.e., no information) and is positive otherwise.

9.3.2.2 Entropy Decomposition Theorem

A powerful and attractive property of entropy statistics is how it handles problems of aggregation and disaggregation (Theil, 1972, pp. 20–22; Zadjenweber, 1972) because of the additivity of the entropy formula.

Let E_i stand for an event, and let there be n events $E_1,\ldots, E_n$ with probabilities $p_1,\ldots, p_n$. Assume all events can be aggregated into a smaller number of sets of events $S_1,\ldots, S_G$ in such a way that each event exclusively falls under one set S_g, where $g = 1,\ldots, G$. The probability of an event falling under S_g is obtained by summation

$$P_g = \sum_{i \in S_g} p_i \tag{9.16}$$

The entropy at the level of sets of events is

$$H_0 = \sum_{g=1}^{G} P_g \log_2\left(\frac{1}{P_g}\right) \tag{9.17}$$

---

[*] In physics, maximum entropy characterizes distributions of randomly moving particles with equal probability of being present in any state (like a perfect gas). When behavior is nonrandom, for example, if particles move toward already crowded regions, the resulting distribution is skewed, and entropy is lower than its maximum value (Prigogine and Stengers, 1984). In biology, meanwhile, maximum entropy refers to a population of genotypes in which the frequency of all possible genotypes is equal. In this case, minimum entropy reflects one genotype's total dominance (that is, resulting when selection is instantaneous [cf. Fisher, 1930, pp. 39–40]).

where H_0 is called the between-group entropy. The entropy decomposition theorem specifies the relationship between the between-group entropy H_0 at the level of sets and the entropy H at the level of events as defined in Equation 9.11. Now, we can write entropy H as

$$
\begin{aligned}
H &= \sum_{i=1}^{n} p_i \log_2 \left(\frac{1}{p_i} \right) = \sum_{g=1}^{G} \sum_{i \in S_g} p_i \log_2 \left(\frac{1}{p_i} \right) \\
&= \sum_{g=1}^{G} P_g \sum_{i \in S_g} \frac{p_i}{P_g} \left(\log_2 \left(\frac{1}{P_g} \right) + \log_2 \left(\frac{P_g}{p_i} \right) \right) \\
&= \sum_{g=1}^{G} P_g \left(\sum_{i \in S_g} \frac{p_i}{P_g} \right) \log_2 \left(\frac{1}{P_g} \right) + \sum_{g=1}^{G} P_g \left(\sum_{i \in S_g} \frac{p_i}{P_g} \log_2 \left(\frac{P_g}{p_i} \right) \right) \\
&= \sum_{g=1}^{G} P_g \log_2 \left(\frac{1}{P_g} \right) + \sum_{g=1}^{G} P_g \left(\sum_{i \in S_g} \frac{p_i}{P_g} \log_2 \left(\frac{1}{p_i/P_g} \right) \right)
\end{aligned}
$$

The first right-hand term in the last line is H_0. Hence,

$$
H = H_0 + \sum_{g=1}^{G} P_g H_g \tag{9.18}
$$

where

$$
H_g = \sum_{i \in S_g} \frac{p_i}{P_g} \log_2 \left(\frac{1}{p_i/P_g} \right) \quad g = 1, \ldots, G \tag{9.19}
$$

The probability p_i/P_g, $i \in S_g$ is the conditional probability of E_i given the knowledge that one of the events falling under Sg is bound to occur. H_g thus stands for the entropy within the set S_g and the term $\Sigma P_g H_g$ in Equation 9.18 is the average within-group entropy. Therefore, entropy equals the between-group entropy plus the average within-group entropy. Two properties of this relationship follow (Theil, 1972, p. 22):

$H \geq H_0$ because both P_g and H_g are nonnegative. This means after grouping, there cannot be more entropy (uncertainty) than there was before grouping.

$H = H_0$ if and only if the term $\Sigma P_g H_g = 0$ and $\Sigma P_g H_g = 0$ if and only if $H_g = 0$ for each set S_g. This means entropy equals between-group entropy if and only if the grouping is such that there is at most one event with nonzero probability.

In informational terms, the decomposition theorem has the following interpretation. Consider the first signal arrives that one of the sets of events occurred. Its expected information content is H_0. Now consider the subsequent message arrives saying that one of the events falling under this set has occurred. Its expected information content is H_g. The total information content becomes $H_0 + \Sigma P_g H_g$.

9.3.2.3 Multidimensional Extensions

Consider a pair of events (Xi, Yj) and the probability of co-occurrence of both events. The probabilities of the two marginal contributions are

$$p_{i.} = \sum_{j=1}^{n} p_{ij} \quad (i = 1,\ldots,m) \tag{9.20}$$

$$p_{.j} = \sum_{i=1}^{m} p_{ij} \quad (j = 1,\ldots,n) \tag{9.21}$$

Now, the marginal entropy values are given by

$$H(X) = \sum_{i=1}^{m} p_{i.} \log_2\left(\frac{1}{p_{i.}}\right) \tag{9.22}$$

$$H(Y) = \sum_{j=1}^{n} p_{.j} \log_2\left(\frac{1}{p_{.j}}\right) \tag{9.23}$$

and two-dimensional entropy is expressed as

$$H(X,Y) = \sum_{i=1}^{m} \sum_{j=1}^{n} p_{ij} \log_2\left(\frac{1}{p_{ij}}\right) \tag{9.24}$$

The conditional entropy value measures the amount of uncertainty in one dimension (e.g., X); this remains even when we know event Yj has occurred, and is given by (Theil, 1972, pp. 116–117)

$$H_{Y_j}(X) = \sum_{i=1}^{m} \frac{p_{ij}}{p_{.j}} \log_2\left(\frac{p_{.j}}{p_{ij}}\right) \tag{9.25}$$

$$H_{X_i}(Y) = \sum_{j=1}^{n} \frac{p_{ij}}{p_{i.}} \log_2 \left(\frac{p_{i.}}{p_{ij}} \right) \tag{9.26}$$

The average conditional entropy is derived as the weighted average of conditional entropies and is expressed as

$$H_Y(X) = \sum_{j=1}^{n} p_{.j} H_{Y_j}(X) = \sum_{i=1}^{m} \sum_{j=1}^{n} p_{ij} \log_2 \left(\frac{p_{.j}}{p_{ij}} \right) \tag{9.27}$$

$$H_X(Y) = \sum_{i=1}^{m} p_{i.} H_{X_i}(Y) = \sum_{i=1}^{m} \sum_{j=1}^{n} p_{ij} \log_2 \left(\frac{p_{i.}}{p_{ij}} \right) \tag{9.28}$$

To sum up, the average conditional entropy is never greater than the unconditional entropy, that is, $H_X(Y) \leq H(Y)$ and $H_Y(X) \leq H(X)$, and the average conditional entropy and the unconditional entropy are equal if and only if the two events are stochastically independent (Theil, 1972, pp. 118–119).

The expected mutual information is a measure of dependence between two dimensions, that is, to what extent events tend to co-occur in particular combinations. In this respect, it is comparable with the product–moment correlation coefficient in the way entropy is comparable to the variance. Mutual information is given by

$$J(X,Y) = \sum_{i=1}^{m} \sum_{j=1}^{n} p_{ij} \log_2 \left(\frac{p_{ij}}{p_{i.} \cdot p_j} \right) \tag{9.29}$$

sometimes also denoted by $M(X,Y)$ or $T(X,Y)$. It can be shown that $J(X,Y) \geq 0$ and $J(X,Y) = H(Y) - H_X(Y)$ and $J(X,Y) = H(X) - H_Y(X)$ (Theil, 1972, pp. 125–131).

It can be further derived that the multidimensional entropy equals the sum of marginal entropies minus the mutual information (Theil, 1972, p. 126):

$$H(X, Y) = H(X) + H(Y) - J(X, Y) \tag{9.30}$$

The interpretation is that when mutual information is absent, marginal distributions are independent, and their entropies add up to the total entropy. When mutual information is positive, marginal distributions are dependent as some combinations occur relatively more often than other combinations, and marginal entropies exceed total entropy by an amount equal to the mutual information (Frenken, 2013).

9.3.3 Applications

Applications of entropy statistics were developed in the late 1960s and the 1970s. Today, tools of entropy statistics are applied in empirical research in industrial organization, regional science, economics of innovation, economics of inequality, and organization theory (Frenken, 2013).

9.4 Detection of Alterations in Information Content

In our technological era, communication plays a major role in everyday life. Methods of communication are changing and developing at an accelerating pace. We have moved from more primitive forms of communication, such as the smoke signals of Native Americans, to handwritten letters, to today's SMS, MMS, and emails (Ercan, 2006).

Communication became faster and easier after the invention of radio, TV, and other mass-media devices. In this book, we mostly consider the exchange of information stored in computers, an exchange based on information theory.

Information theory dates back to the seminal work of Claude Elwood Shannon in 1948. It enables the storage of information in computers and their transfer in a faster and easier manner than previously. The theory is mostly based on mathematics but includes fundamental concepts of source coding and channel coding. But before dealing with those topics, we need to understand some basics concepts of physics on which the mathematical model of the theory is constructed. Let us begin by considering what information is and whether it is physical.

9.4.1 What Is the Information Content of an Object?

Say we are holding an object, cards, geometric shapes, or a complex molecule. We ask the following question: what is the information content of this object? To answer this question, we introduce another party, say a friend, who shares some background knowledge (e.g., the same language or other sets of prior agreements that make communication possible), but who does not know the state of the object. We define the information content of the object as the size of the set of instructions that our friend requires to be able to reconstruct the object or its state (Plenio and Vitelli, 2001) (Figure 9.18).

An example of a decision tree is shown on the right side of Figure 9.18. Two binary choices have to be made to identify the shape (triangle or square) and the orientation (horizontal or rotated). In sending with equal probability one of the four objects, we transmit 2 bits of information (Plenio and Vitelli, 2001).

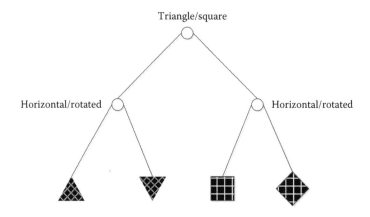

Triangle/square

Horizontal/rotated   Horizontal/rotated

FIGURE 9.18
Information content of an object. (From Plenio, M. and Vitelli, V., 2001. *Contemporary Physics*, 42(1), 25–60.)

The information content of an object can easily be obtained by counting the number of binary choices. In classical information theory, a variable that can assume only the values 0 or 1 is called a bit. Instructions to make a binary choice can be given by transmitting 1 to suggest one of the alternatives and 0 for the other. To sum up, we say that n bits of information can be encoded in a system when instructions in the form of n binary choices need to be transmitted to identify or recreate the state of the system (Plenio and Vitelli, 2001).

9.4.1.1 Is Information Physical?

In digital computers, the voltage between the plates in a capacitor represents a bit of information: a charged capacitor denotes a bit value of 1, whereas an uncharged capacitor shows a bit value of 0. One bit of information can also be encoded using two different polarizations of light or two different electronic states of an atom (Steane, 1998).

For example, let us consider the analysis of a 1-GB hard disk drive. At the core of this device is a metal disk coated with a film of magnetic material. Like any macroscopic object, the disk has many degrees of freedom. Of these, a tiny fraction, about $8 * 10^9$ (corresponding to 1 GB), represent information-bearing degrees of freedom that can be read or modified. The information-bearing degrees of freedom are collective variables, as many electrons in a specific region on the disk can be magnetized. Reading (writing) is done by a head riding over the disk's surface measuring (changing) the magnetization. The information-bearing degrees of freedom contribute to the algorithmic entropy in the same fashion as all other degrees of freedom (Machta, 1999).

Information is stored by encoding it in the physical systems. The laws of classical mechanics, electrodynamics, or quantum mechanics indicate the properties of these systems and limit our capabilities for information processing. These rather obvious-looking statements, however, have significant implications for our understanding of the concept of information, as they emphasize that the theory of information is not purely a mathematical concept, but the properties of its basic units are dictated by the laws of physics.

9.4.1.2 Shannon and the Complexity of the Data

Shannon's definition of information can also be thought of as complexity. Intuitively, we think a system is more complex when there are more numbers of possible configurations. A mechanical system with two parts where each part can be adjusted to six different settings will have 36 different configurations. A biological system such as a cell with an enormous number of parts (organelles, molecules) will have an exponentially larger amount of configurations.

Let us take a simple system of a two-sided coin. The outcome of flipping an unbiased coin is either a head or a tail. There are two possible configurations of this system. On the other hand, if we toss a coin where both sides are heads, we will always get head as the outcome. There is only one configuration of this two-headed coin system. Tossing this two-headed coin will tell us nothing new because we know it will always be heads.

The information content of a system is quantified by the complexity or entropy of the system. The more entropic a system is, the more possibilities there are for the outcome. More possibilities for the outcome make a message more significant and contain more information. Conversely, a less entropic system where there are fewer variations and fewer possibilities makes the message less significant and contain less information.

It is important to note that information as defined by Shannon is independent of the meaning or interpretation of the message. Two pieces of data describing complete disparate ideas can have the same quantity of information content (Kao, 2012).

9.4.2 Entropy as a Measure of Information Integrity

An issue that arose in the first explorations of multiclustering is that the cut-plot gives less-than-detailed information about the size of clusters forming as the cut-value is increased. The entropy plot was developed to address this limitation. In brief, the entropy-plot replaces integer-valued jumps that appear each time a cluster is divided with a real-valued jump that reflects the information present in the cluster division. The entropy-plot is based on the information-theoretic entropy of the distribution of cluster sizes. Entropy is a standard method of measuring the information present in a probability distribution. In order to compute the entropy of a set of cluster sizes, the sizes are normalized by dividing by the total number of points. This gives a probability distribution that measures the empirical probability P_i of a point belonging to a given cluster. The entropy used in the entropy-plot is the entropy of this distribution, given by Equation 9.10 as

$$E = \sum_i p_i \log(p_i)$$

Entropy is a measure of information content, so the entropy-plot displays how the information in the cluster structure changes as new clusters form. A simple example of the value of entropy comes from recording many experiments in which a coin is flipped. In this case, the coin has probability p of heads and probability $1 - p$ of tails. In this situation, simplify Equation 9.10 to Equation 9.11. Note that the maximum information content for a coin is 1 bit and this occurs for a fair coin, one with a 50/50 chance of producing a head or tail. This type of coin corresponds, in an information-theoretic sense, to dividing a cluster in half and can be expressed as

$$E = -(p\log(p) + (1 - p)\log(1 - p))$$

Entropy measures the information content of a coin with probability p of heads in the following sense. A fair coin, flipped many times, will generate a random string without bias or pattern. One bit of information is needed to report a flip of this fair coin and compression of the pattern of results produced by this coin is negligible. A coin with a high probability of heads will generate a string that is mostly heads. This string has low-information content and so is easy to compress. In an information-theoretic sense, the flips of the biased coin contain far less information than those of the fair coin. Likewise, even division of a large cluster is a more informative event than uneven division. It is worth making the notion of information content precise. The entropy of the coin, given in Equation 9.11, measures the number of bits required to store the outcomes of flipping the coin (so long as we are storing many flips of the coin). If a coin with probability p of heads is flipped many times, Equation 9.11 gives a close estimate of how much a long string of flips can be compressed (Figure 9.19).

We return now to the entropy-plot. If, as we change the cut-value, a cluster divides in half, this creates the maximum number of new relationships between data points and, hence, the greatest possible increase in information present in the cluster structure of the data. If, on the other hand, a single point splits off of a cluster, this represents the smallest possible change in the information present in the cluster structure. Using the entropy-plot permits the user to see the relative importance of different cluster divisions, which actually looked the same in the original cut-plot. As a convenience to the user, we make the process of using the entropy-plot automatic in one simple way by declining to report tiny clusters. Such clusters are often artifactual and unimportant, cluttering the output with insignificant information (BioMed Central, 2009).

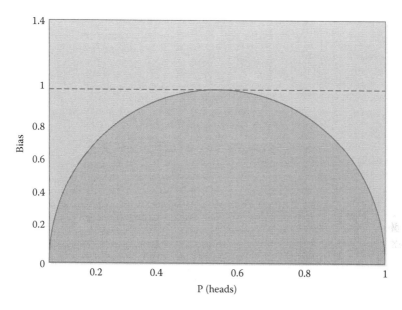

FIGURE 9.19

Entropy of a coin as a function of the coin's bias. (From BioMed Central, 2009. *Supplementary Material.* [Online] Available at: http://www.biomedcentral.com/content/supplementary/1471-2105-10-260-s1.doc_[Använd 2014].)

9.5 Advantages of Information Theory as an Unsupervised System

Information theory plays an important role in the study of learning systems. Just as information theory deals with quantifying information regardless of its physical medium of transmission, learning theory deals with understanding systems that learn irrespective of whether they are biological or artificial.

Learning systems can be broadly categorized by the amount of information they receive from the environment in their supervision signal. In *unsupervised learning*, the goal of the system is to learn from sensory data with no supervision. This can be achieved by casting the unsupervised learning problem as one of discovering a code for the system's sensory data; this code should be as efficient as possible. Thus, the family of concepts—entropy, Kolmogorov complexity, and the general notion of description length—can be used to formalize unsupervised learning problems.

We know from the source coding theorem that the most efficient code for a data source uses $-\log_2 p(x)$ bits per symbol x. Therefore, discovering the optimal coding scheme for a set of sensory data is equivalent to the problem of learning what the true probability distribution $p(x)$ of the data is. If at some stage we have an estimate $q(x)$ of this distribution, we can use this estimate instead of the true probabilities to code the data. However, we incur a loss in efficiency measured by the *relative entropy* between the two probability distributions p and q

$$D(p \parallel q) = \sum_{\text{all } x} p(x) \log_2 \frac{p(x)}{q(x)},$$

also known as the Kullback–Leibler divergence. This measure is the inefficiency in bits of coding messages with respect to a probability distribution q instead of the true probability distribution p, and is zero if and only if $p = q$. Many unsupervised learning systems can be designed from the principle of minimizing this relative entropy (Ghahramani, 2006).

9.5.1 Information and Learning Process

Learning is a broad topic in AI. What does it mean for a machine to learn? Machines do not *understand* programs, data, or knowledge; we simply program them to process data and apply knowledge. For a machine to learn, this does not imply the machine is physically learning something new or different. Rather, through the learning algorithm(s), it is refining its problem-solving ability (its program and/or its stored knowledge) so that it is more capable of solving that particular class of problem in the future.

References

Abbasion, S., Rafsanjani, A., Farshidianfar, A., and Irani, N., 2007. Rolling element bearings multi-fault classification based on the wavelet denoising and support vector machine. *Mechanical Systems and Signal Processing*, 21, 2933–2945.

Abele, E., Altintas, Y., and Brecher, C., 2010. Machine tool spindle units. *CIRP Annals—Manufacturing Technology*, 59(2), 781–802.

Aha, D. W., Kibler, D., and Albert, M. K., 1991. Instance based learning algorithms. *Machine Learning*, 6, 37–66.

Alzaga, A., Konde, E., Bravo, I., Arana, R., Prado, A., Yurre, C., Monnin, M., Medina-Oliva, G. 2014. New technologies to optimize reliability, operation and maintenance in the use of Machine-Tools. *Euro-Maintenance Conference*, 5–8 May 2014, Helsinki, Finland.

Baeyer, V. H. C., 2003. *Information: The New Language of Science*. Harvard University Press, London: Weidenfeld and Nicolson.

Bar-Hillel, M., 1980. The base-rate fallacy in probability judgments. *Acta Psychologica*, 44(3), 211–233.

Barwise, J. and Perry, J., 1983. *Situations and Attitudes*. Cambridge, MA: MIT Press.

Bateson, G., 1972. *Steps to an Ecology of Mind: Collected Essays in Anthropology, Psychiatry, Evolution, and Epistemology.* Chicago: University of Chicago Press.

Bateson, G., 1973. *Steps to an Ecology of Mind.* Frogmore, St. Albans: Paladin.

Benbouzid, E. H. M., 2000. A review of induction motors signature analysis as a medium for faults detection. *IEEE Transactions on Industrial Electronics,* 47(5), 984–993.

Benthem, V., 2005. Guards, bounds and generalized semantics. *Journal of Logic, Language and Information,* 14, 263–279.

Bickhard, M. H. and Terveen, L., 1995. *Foundational Issues in Artificial Intelligence and Cognitive Science—Impasse and Solution.* Vol. 109. Amsterdam: Elsevier Scientific.

BioMed Central, 2009. *Supplementary Material.* [Online] Available at: http://www.biomedcentral.com/content/supplementary/1471-2105-10-260-s1.doc_ [Använd 2014].

Boltzmann, L., 1877. Ueber die Beziehung eines allgemeine mechanischen Satzes zum zweiten Hauptsatzes der Warmetheorie. *Akademie der Wissenschaften in Wien, Sitzungsberichte, Mathematisch-naturwissenschaftliche,* 75, 67–73.

Butler, S., 2012. *Prognostic Algorithms for Condition Monitoring and Remaining Useful Life Estimation.* Doctoral dissertation. Maynooth: National University of Ireland.

Capurro, R., 2002. *Philosophical Presuppositions of Producing and Patenting Organic Life.* [Online] Available at: http://www.capurro.de/patent.html.

Capurro, R., Fleissner, P. and Hofkirchner, W., 1999. Is a unified theory of information feasible? A trialogue. *World Futures: Journal of General Evolution,* 49(3-4), 213–234.

Capurro, R. and Hjørland, B., 2003. The concept of information. In: Cronin, B. (Ed.), *Annual Review of Information Science and Technology (ARIST).* Medford, NJ: Information Today, Inc.

Carnap, R. and Bar-Hillel, Y., 1964. An outline of semantic information. In: *Language and Information: Selected Essays on Their Theory and Application.* Reading, MA: Addison-Wesley.

Chandran, P., Lokesha, M., Majumder, M. C., and Raheem, K. F. A., 2012. Application of laplace wavelet Kurtosis and wavelet statistical parameters for gear fault diagnosis. *International Journal of Multidisciplinary Sciences and Engineering,* 3(9), 1–7.

Collier, J., 1990. Intrinsic information in Philip Hanson. In: Hanson, P. P. (Ed.), *Information Language and Cognition: Vancouver Studies in Cognitive Science.* Oxford, UK: University of Oxford Press.

Cusido, J., 2008. Fault detection in induction machines using power spectral density in wavelet decomposition. *IEEE Transactions on Industrial Electronics,* 55(2), 633–643.

Dash, M. and Liu, H., 1997. Feature selection for classification. *Intelligent Data Analysis* 1(3), 131–156.

Devijver, P. and Kittler, J., 1982. *Pattern Recognition: A Statistical Approach.* Prentice Hall.

Devlin, K., 1991. *Logic and Information.* Cambridge: Cambridge University Press.

Dodig-Crnkovic, G., 2005. Philosophy of information, A new renaissance and the discreet charm of the computational paradigm. In: Magnani, L. (Ed.), *Computing, Philosophy and Cognition.* London: King's College Publications.

Dodig-Crnkovic, G., 2006. *Investigations into Information Semantics and Ethics of Computing.* Västerås, Sweden: Mälarden University Press.

Dretske, F. I., 1981. *Knowledge and the Flow of Information.* Cambridge: MIT Press.

Ducker, C., 2009. *Force, Work & Energy with the Ropes and Pulleys*. [Online] Available at: http://www.cdrucker.com/files/labsphys/forceworkenergy.html.

Ercan, I., 2006. *An Introduction to the Information Theory and Shannon Theory*. [Online] Available at: http://www.physics.metu.edu.tr/~ozpineci/dersler/20052/phys430/presentations/PHYS_430_Rapor_ilke_Ercan.doc.

Finney, C. E. A., Green, J. B., and Daw, C. S. 1998. Symbolic time-series analysis of engine combustion measurements (No. 980624). SAE Technical Paper. Chicago.

Fisher, R. A., 1930. *The Genetic Theory of Natural Selection*. Oxford: Clarendon Press.

Floridi, L., 2004. Open problems in the philosophy of information. *Metaphilosophy*, 35(4), 554–582.

Floridi, L., 2005. *Semantic Conceptions of Information, The Stanford Encyclopedia of Philosophy (Winter Edition)*. [Online] Available at: http://plato.stanford.edu/entries/information-semantic.

Frenken, K., 2013. *Entropy Statistics and Information Theory*. [Online] Available at: http://econ.geo.uu.nl/frenken/Entropy_statistics_final.doc.

Frieden, B. R., 2014. *Science from Fisher Information: A Unification*, 2nd edition. Cambridge University Press.

Fuqing, Y., Kumar, U., and Galar, D., u.d. Fault diagnosis on time domain for rolling element bearings using support vector machine. *Reliability Engineering and System Safety*.

Galar, D., Kumar, U., Lee, J., and Zhao, W., 2012. Remaining useful life estimation using time trajectory tracking and support vector machines. *International Journal of COMADEM*, 15(3), 2–8.

Gamberger, D. and Lavrač, N. 1997. *Conditions for Occam's Razor Applicability and Noise Elimination*. Heidelberg, Berlin: Springer, pp. 108–123.

Gennari, J., Langley, P., and Fisher, D., 1989. Models of incremental concept formation. *Artificial Intelligence*, 40, 11–61.

Ghahramani, Z., 2006. *Information Theory. Encyclopedia of Cognitive Science*. New York: John Wiley & Sons, Ltd.

Ghiselli, E. E., 1964. *Theory of Psychological Measurement*. New York: McGraw-Hill.

Goertzel, B. 1994. Chaotic Logic: Language, Thought, and Reality from the Perspective of Complex Systems Science. New York: Plenum Press.

Hall, M. A. and Smith, L. A., 1999. Feature selection for machine learning: comparing a correlation-based filter approach to the wrapper. In: *FLAIRS Conference*, pp. 235–239.

Haugeland, J., 1995. *Artificial Intelligence: The Very Idea*. Cambridge, MA: MIT Press.

Heng, R. and Nor, M., 1998. Statistical analysis of sound and vibration signals for monitoring rolling element bearing condition. *Applied Acoustics*, 53, 211–226.

Heylighen, F., 2003. *Entropy Increase: The Box with Two Compartments*. [Online] Available at: http://pespmc1.vub.ac.be/ENTROBOX.html.

Hogarth, R. M., 1977. Methods for aggregating opinions. In: Jungermann, H. and De Zeeuw, G. (Eds.), *Decision Making and Change in Human Affairs*. Dordrecht-Holland: D. Reidel Publishing.

Holgate, J., 2002. *The Phantom of Information—Peripatetic thoughts on Information Language and Knowing*. [Online] Available at: http://www.mdpi.net/ec/papers/fis2002/183/phantom-holgate.htm.

Javorszky, K., 2003. Information processing in auto-regulated systems. *Entropy*, 5, 161–192.

Johansson, C.-A., 2014. *The Importance of Operation Context for Proper Remaining Useful Life Estimation.*

Kao, D., 2012. *Concept: Information Theory and Evolution.* [Online] Available at: http://blog.nextgenetics.net/?e=38.

Kliman, G. B. and Stein, J., 1992. Methods of motor current signature analysis. *Electric Machines and Power System*, 20(5), 463–474.

Kohavi, R., 1995. Wrappers for performance enhancement and oblivious decision graphs. Doctoral dissertation, Stanford University.

Kohavi, R. and John, G., 1996. Wrappers for feature subset selection. *Artificial Intelligence, Special Issue on Relevance*, 97(1–2), 273–324.

Kononenko, I., 1994. *Estimating Attributes: Analysis and Extensions of RELIEF.* pp. 171–182.

Landauer, R., 1996. The physical nature of information. *Physics Letters*, 217, 188.

Langley, P. and Sage, S., 1994a. Induction of selective Bayesian classifiers. In: *Proceedings of the Tenth Conference on Uncertainty in Artificial Intelligence.* Seattle, W.A: Morgan Kaufmann.

Langley, P. and Sage, S., 1994b. Oblivious decision trees and abstract cases. In: *Working Notes of the AAAI-94 Workshop on Case-Based Reasoning.* Seattle, W.A: AAAI Press.

Langley, P. and Sage, S. 1997. Scaling to domains with irrelevant features. *Computational Learning Theory and Natural Learning Systems*, 4, 51–63.

Lee, S.-K., 2011. *A Survey of Context-Aware System—The Past, Present, and Future.* [Online] Available at: ids.snu.ac.kr/w/images/d/d7/Survey(draft).doc.

Lee, S., Chang, J., and Lee, S.-G., 2011. Survey and trend analysis of context-aware systems. *Information-An International Interdisciplinary Journal*, 14(2), 527–548.

Leslie, A., 2011. *Entropy.* [Online] Available at: http://media.wix.com/ugd/b98542 _8a0b856e6e949ae49f8823e7a083a0af.docx?dn=Physics%20final%20paper.docx.

Leyton, M., 1992. *Symmetry Causality Mind.* Cambridge, MA: MIT Press.

Machlup, F., 1983. *The Study of Information: Interdisciplinary Messages.* New York: John Wiley & Sons.

Machta, J., 1999. *Entropy, Information, and Computation.* Amherst, Massachusetts.

Mahler, S., 1996. *Bringing Gender to a Transnational Focus: Theoretical and Empirical Ideas.* Burlington. University of Vermont (Unpublished Manuscript).

Marijuan, P. C., 2003. Foundations of information science: Selected papers from FIS 2002. *Entropy*, 5(2), 214–219.

McCarthy, J. and Buvac, S., 1997. *Formalizing Context (Expanded Notes).* [Online] Available at: http://cogprints.org/419/2/formalizing-context.ps.

Medina-Oliva, G., Weber, P., and Iung, B., 2013. PRM-based patterns for knowledge for-malisation of industrial systems to support maintenance strategies assessment. *Reliability Engineering & System Safety*, 116, 38–56.

Miller, A. J., 1990. *Subset Selection in Regression.* New York: Chapman and Hall.

Newell, A. and Simon, H. A., 1972. *Human Problem Solving.* Englewood Cliffs, NJ: Prentice Hall.

Peng, Z. K. and Chu, F. L., 2004. Application of the wavelet transform in machine condition monitoring and fault diagnostics: A review with bibliography. *Mechanical Systems and Signal Processing*, 18(2), 199–221.

Plenio, M. and Vitelli, V., 2001. The physics of forgetting: Landauer's erasure principle and information theory. *Contemporary Physics*, 42(1), 25–60.

Prigogine, I. and Stengers, I., 1984. *Order Out of Chaos: Man's New Dialogue with Nature.* London: Fontana.

Quinlan, J., 1986. Induction of decision trees. *Machine Learning*, 1, 81–106.

Quinlan, J., 1993. *C4.5 Programs for Machine Learning.* Los Altos, California: Morgan Kaufmann.

Salber, D., Dey, A. K., and Abowd, G. D., 1999. *The Context Toolkit: Aiding the Development of Context-Aware Applications.*

Sankararaman, S. and Goebel, K., 2013. *Why Is the Remaining Useful Life Prediction Uncertain?* New Orleans, Texas, USA.

Schilit, B., Adams, N., and Want, R., 1994. *Context-Aware Computing Applications.* pp. 85–90.

Shannon, C., 1948. The mathematical theory of communication. *Bell System Technical Journal*, 27, 323–332; 379–423.

Smoli, L., 2007. *The Trouble with Physics: The Rise of String Theory, The Fall of a Science, and What Comes Next.* Mariner Books.

Sreejith, B., Verma, A., and Srividya, A., 2008. *Fault Diagnosis of Rolling Element Bearing Using Time-Domain Features and Neural Network.* Kharagpur, India.

Steane, A., 1998. Quantum computing. *Reports on Progress in Physics*, 61, 117–173.

Stonier, T., 1997. *Information and Meaning. An Evolutionary Perspective.* Berlin, New York: Springer.

Subasi, A., 2007. EEG signal classification using wavelet feature extraction and a mixture of expert model. *Expert Systems with Applications*, 32(4), 1084–1093.

Tang, X., 1995. Symbol sequence statistics in noisy chaotic signal reconstruction. *Physical Review E*, 1995(51), 3871–3889.

Tato, R., Santos, R., Kompe, R., and Pardo, J. M., 2002. *Emotion Recognition in Speech Signal.* Sony, Interspeech.

Theil, H., 1967. *Economics and Information Theory.* Amsterdam: North-Holland.

Theil, H., 1972. *Statistical Decomposition Analysis.* Amsterdam: North-Holland.

Van, B. J. and Adriaans, P., 2006. Handbook on the Philosophy of Information. In: Gabbay, D. M., Thagard, P., and Woods, J. (Eds.), *Handbook of the Philosophy of Science.* Amsterdam: North Holland.

Vass, J., Šmíd, R., Randall, R. B., Sovka, P., Cristalli, C., and Torcianti, B., 2008. Avoidance of speckle noise in laser vibrometry by the use of kurtosis ratio: Application to mechanical fault diagnostics. *Mechanical Systems and Signal Processing*, 22(3), 647–671.

Vedral, V., 2012. *Decoding Reality: The Universe as Quantum Information.* Oxford University Press.

Walker, J. S., 2009. *Physics*, 4th edition. Pearson College Division.

Wang, W., Ismail, F., and Golnaraghi, M., 2001. Assessment of gear damage monitoring techniques using vibration measurements. *Mechanical Systems Signal Processing*, 15, 905–922.

Wiener, N., 1948. *Cybernetics.* New York: J. Wiley.

Zadjenweber, D., 1972. Une application de la théorie de l'information à l'économie: la mesure de la concentration. *Revue d'Economie Politique*, 82, 486–510.

Zajonic, R. B., 1962. A note on group judgements and group size. *Human Relations*, 15, 177–180.

10

Uncertainty Management

10.1 Classical Logic and Fuzzy Logic

In our daily lives, we can usually give reasons for the things we do. Moreover, when we want others to do something, we give them reasons.

This may be considered using logic.

Every day, we have options, from what to wear, to what to eat, to what bus to take to work, and so on. We must choose between options and decide what is best for us, based on the situation. Here logic, or reasoning, defines what is the best among all our possible options.

Other people may try to convince to think a certain way by giving us reasons. Logic allows us to distinguish between valid and invalid arguments. From several premises offered to us, we ultimately reach a single conclusion. To reach the conclusion, we must accept one of the premises using logic, a process that may be intuitive.

10.1.1 Classical Logic

Classical logic is the simplest of all major logics. In classical logic, we can only find two truth values for any proposition or statement (Khaliq and Ahmad, 2010). These are:

1. True (1, yes)
2. False (0, no)

A proposition can be true or false but it cannot be both simultaneously.
For example,
"The sun rises in the east" is a true statement and has truth value 1. Normally, we use 1s and 0s in mathematical classical logic and true and false in propositional classical logic (Restall, 2006).

10.1.1.1 Basic Operations on Classical Truth Values

Operations in propositional classical logic can be described in terms of tables of 0s and 1s and are called truth tables. Truth tables for classical logic are based on the following basic operations:

1. $\neg p = 1 - p$
2. $p \vee q = \max(p, q)$
3. $p \wedge q = \min(p, q)$
4. $p \rightarrow q = \min\{1, 1 - p + q\}$

where p and q are two propositions. Their truthfulness is used as inputs in Table 10.1.

10.1.1.2 Example of Basic Classical Logical Operations

The following are four fuzzy operations that are significant for the example presented in this book:

p = "We are sitting in a restaurant."
q = "We are drinking tea."

Now, according to the above operations and truth table, we have

1. $\neg p$ = "We are not sitting in a restaurant."
   $\neg q$ = "We are not drinking tea."
2. $p \vee q$ = "We are sitting in a restaurant or we are drinking tea."
   This means the compound statement is true when one of p and q is true or both p and q are true.
3. $p \wedge q$ = "We are sitting in a restaurant and we are drinking tea."
   This means the compound statement is true only when both p and q are true.
4. $p \rightarrow q$ = "If we are sitting in a restaurant, then we are drinking tea."

TABLE 10.1

Truth Table

p	q	$\neg p$	$p \wedge q$	$p \vee q$	$p \rightarrow q$
0	0	1	0	0	1
0	1	1	1	0	1
1	0	0	1	0	0
1	1	0	1	1	1

Source: Data from Khaliq, A. and Ahmad, A., 2010. *Fuzzy Logic and Approximate Reasoning*, Master's Thesis, Blekinge Institute of Technology, Karlskrona.

This means the compound statement will be incorrect when the true statement implies the wrong one. We can also say p is a sufficient condition for q, or alternatively, q is a necessary condition for p.

The implication should clarify a relationship between the premise and the conclusion, as for example, the following: if I fall into the lake, then I will get wet. We also use implication in theorems, as for example: if ABC is the correct triangle with the right angle at B, then $AC^2 = AB^2 + BC^2$ (Khaliq and Ahmad, 2010).

10.1.2 Fuzzy Logic

During the past few years, the use of fuzzy logic has been introduced widely in the area of control systems. It has received major acceptance in Japan, and the Japanese are moving from a theoretical point of view into technological realization. Today, we can find a huge variety of products that use fuzzy logic-based control systems or simple fuzzy logic controllers.

The key reason for the success of fuzzy logic controllers is their ability to cope with knowledge presented in a linguistic form. For the control engineers, knowledge has traditionally been represented by conventional mathematical frameworks in their designs.

But fuzzy logic controllers can add experience, intuition, and heuristics into a system instead of relying on mathematical models, making them more effective in applications where the existing models are poorly defined and not reliable enough (Cristea et al., 2002).

10.1.2.1 Historical Review

The term "fuzzy" was first applied to logic in 1965 by Professor Lofti Zadeh, Chair of University of California, Berkeley's Electrical Engineering Department. He used the term to describe multivalued sets in the seminal paper, "Fuzzy Sets" (Zadeh, 1965). The work in his paper is derived from multivalued logic, a concept that emerged in the 1920s to deal with Heisenberg's Uncertainty Principle in quantum mechanics. Multivalued logic was further developed by distinguished logicians such as Jan Lukasiewicz, Bertrand Russell, and Max Black. At the time, multivalence was usually described by the term "vagueness."

Zadeh applied Lukasiewicz's multivalued logic to set theory and created what he called fuzzy sets—sets whose elements belong to it in different degrees (Zadeh, 1973). The fuzzy principle says that "everything is a matter of degree." Conventional logic is bivalence (TRUE or FALSE, 1 or 0), but fuzzy logic is multivalence (from 0 through 1). It represents a shift from conventional mathematics to philosophy and language. At the start, fuzzy logic was a theoretical concept with little practical application. Zadeh was mainly involved in computer simulations of mathematical ideas. In the 1970s, Professor Edrahim Mamdani of Queen Mary College, London, built the first fuzzy system, a steam-engine controller, and he later designed the first fuzzy traffic lights. His work led to an extensive development of fuzzy control applications and products (Cristea et al., 2002).

10.1.2.2 Fuzzy Sets and Fuzzy Logic

Classical set theory was formulated by German mathematician Georg Cantor (1845–1918). The theory defines a universe of discourse, U, as a collection of objects with the same characteristics. A classical set is a collection of some of those elements; moreover, the classical set's member elements belong to the set 100%. Those elements in the universe of discourse which are nonmember elements of the set are, in fact, not related to the set at all. Therefore, we can draw a definitive boundary for the set, as shown in Figure 10.1.

A classical set may be denoted by $A = \{x \in U | P(x)\}$ where the elements of A have the property P, and U is the universe of discourse. The characteristic function $\mu A(x):U \rightarrow \{0, 1\}$ is defined as 0 if x is not an element of A and as 1 if x is an element of A. Here, U contains two elements: 1 and 0. Hence, an element x, in the universe of discourse is either a member of set A or not a member of set A. Membership has a certain ambiguity, however. For example, consider a set ADULT, which contains elements classified by the variable AGE. An element with AGE = 5 would not be a member of the set whereas an element with AGE = 45 would be. That is all very well and good, but where can a sharp and discrete line be drawn to separate members from nonmembers? Will we pick AGE = 18? If we do this, then elements with AGE = 17.9 are not members of the set ADULT but those with AGE = 18.1 are. Obviously the system cannot realistically model the definition of an adult human. This kind of simple problem embodies the notion behind Zadeh's Principle of Incompatibility (Cristea et al., 2002).

10.1.3 Fuzzy Set Concept

The difference between crisp, or classical, and fuzzy sets can be established by introducing a membership function.

(a)

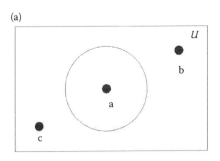

(b)

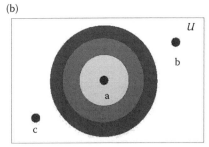

FIGURE 10.1
(a) Classical/crisp set boundary; and (b) fuzzy set boundary.

Let us begin with a finite set $X = \{x_1, x_2,..., x_n\}$ that will be considered the universal set in the following explanation. The subset A of X consisting of the single element x_1 can be described by the n-dimensional membership vector $Z(A) = (1, 0, 0,..., 0)$, where a convention has been adopted whereby a 1 at the ith position indicates x_i belongs to A. The set B, composed of the elements x_1 and x_n, is described by the vector $Z(B) = (1, 0, 0,..., 1)$. Any other crisp (i.e., classical) subset of X can be represented in the same way by an n-dimensional binary vector. Will this change if we lift the restriction on binary vectors? In that case, we can define the fuzzy set C with the following vector description.

In classical set theory, such a set cannot be defined. An element either does or does not belong to a subset. In the theory of fuzzy sets, we make a generalization and then accept descriptions of this type. In our example, the element x_1 belongs to the set C only to a certain extent. The degree of membership is expressed by a real number in the interval $[0, 1]$, here 0.5. This interpretation of the degree of membership is much like the meaning we assign to statements such as "person x_1 is an adult." It is obviously not possible to define an exact age representing the absolute threshold to enter adulthood. The act of becoming mature is more likely to be interpreted as a continuous process where the membership of a person in the set of adults goes slowly from 0 through 1.

Other examples of such diffuse statements abound. For example, the concepts "old" and "young" or "fast" and "slow" are imprecise but easy to interpret in a given context. In some applications, such as expert systems, it is necessary to introduce formal methods capable of dealing with such imprecise expressions to permit a computer using rigid Boolean logic to process them. This is where the theory of fuzzy sets and fuzzy logic comes in. Figure 10.2 shows three examples of a membership function in the interval 0–70 years. These three define the degree of membership of any given age in one of the three sets: young, adult, and old age. In this example, if a person is 20 years old, his/her degree of membership in the set of young persons is 1.0, in the set of adults 0.35, and in the set of old persons 0.0. If a person is 50 years old, the degrees of membership are 0.0, 1.0, and 0.3 in the respective sets (Rojas, 1996).

Definition 10.1

Let X be a classical universal set. A real function $\mu A: X \rightarrow [0, 1]$ is termed the membership function of A. It defines the fuzzy set A of X. This is the set of all pairs $(x, \mu A(x))$ with $x \in X$.

A fuzzy set is completely determined by its membership function. *Note:* The above definition covers the case where X is not a finite set.

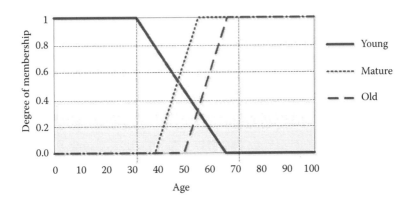

FIGURE 10.2
Membership functions for the concepts young, mature, and old. (Redrawn from Rojas, R., 1996. *Neural Networks: A Systematic Introduction.* Berlin, Heidelberg: Springer-Verlag.)

The set supporting a fuzzy set A is the set of all elements x of X for which $(x, \mu A(x)) \in A$ and $\mu A(x) > 0$ holds. A fuzzy set A with the finite set of support $\{a_1, a_2,..., a_m\}$ can be described as

$$A = \mu_1/a_1 + \mu_2/a_2 + \cdots + \mu_m/a_m$$

where $\mu i = \mu A(ai)$ for $i = 1,..., m$. *Note:* The symbols "/" and " + " are used only as syntactical constructors.

Crisp sets are a special case of fuzzy sets, as the range of the function is restricted to the values 0 and 1. But operations defined over crisp sets, such as union or intersection, can be generalized to cover fuzzy sets.

Assume $X = \{x_1, x_2, x_3\}$. In the classical subsets $A = \{x_1, x_2\}$ and $B = \{x_2, x_3\}$, the union of A and B is computed by taking for each element xi, the maximum of its membership in both sets

$$A \cup B = 1/x_1 + 1/x_2 + 1/x_3$$

The fuzzy union of two fuzzy sets can be computed in the same way. Accordingly, the union of the two fuzzy sets is

$$C \cup D = 0.7/x_1 + 0.6/x_2 + 0.8/x_3$$

The fuzzy intersection of two sets A and B can be defined in a similar way, but instead of taking the maximum, we can compute the minimum of the membership of each element xi to A and B. The maximum or minimum of the membership values represent only one pair of possible definitions of union and intersection operations for fuzzy sets. As we go on to show, there are alternative definitions (Rojas, 1996).

10.1.3.1 Fuzzy Logic

Logical connectives are also defined for fuzzy logic operations. They are closely related to Zadeh's definitions of fuzzy set operations. The following are four fuzzy operations that are significant for the second example presented here. R denotes the relation between the fuzzy sets A and B.

Negation: $\mu A(x) = 1 - \mu A(x)$
Disjunction: $R: A$ OR B $\mu_R(x) = \max[\mu_A(x), \mu_B(x)]$
Conjunction: $R: A$ AND B $\mu_R(x) = \min[\mu_A(x), \mu_B(x)]$
Implication: $R: (x = A) \rightarrow (y = B)$ IF x is A THEN y is B

Fuzzy implication is an important connective in fuzzy control systems because the control strategies are embodied by sets of IF-THEN rules. There are various techniques involving fuzzy implication. These relationships are mostly derived from multivalued logic theory. The following are some of the common techniques of fuzzy implication found in the literature:

Zadeh's classical implication: $\mu_R(x, y) = \max\{\min[\mu_A(x), \mu_B(y)], 1 - \mu_A(x)\}$
Mamdani's implication: $\mu_R(x, y) = \min[\mu_A(x), \mu_B(y)]$

Note that Mamdani's implication is equivalent to Zadeh's classical implication when $\mu_A(x) \geq 0.5$ and $\mu_B(y) \geq 0.5$.

$$\text{Godel's implication :} \begin{cases} 1 & \mu_A(x) \leq \mu_B(y) \\ \mu_B(y) & \text{otherwise} \end{cases}$$

Lukasiewicz' implication: $\mu R(x, y) = \min\{1, [1 - \mu A(x) + \mu B(y)]\}$

10.1.3.2 Control with Fuzzy Logic

A fuzzy controller is a regulating system whose modus operandi is specified with fuzzy rules, but it generally uses a small set of rules. The measurements are processed in their fuzzified form, fuzzy inferences are computed, and the result is defuzzified, that is, it is transformed back into a specific number (Cristea et al., 2002).

10.1.3.3 Fuzzy Controllers

The example of the electrical heater will be explained in this section. We must first determine the domain of variable definitions used in the problem. Assume the room temperature is a number between 0°C and 40°C. The controller can vary the electrical power consumed between 0 and 100 (in some suitable units), whereby 50 is the normal standby value.

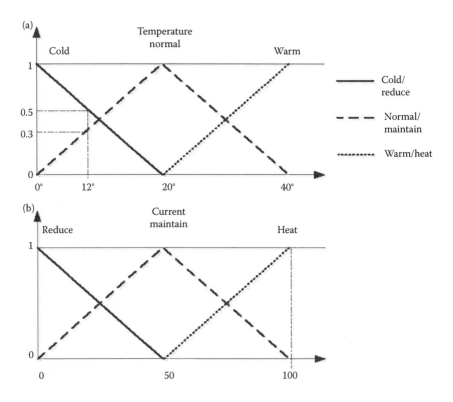

FIGURE 10.3
Membership functions for temperature (a) and electric current (b) categories in cold, normal, and warm conditions. (Redrawn from Rojas, R., 1996. *Neural Networks: A Systematic Introduction.* Berlin, Heidelberg: Springer-Verlag.)

Figure 10.3 shows the membership functions for the temperature categories "cold," "normal," and "warm" and the control categories "reduce," "maintain," and "heat."

A temperature of 12°C corresponds to the fuzzy number T = cold/0.5 + normal/0.3 + warm/0.0, values leading to the previously computed inference action = heat/0.5 + maintain/0.3 + reduce/0.0. The controller must now transform these fuzzy inference results into a definite value. In the first step, the controller calculates the surfaces of the membership triangles below the inferred degree of membership. The action "heat," which is valid to 50%, appears as the lighter surface in Figure 10.4. The darker region represents "maintain," valid to 30%. The centroid of the two shaded regions is somewhere around 70, so this value for the power consumption value is selected by the controller to heat the room.

Of course, we can formulate more complex rules using more than two variables, but in all instances, we must evaluate all rules simultaneously (Rojas, 1996).

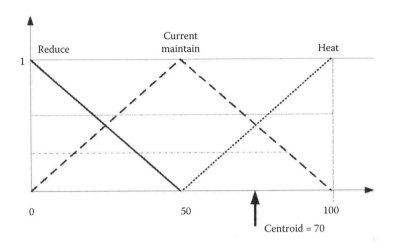

FIGURE 10.4
Centroid computation. (Redrawn from Rojas, R., 1996. *Neural Networks: A Systematic Introduction.* Berlin, Heidelberg: Springer-Verlag.)

10.1.4 Fuzzy Networks

We can represent fuzzy systems as networks with the computing imple-menting fuzzy operators. Figure 10.5 depicts a network with four hidden units. Each receives the inputs x_1, x_2, and x_3, with each of these correspond-ing to the fuzzy categorization of a specific number. The fuzzy operators are evaluated in parallel in the hidden layer of the network, with the latter cor-responding to the set of inference rules. The defuzzifier, the final part of the network, transforms the fuzzy inferences into a specific control variable. To

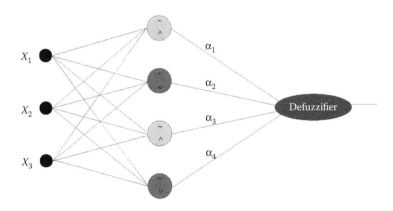

FIGURE 10.5
Example of a fuzzy network. (Redrawn from Rojas, R., 1996. *Neural Networks: A Systematic Introduction.* Berlin, Heidelberg: Springer-Verlag.)

assign importance, each fuzzy inference rule is weighted by the numbers α_1, α_2, and α_3 as in weighted centroid computations.

Obviously, if we implement more complex rules, we will have networks with several layers, but fuzzy systems do not usually have deep networks. As every fuzzy inference step reduces the precision of the conclusion, building a long inference chain may not advisable.

Fuzzy operators cannot be computed exactly by sigmoidal units, but a relatively good approximation is possible for some. For example, a fuzzy inference chain using the bounded sum or bounded difference can be approximated by a neural network.

Standard units can be used to approximate the defuzzifier operator in the last layer. If the membership functions are triangles, their surface grows quadratically with the height. A quadratic function of this form can be approximated in the relevant interval using sigmoids, using a learning algorithm to help set the parameters of the approximation (Rojas, 1996).

10.1.4.1 Function Approximation with Fuzzy Methods

A fuzzy controller is simply a system to rapidly compute an approximation of a coarsely defined control surface; an example is shown in Figure 10.6.

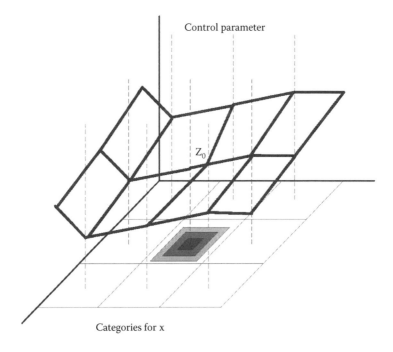

FIGURE 10.6
Approximation of a control surface. (Redrawn from Rojas, R., 1996. *Neural Networks: A Systematic Introduction.* Berlin, Heidelberg: Springer-Verlag.)

The fuzzy controller computes a control variable according to the values of the variables x and y, with both variables transformed into fuzzy categories. Assume each variable is transformed into a combination of three categories, with nine different combinations of the categories for x and y. The value of the control variable is defined for each of these nine combinations, thus fixing nine points of the control surface.

Arbitrary values of x and y belong to the nine combined categories to different degrees. Therefore, for arbitrary combinations of x and y, an interpolation of the known function values of the control variable is required. The computation is done by a fuzzy controller according to the degree of membership of (x, y) in each combined category. The various shadings of the quadratic regions in the xy plane in Figure 10.6 represent the membership of the input in the category for which the control variable assumes the value $z0$. Other values correspond to the lighter-shaded regions and receive a value for the control variable interpolating the neighboring z-values. The control surface can be defined using points; if the control function is smooth, we will obtain a good approximation to other values with simple interpolation. A reduced number of given points corresponds to a reduced number of inference rules in the fuzzy controller. The economic use of rules constitutes a main advantage of such an approach. The definition of inference rules is straightforward with fuzzy formalism, as the interpolation mechanism is taken as given. However, the approach only works when the control function has an adequate degree of smoothness (Rojas, 1996).

10.2 Using Fuzzy Logic to Solve Diagnosis Problems

We are seeing an increasing demand for safer and more reliable man-made dynamical systems. This demand is extending into the processes of industrial plants using servo-actuated flow control valves.

To avoid expensive damage or low efficiency and productivity, we must detect and isolate a malfunction quickly. A fault diagnosis and isolation (FDI) system includes the capacity of detecting, isolating, and identifying faults. Some research has involved analytical approaches based on quantitative models. The main aim is to get different signals when there are inconsistencies between a normal and faulty system operation. These signals are usually generated by an analytical approach using observers, parameter estimation, or parity equations, and are the residual signals. Using either qualitative or quantitative modeling, however, we can achieve an early detection and isolation of abrupt and incipient faults.

The most common method of fault detection is process-variable monitoring and model-based methods (usually more complex). Simple faults can be detected with single measurements. For example, an appropriate threshold check can be defined fairly simply in systems with low complexity, but in industrial systems, it is not so easy, and we need more sophisticated solutions to get a reliable indicator. A model-based approach is more suitable for the former, whereas process modeling is more suitable in the latter. We must take this into consideration especially if we are dealing with a nonlinear process.

The idea for model-based fault detection is to compare different signals of one model with the real measurements taken within the process. From this, we can derive residuals (fault indicators) giving locations and the time of the fault. Having a precise mathematical relationship between the model and the process allows the detection of sudden faults quickly and reliably.

Today, there are numerous methods for model estimation. The most popular are analytical, for example, the Kalman filter and the Luenberger observer, among others (Chen and Patton, 1999; Patton, et.al., 1999). The performance of the resulting FDI systems will depend on creating precise and accurate analytical models, something that is especially important if we are talking about dynamically nonlinear and uncertain systems, which comprise the majority. Using a model-based FDI approach, we can make a precise mathematical model of a plant. We can apply complicated quantitative model-based approaches in real systems, as no unmodeled dynamics can affect the FDI performance. We can design robust algorithms in which the disturbances are reduced and the faults maximized.

Many approaches have been developed, including the unknown input observers and eigenstructure assignment observers, as well as frequency-domain techniques for robust FDI filters, such as the minimization of multiobjective functions, although they have little success in nonlinear cases. Recently, other methods such as neural networks, expert systems, fuzzy systems, and neuro-fuzzy systems have been used with relative success (Calado et al., 2001).

Fuzzy techniques are widely accepted due to their fast and robust implementation, capacity to embed knowledge, performance in reproducing nonlinear mappings, and ability to generalize. This has captured the attention of FDI researchers who are now investigating it as a powerful modeling and decision-making tool, along with neural networks and other more traditional techniques such as nonlinear and robust observers, parity space methods and hypothesis-testing theory.

To help solve the problem of precision in modeling, we can use abstract models based on qualitative approaches. Alternatively, we can use fuzzy logic rules to either assist or replace a model in diagnosis. Using fuzzy logic, we can describe "if-then" relations, making it extremely advantageous in describing system behavior. In the past few years, several research groups have focused on developing residuals using either parameter estimation or observers in fuzzy FDI systems, thereby allocating decision making to a fuzzy logic inference engine. When symbolic knowledge is combined with quantitative information, the false alarm rate goes down.

As noted above, the key advantage to using fuzzy logic techniques is that the operator can describe the system behavior or the fault symptom relationship using simple if-then rules. If we introduce fuzzy observers and plant measurements, we will derive symptoms. The goal is to predict the system outputs from the available process inputs and outputs. The residual is a weighted difference between the predicted and the actual outputs. Simply stated, fuzzy observers compare normal and faulty operations, allowing the detection and isolation of faults (Mendonça et al., 2003).

10.2.1 Architecture for Fault Detection and Diagnosis

The FDI system provides a simple architecture to detect, isolate, and identify faults. It is based on fuzzy observers (models) identified directly from data. This model-based technique uses a fuzzy model for the relevant process, running in normal operation, and one observer (model) for each of the faults to be detected. Suppose a process is running, and there are n possible faults. The fault detection and isolation system proposed for these n faults is shown in Figure 10.7. The multidimensional input, u, of the system enters both the process model and the observer model in normal operation. The vector of residuals ε can be defined as

$$\varepsilon = y - \hat{y}$$

where y is the output of the system and $\hat{y}$ is the output of the model during normal operation. When any component of ε is larger than a certain

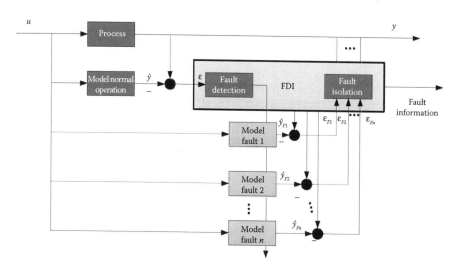

FIGURE 10.7
Fault detection and identification scheme. (Redrawn from Mendonça, L. F., Sá da Costa, J. M. G. and Sousa, J. M., 2003. *Fault Detection and Diagnosis Using Fuzzy Models.* pp. 1–6.)

threshold δ, faults are detected by the system. In this case, n observer models, one for each fault, are activated, and n vectors of residuals computed. Each residual i, with $i = 1,\ldots, n$ is computed as

$$\varepsilon_{F_i} = y - \hat{y}_{F_i}$$

where $\hat{y}Fi$ is the output of the observer for the fault i. Residuals $\varepsilon F_1,\ldots, \varepsilon F_n$ are evaluated, and the detected faults are the outputs of the FDI system. In this chapter, all models, that is, the observer models for normal operation as well as the observers for the n faults, are fuzzy models; they reproduce the dynamic behavior of the process for each condition considered. This technique can identify models extracted from real data (Mendonça et al., 2003).

10.2.2 A Fuzzy Filter for Residual Evaluation

In practice, analytical models often exist for only parts of the plant, with no analytical connections between the models; therefore, analytical model-based methods fail to serve as useful fault diagnosis concepts for the whole plant. However, there is always some qualitative or heuristic knowledge of the connections between the existing analytical submodels that may not be very detailed, but can be expressed linguistically. Such knowledge can be expressed using fuzzy rules, thus describing the normal and faulty behavior of the system in a fuzzy manner.

This means we can use quantitative model-based techniques for the submodels, and qualitative and heuristic knowledge of the connections can be used for the fault symptom generation of the complete system.

The advantages of using such a combined quantitative/knowledge-based approach can be summarized as follows (Jain and Martin, 1998):

1. It is not necessary to build an analytical model of the complete process. It is sufficient to have analytical models of the subparts.
2. The connections between the submodels can be described by qualitative or heuristic knowledge. This is often easier because some qualitative or heuristic description of the plant or the interconnections between the submodels is normally known.
3. The mathematical effort, compared to using a model of the complete plant, is significantly reduced.
4. The causes and effects of the faults can be transferred more easily into the fault diagnosis concept.

10.2.2.1 Structure of the Fuzzy Filter

Fuzzy residual evaluation is a process that transforms quantitative knowledge (residuals) into qualitative knowledge (fault indications). Residuals generated by analytical submodels represent the inputs of a fuzzy filter that consists of the three basic components:

1. Fuzzifiation
2. Inference
3. Presentation of the fault indication

As a first step, a knowledge base has to be built that includes the definition of the faults of interest, the measurable residuals (symptoms), the relationship between the residuals and the faults in terms of IF/THEN rules, and the representation of the residuals in terms of fuzzy sets, for example, "normal" and "not normal."

The process of fuzzification includes the proper choice of the membership functions for the fuzzy sets. This is defined as the assignment of a suitable number of fuzzy sets to each residual component r_i with

$$r = \{r_1, ...r_i, ...r_n\}$$

but not for the fault symptoms fi. This procedure can be mathematically described for the residuals as

$$r_i \rightarrow r_{i1} \, a \, r_{i2} \, a \, ... \, r_{in} \quad r \rightarrow [0,1]$$

where rij describes the jth fuzzy set of the ith residual and a describes the fuzzy composition operator. This part is very important because the coupling or decoupling of the faults will be significantly influenced by this procedure.

The task of the FDI system is now to determine, from the given rule base, indication signals for the faults with the aid of an inference mechanism. The inference can be appropriately carried out by using so-called fuzzy conditional statements

$$IF(effect = r_{11})AND\ IF(effect = r_{12})...THEN(cause = f_m)$$

where f_m denotes the mth fault of the system. The result of this fuzzy inference is a fault indication signal determined from a corresponding combination of residuals as characterized by the rules. Note that this fault indication signal is still in a fuzzified format. Therefore, the signal is called a fuzzy fault indication signal (FFIS). The final task of the proposed FDI concept is the proper presentation of the fault situation to the operator who has to make the final decision about the appropriate fault handling. Typical of the fault detection problem is that the output consists of a number of fault indication signals, one for each fault, where these signals can take only the values 1 or 0 (yes or no). For a fuzzy representation, this means it is not necessary to have a number of fuzzy sets to represent the output, as in control. Rather, each FFIS is, by its nature, a singleton whose amplitude characterizes the degree of membership in only one preassigned fuzzy set "faultm." This degree is characterized by the FFIS, that is, the signal obtained as a result of the inference. Specific to this approach is that it refrains from defuzzification and represents the fault indication signal for each fault to the operator in the fuzzy format, that is, in terms of the FFIS, which represents the desired degree of membership in the set "faultm."

There are some advantages to this procedure in terms of computational expenditure. There is no need to represent the output using a number of fuzzy sets. All available information about the appearance of a fault can be incorporated into the definition of the fuzzy sets of the inputs. We can also dispense with the defuzzification of the signals obtained after the inference has been performed. To be more specific, instead of using the standard format of the statement

$$\text{THEN fault} = \text{big}$$

where big is defined as one of a number of fuzzy sets characterizing the output, we can use the following format:

$$\text{THEN fault} = \text{fault}_1$$

where "fault_1" is the only existing fuzzy set of fault f_1. This applies in a similar way to all faults under consideration. Note that the fuzzy set "$\text{fault}i$" has a degree of membership that is identical to the aggregated output of the evaluated residuals. As a result, one of the key issues of the fuzzy inference approach is that the representation of the result of the residual evaluation concept is different from the conventional concepts in that it directly provides the human operator with the FFIS, leaving him or her to decide whether or not a fault has occurred. This combination of a human expert with a fuzzy FDI toolbox allows us to avoid false alarms, because the fault situation can be assessed on the basis of a fuzzy characterization of the fault situation, together with human expertise and experience. The key issue of this kind of residual evaluation approach is the design of the fuzzy filter. To simplify this design problem, in the following section we present an algorithm that provides systematic support by efficiently reducing the degrees of freedom in the design process (Jain and Martin, 1998).

10.2.3 Identification by Fuzzy Clustering

There are two steps involved in solving the nonlinear identification problem: structure identification and parameter estimation.

10.2.3.1 Structure Identification

In structure identification, the designer must first choose the order of the model and its significant state variables x. This step is extremely important in identifying fuzzy observers for FDI, because the smaller the vector x, the faster the model. *Note*: Fuzzy observers for FDI must be simple but accurate, detecting faults as quickly possible. To identify the model, we construct regression matrix X and output vector y using the available data:

$$X^T = [x_1, ...x_N], \quad y^T = [y_1, ..., y_N]$$

where $N \gg n$ is the number of samples used for identification. The objective is to construct the unknown nonlinear function $y = f(x)$, where f is the Takagi-Sugeno (TS) model (Mendonça et al., 2003).

10.2.3.2 Parameter Estimation

Parameter estimation determines the number of rules, K, the antecedent fuzzy sets, A_{ij}, and the consequent parameters, ai, bi, using fuzzy clustering in the product space of $X \times Y$. Hence, the data set Z to be clustered is composed from X and Y as

$$Z^T = [X, Y]$$

Given Z and an estimated number of clusters K, we can apply the Gustafson–Kessel fuzzy-clustering algorithm to compute the fuzzy partition matrix U. This describes the system in terms of its local characteristic behavior in regions of the data identified by the clustering algorithm, wherein each cluster defines a rule. Unlike the popular fuzzy c-means algorithm, the Gustafson–Kessel algorithm uses an adaptive distance measure, allowing it to find hyperellipsoid regions in the data that can be efficiently approximated by the hyperplanes described by the consequents in the TS model. The fuzzy sets in the antecedent of the rules are obtained from partition matrix U, whose ikth element $\mu_{ik} \in [0, 1]$ represents the membership degree of data object z_k in cluster i. One-dimensional fuzzy sets A_{ij} are obtained from the multidimensional fuzzy sets defined pointwise in the ith row of the partition matrix by projections onto the space of the input variables x_j

$$\mu_{Aij}(x_{jk}) = \text{proj}_j^{N_{n+1}}(\mu_{ik})$$

where proj is the pointwise projection operator (Kruse et al., 1994). The point-wise-defined fuzzy sets A_{ij} are approximated by suitable parametric functions to compute $\mu_{Aij}(x_j)$ for any value of x_j. After clustering, the consequent parameters for each rule are obtained as a weighted ordinary least-square [1] estimate. Let $\theta^T = [a^T; b]$, let X denote the matrix $[X; 1]$, and let W denote a diagonal matrix in R $N \times N$ having the degree of activation, $\beta i(xk)$, as its kth diagonal element as defined as $\beta i = \Pi_{j=1}^n \mu A_{ij}(x_j)$, $i = 1,2,...,k$. Assuming the columns of X_e are linearly independent and $\beta i(xk) > 0$ for $1 \leq k \leq N$, the weighted least-squares solution of $\mathbf{y} = X_e\theta + \varepsilon$ becomes

$$\theta_i = [X_e^T W_i X_e]^{-1} X_e^T W_i y$$

Rule bases constructed from clusters are often unnecessarily redundant because the rules defined in the multidimensional premise overlap in one or more dimension (Mendonça et al., 2003).

10.3 Defuzzification

Defuzzification means fuzzy-to-crisp conversions. The fuzzy results generated cannot be used as such in applications; hence, it is necessary to convert the fuzzy quantities into crisp quantities for further processing. This can be achieved by using a defuzzification process. Defuzzification has the ability to reduce a fuzzy to a crisp single-valued quantity or set, or convert it into a form in which the fuzzy quantity is present. Defuzzification can also be called the "rounding off" method, as it reduces the collection of membership function values into a single-sealer quantity. In what follows, we will discuss several methods of obtaining defuzzified values (Sivanandam et al., 2007).

10.3.1 Lambda Cuts for Fuzzy Sets

Consider a fuzzy set $A\sim$; then, the λ-cut set can be denoted by $A\lambda$, where λ ranges between 0 and 1 ($0 \leq \lambda \leq 1$). The set $A\lambda$ is will be a crisp set. This crisp set is called the λ-cut set of the fuzzy set $A\sim$, where

$$A\lambda = \{x/\mu_A(x) \geq \lambda\}$$

That is, the value of λ-cut set is x, when the membership value corresponding to x is greater than or equal to the specified λ. This λ-cut set can also be called an α-cut set. The λ-cut set $A\lambda$ does not have a title underscore because it is derived from the parent fuzzy set $A\sim$. Since the λ ranges in the interval [0, 1], the fuzzy set $A\sim$ can be transformed into an infinite number of λ-cut sets.

Properties of λ-cut sets:
Λ-cut sets have four main properties:

1. $(A \cup B)_\lambda = A_\lambda \cup B_\lambda$
2. $(A \cap B)_\lambda = A_\lambda \cap B_\lambda$
3. $(A)_\lambda \neq A_\lambda$ except for a value of $\lambda = 0.5$
4. For any $\lambda \leq \alpha$, where α varies between 0 and 1, $A\alpha \subseteq A\lambda$,

where the value of $A0$ will be the universe defined.

As indicated by the properties, the standard set of operations on fuzzy sets is similar to the standard set of operations on λ-cut sets (Sivanandam et al., 2007).

10.3.2 Lambda Cuts for Fuzzy Relations

The λ-cut procedure for relations is similar to that used for the λ-cut sets. Consider a fuzzy relation $R\sim$, in which some of the relational matrix represents a fuzzy set. A fuzzy relation can be converted into a crisp relation depending on the λ-cut relation to the fuzzy relation (Sivanandam et al., 2007)

$$R_\lambda = \left\{ x, y / \mu R(x, y) \geq \lambda \right\}$$

Properties of λ-cut relations:
λ-cut relations satisfy some properties similar to those of λ-cut sets

1. $(R \cup S)_\lambda = R_\lambda \cup S_\lambda$
2. $(R \cap S)_\lambda = R_\lambda \cap S_\lambda$
3. $(R)_\lambda \neq R_\lambda$
4. For $\lambda \leq \alpha$, where α is between 0 and 1, $R\alpha \subseteq R\lambda$.

10.3.3 Defuzzification Methods

Basically, defuzzification maps the output of fuzzy sets defined over an output universe of discourse to crisp outputs. It is employed because in many practical applications, a crisp output is required. A defuzzification strategy is aimed at producing the nonfuzzy output that best represents the possibility distribution of an inferred fuzzy output. Currently, the commonly used strategies are as follows (Jain and Martin, 1998).

10.3.3.1 Max Criterion Method

The max criterion method produces the point at which the possibility distribution of the fuzzy output reaches a maximum value.

10.3.3.2 Mean of Maximum Method

The mean of maximum generates an output that represents the mean value of all local inferred fuzzy outputs whose membership functions reach the maximum. In the case of a discrete universe, the inferred fuzzy output may be expressed as

$$Z_0 = \sum_{j=1}^{I} \frac{w_j}{I}$$

where w_j is the support value at which the membership function reaches the maximum value $1/4z(w_j)$ and l is the number of such support values.

10.3.3.3 Center of Area Method

The center of area generates the center of gravity of the possibility distribution of the inferred fuzzy output. In the case of a discrete universe, this method yields

$$Z_0 = \frac{\sum_{j=1}^{n} \mu_z(w_j)w_j}{\sum_{j=1}^{n} \mu_z(w_j)}$$

where n is the number of quantization levels of the output (Jain and Martin, 1998).

10.4 Need for Complex Relations in Contextual Decision Making

As the key objective of applied fuzzy systems is to transfer ambiguity into value, the main application effort is the effective translation of the vague information from the natural language of the experts into the precise and highly interpretable language of fuzzy sets and rules. Experts play a key role in this process even in the case of data-driven fuzzy systems where the rules are automatically discovered by the clustering algorithms, as they must be interpreted and blessed by the experts (Kordon, 2010).

10.4.1 When Do We Need Fuzzy Systems?

Fuzzy systems are useful for the following reasons:

- Vague knowledge can be included in the solution.
- Interpretation is as important as performance and with fuzzy systems, the solution is interpretable in the form of linguistic rules; that is, we can learn about our data/problem.
- The solution is easy to implement, use, and understand.

10.4.2 Applying Expert-Based Fuzzy Systems

There is no fixed order for the design of a fuzzy system, but an attempt to define an application sequence for classical expert-based systems is given in Figure 10.8.

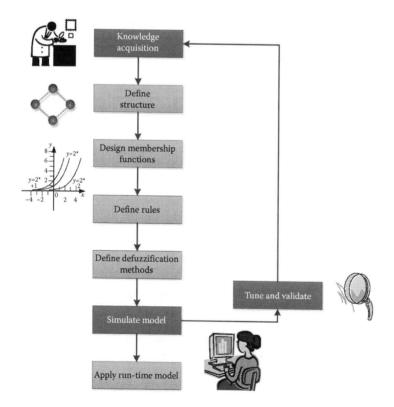

FIGURE 10.8

Application sequence for classical expert-based systems. (Redrawn from Kordon, A. K., 2010. *Applying Computational Intelligence: How to Create Value*. Berlin, Heidelberg: Springer-Verlag.)

Probably 80% of the application's success depends on the efficiency and quality of knowledge acquisition. This is the process of extracting useful information from the experts, data sets, known documents, and common sense reasoning, and then applied to a specific objective. It includes interviewing the experts and defining key features of the fuzzy system, such as identifying input and output variables, separating crisp and fuzzy variables, formulating the protorules using the defined variables, ranking the rules according to their importance, identifying operational constraints, and defining expected performance.

As a result of knowledge acquisition, the structure of the fuzzy system is defined by its functional and operational characteristics, key inputs and outputs, and performance metric.

The next phase of the application sequence is applying the defined structure into a specific software tool, such as the Fuzzy Logic toolbox in MATLAB. The development process includes designing the membership functions, defining the rules, and creating the corresponding defuzzification methods. The aggregated model is simulated and validated with independent data in several iterations until the defined performance is achieved, mostly by tuning the membership functions. A run-time version of the model can be applied in a separate software environment such as Excel (Kordon, 2010).

10.4.3 Applying Data-Based Fuzzy Systems

If the success of expert-based fuzzy systems depends mostly on the quality of knowledge acquisition, the success of data-based fuzzy systems is heavily dependent on the quality of the available data.

The defined structure at the beginning of the application mostly includes data-related issues for the selection of process inputs and outputs from which we expect to find potential rules. Data collection is the critical part in the whole process and could be a significant obstacle if the data have a very narrow range and the process behavior cannot be represented adequately. The possibility of appropriate fuzzy rule discovery is very low in this case.

The data-processing part of the process includes the discovery of protoclusters from the data and defining the corresponding rules. The most interesting step of the design is the decision about the size of the granule of the protoclusters. In principle, the broader the cluster space, the more generic the defined rule. However, some important nonlinear behaviors of the process could be lost. It is therefore recommended that the proper size of the fuzzy clusters be decided by domain experts.

The final result of the development process is a fuzzy system model based on the generalized rules. Notably, there is no difference in the run-time application of either approach (Kordon, 2010).

10.5 Bayesian Analysis versus Classical Statistical Analysis

Bayesian and classical methods are similar in some respects, and both have specific advantages and disadvantages. When the sample size is large, Bayesian inference often provides results for parametric models that are very similar to the results produced by frequentist methods. Some advantages of using Bayesian analysis include the following (SAS/STAT, 2009):

- It provides a natural and principled way to combine prior information with data, within a solid decision-theoretical framework. We can incorporate past information about a parameter and formulate a prior distribution for future analysis. When new observations become available, the previous posterior distribution can be used as a prior one. All inferences logically follow from Bayes' theorem.

- It provides inferences that are conditional on the data and are exact, without relying on asymptotic approximation. Small sample inference proceeds in the same manner as if we have a large sample. Bayesian analysis can also estimate any functions of parameters directly, without using the "plug-in" method (a way to estimate functionals by plugging the estimated parameters into them).

- It obeys the likelihood principle. If two distinct sampling designs yield proportional likelihood functions, then all inferences about them should be identical. Classical inference does not generally obey the likelihood principle.

- It provides interpretable answers, such as "the true parameter has a probability of 0.95 of falling in a 95% credible interval."

- It provides a convenient setting for a wide range of models, such as hierarchical models or missing data problems. The use of Monte Carlo Markov chain (MCMC), along with other numerical methods, makes computations possible for virtually all parametric models.

There are also disadvantages to using Bayesian analysis:

- It does not tell you how to select a prior, and there is no correct way to do so. Bayesian inferences require skills to translate subjective prior beliefs into mathematically formulated priors. If we do not proceed with caution, we can generate misleading results.

- It can produce posterior distributions that are heavily influenced by the priors. From a practical point of view, it might sometimes be difficult to convince subject-matter experts who do not agree with the validity of the chosen prior.

- It often comes with a high computational cost, especially in models with a large number of parameters. In addition, simulations provide slightly different answers unless the same random seed is used. Note that slight variations in simulation results do not contradict the early claim that Bayesian inferences are exact. The posterior distribution of a parameter is exact, given the likelihood function and the priors, while simulation-based estimates of posterior quantities can vary due to the random number generator used in the procedures.

References

Calado, J. M. F., Korbicz, J., Patan, K., Patton, R. J., and Sa da Costa, J. M. G. 2001. Soft computing approaches to fault diagnosis for dynamic systems. *European Journal of Control*, 7(2), 248–286.

Chen, R. and Patton, R., 1999. *Robust Model-Based Fault Diagnosis for Dynamic Systems.* Boston, MA: Kluwer Academic Publishers.

Cristea, M., Dinu, A., Khor, J. and McCormic, M., 2002. *Neural and Fuzzy Logic Control of Drives and Power Systems.* Woburn, MA: Elsevier Science.

Jain, C. L. and Martin, N., 1998. *Fusion of Neural Networks, Fuzzy Systems and Genetic Algorithms: Industrial Applications.* Florida: CRC Press.

Khaliq, A. and Ahmad, A., 2010. *Fuzzy Logic and Approximate Reasoning.* Master's Thesis, Blekinge Institute of Technology, Karlskrona.

Kordon, A. K., 2010. *Applying Computational Intelligence: How to Create Value.* Berlin, Heidelberg: Springer-Verlag.

Kruse, R., Gebhardt, J. and Klawonn, F., 1994. *Foundations of Fuzzy Systems.* Chichester, UK: John Wiley and Sons.

Mendonça, L. F., Sá da Costa, J. M. G. and Sousa, J. M., 2003. *Fault Detection and Diagnosis Using Fuzzy Models.* pp. 1–6.

Patton, R. J., Lopez-Toribio, C. J. and Uppal, F. J., 1999. Artificial intelligence approaches to fault diagnosis. *IEE Colloquium on Condition Monitoring: Machinery, External Structures and Health (Ref. No. 1999/034)*, Birmingham, UK, p. 15.

Restall, G., 2006. *Logic*, Vol. 8. New York: Routledge.

Rojas, R., 1996. *Neural Networks: A Systematic Introduction.* Berlin, Heidelberg: Springer-Verlag.

SAS/STAT, 2009. *Bayesian Analysis versus Classical Statistical Analysis.* [Online] Available at: https://support.sas.com/documentation/cdl/en/statug/63033/HTML/default/viewer.htm#statug_introbayes_sect006.htm.

Sivanandam, S., Sumathi, S. and Deepa, S., 2007. *Introduction to Fuzzy Logic Using MATLAB.* Berlin, Heidelberg: Springer-Verlag.

Zadeh, L., 1965. Fuzzy sets. *Information and Control*, 8, 338–353.

Zadeh, L., 1973. Outline of a new approach to the analysis of complex systems and decision processes. *IEEE Transactions on Systems, Man and Cybernetics*, 1(1), 28–44.

Index

A

Printed and bound by CPI Group (UK) Ltd, Croydon, CR0 4YY

24/10/2024

01778281-0017